Concise Optics
Concepts, Examples, and Problems

TEXTBOOK SERIES IN PHYSICAL SCIENCES

PUBLISHED TITLE

Concise Optics: Concepts, Examples, and Problems
Ajawad I. Haija, M. Z. Numan, and W. Larry Freeman
ISBN: 978-1-138-10702-1 • 2018

Concise Optics
Concepts, Examples, and Problems

Ajawad I. Haija, PhD
Professor of Physics, Physics Department,
Indiana University of Pennsylvania

M. Z. Numan, PhD
Professor of Physics, Physics Department,
Indiana University of Pennsylvania

W. Larry Freeman, PhD
Emeritus Professor of Physics,
Indiana University of Pennsylvania

CRC Press
Taylor & Francis Group
Boca Raton London New York

CRC Press is an imprint of the
Taylor & Francis Group, an **informa** business

CRC Press
Taylor & Francis Group
6000 Broken Sound Parkway NW, Suite 300
Boca Raton, FL 33487-2742

Printed on acid-free paper

International Standard Book Number-13: 978-1-138-10712-0 (Hardback), 978-1-138-10702-1 (Paperback)

Visit the Taylor & Francis Web site at
http://www.taylorandfrancis.com

and the CRC Press Web site at
http://www.crcpress.com

Dedicated to my parents, in whose hands I had received my real education, and to my wife, Samira, a family pillar of love and sacrifice, for her continued passionate support.

Ajawad I. Haija

Dedicated to my wife, Cindy, who has been an extraordinary beacon of solace and sensibility.

M. Z. Numan

Dedicated to my wife, Gloria, for her patience and support throughout this endeavor. Her ability to listen and soften the pains of frustration cannot be overstated or properly compensated.

W. Larry Freeman

Contents

SECTION I — Introduction

SECTION II — Geometrical Optics of Light

SECTION III — Wave Optics

Series Preface: Textbook Series in Physical Sciences

This textbook series offers pedagogical resources for the physical sciences. It publishes high-quality, high-impact texts to improve the understanding of fundamental and cutting-edge topics, as well as to facilitate instruction. The authors are encouraged to incorporate numerous problems and worked examples, as well as making available solutions manuals for undergraduate- and graduate-level course adoptions. The format makes these texts useful as professional self-study and refresher guides as well. Subject areas covered in this series include condensed matter physics; quantum sciences; atomic, molecular, and plasma physics; energy science; nanoscience; spectroscopy; mathematical physics; geophysics; environmental physics; and so on, in terms of both theory and experiment.

New books in the series are commissioned by invitation. Authors are also welcome to contact the publisher (Lou M. Han, Executive Editor: lou.han@taylorandfrancis.com) to discuss new title ideas.

Preface

The ignorant is always prejudiced and the prejudiced are always ignorant.

Charles V. Roman (1864–1934)

This book provides an introduction to optics, spanning a broad range of topics in geometrical and physical optics, as well as electromagnetism. We developed it out of our need for an introductory textbook that would not be too elaborate or advanced for our undergraduate students. Specifically, it is designed for sophomores, juniors, and seniors in physics, chemistry, electro-optics, materials science, and pre-electrical engineering programs taking a first course in optics. It is concise in that it illuminates the essentials necessary for understanding the fundamentals, and there is minimum redundancy so that it can be covered in one semester. The book contains an abundance of solved examples and problems in each chapter to help students master the material.

We have assumed a basic background in physics and calculus. The treatment of each unit in this text is presented in a sequence of small steps that employ essential math, smoothly correlated with, and tied to, the physical content. This approach, as developed from teaching this subject for many years, has been found to facilitate gradual familiarity, a steady buildup of confidence, and improved comprehension of the subject. Several appendices are provided to help overcome likely impediments to success due to math anxiety. Those resources cover topics in basic trigonometry, complex variables, mathematical operators, physical constants, and matrices. Solutions for several problems in Chapter 9 using Mathematica are detailed in Appendix F. Appendix G provides solutions for several problems in Chapter 10 using matrices.

The first section begins with a brief account of the historical progress of optics in Chapter 1, where we briefly survey important explorations, measurements, and junctions that marked significant developments in geometric and physical optics. Among these are major contributions such as Fermat's principle, Maxwell's electromagnetic theory, and Michelson's measurement of the speed of light that crowned physics with an accurate value for the speed of light in vacuum, offering it as a universal constant. This was a remarkable feat, which supported establishing the constancy of light in vacuum and served as the second postulate of the special theory of relativity conceived by Einstein in 1905.

The second section addresses geometric optics in Chapters 2 through 4, covering phenomena of reflection and refraction, Snell's laws, mirrors, and lenses, together with their properties and resulting features. This section concludes with a sizeable chapter (Chapter 4) on matrix representation of flat surfaces, mirrors, and lenses. The third section covers wave or physical optics in Chapters 5 through 13, including wave motion, derivation of the wave equation, superposition of electromagnetic waves, interference, diffraction, and polarization. With respect to interference in thin films and layer stacks, the text incorporates matrix optics after presenting the characteristic matrix technique (CMT) in Chapter 10. CMT has proven to be a powerful tool for determining the basic optical properties of thin films and multilayer structures (reflectivity, transmissivity, and absorptivity). CMT simplifies the calculations for complex, multicomponent layer systems.

In all chapters, the International System of Units (SI) has been adopted. A solutions manual containing detailed, worked out solutions to end-of-chapter problems is available from the publisher for instructors adopting the text.

Ajawad I. Haija

M. Z. Numan

W. Larry Freeman

Acknowledgments

The material has been classroom tested by A. J. Haija over the last several years. Student feedback has been incorporated into the content. Special thanks to our students, Ryan A. Perrin and Megan C. McKillop, who were the first to suggest using Mathematica to solve some of the text problems, especially those in Chapter 9, and Sky Semone for following this interest and devising a simple program in Mathematica for generating the Cornu spiral. He applied the program as a starting point to solve many of the examples and problems in Chapter 9. He also participated in reviewing parts of the manuscript. Another student, Nicholas G. Dzuricky, kindly participated in reviewing several chapters of the text.

Illustrations in the text were drawn by A. J. Haija and W. L. Freeman using Adobe Illustrator. Several images in Chapter 9 were plotted using Mathematica and then formatted in Illustrator by A. J. Haija to fit the text. EasyPlot was used to generate sinusoidal functions before transferring them to Illustrator for further production and completion.

A. J. Haija gratefully acknowledges a Sabbatical from the IUP University Senate, which provided an opportunity to launch the manuscript and implement a major portion of its earlier draft.

Authors

Ajawad I. Haija was born in Sirin, Palestine, grew up in Jordan, and attended the University of Alexandria in Egypt, where he received his BSc in 1968 with distinction, first honor. He received his PhD (1971–1977) from Pennsylvania State University in 1977. His dissertation in solid-state physics addressed electrical and optical properties of superlattices. Dr. Haija joined the University of Jordan in Amman, Jordan, in 1977, where he served until 2000. In 2000, he moved to the United States and joined the Indiana University of Pennsylvania. In addition to numerous publications in the field of thin films, multilayers, and superlattices, he has authored two books, both in Arabic: *Classical Physics*, published in 1998, another (in literature), *Satirical Thoughts from the Inspiration of the Twentieth Century*, published in 2000. He coauthored a third book, *Essential Physics*, published in 2014 by CRC Press/Taylor & Francis Group. He has also translated into Arabic, reviewed, and edited numerous translations of the Arabic version of *Scientific American* that is published in Kuwait.

He is currently on the physics faculty of Indiana University of Pennsylvania, where he conducts research on the properties of thin multilayer structures and superlattices. He enjoys teaching physics courses that extend from freshman to senior and graduate levels. In 2014, Dr. Haija was awarded the Distinguished Faculty Award for Teaching, 2013–2014, Indiana University of Pennsylvania (IUP).

Dr. Haija is a former member of New York Academy of Sciences and a current member of the American Physical Society.

M. Z. Numan hails from Bangladesh, where he received his BSc (Hons.) and MSc in physics from Dhaka University. He received his PhD from The College of William and Mary in Williamsburg, Virginia, in 1982. He taught at Virginia Commonwealth University in Richmond, Virginia; University of North Carolina at Chapel Hill; and Indiana University of Pennsylvania, where he is currently the chair of the department of physics. His research focused on materials modification and characterization using ion implantation, back-scattering, and channeling; optical and electrical characterization of metallic multilayers and semiconductor materials; and light harvesting through silicon micro- and nanostructures.

Dr. Numan is very interested in the teaching and learning of physics and other STEM fields. He is actively engaged in curriculum development using Physics Education Research.

W. Larry Freeman was born and grew up in South Carolina, and earned his BSc in physics from Appalachian State University, Boone, North Carolina, in 1969. He received his MSc (1969–1971) from the University of North Carolina, Greensboro, in 1971. His thesis on solid-state physics addressed the physical properties of the gallium molecule at low temperature under neutron excitations. Dr. Freeman received his PhD from Clemson University, Clemson, South Carolina, in 1976, where his dissertation explored quantum size effects in thin bismuth films at low temperatures. After leaving Clemson University and teaching high school and technical college, he returned to Clemson in a postdoctoral position. He joined the U.S. Naval Intelligence service in 1978, where he was employed as an electronic signals analyst working closely with other agencies in analyzing covert signals and photographs to develop computer models of missile hardware developed by potential adversaries of the United States. He moved to the U.S. Army Night Vision and Electro-Optics Laboratory in 1980, where he was heavily involved with the development of solid-state materials used for the detection of radiation in the infrared radiation wavelength range. He was instrumental in developing nondestructive testing and characterization as well as fabrication techniques for the manufacturing of solid-state infrared detector arrays.

Dr. Freeman moved to Indiana University of Pennsylvania in 1984, where he taught graduate and undergraduate courses in physics. He focused on the involvement of undergraduate students in all aspects of his work, including thin film research and developing novel teaching techniques. He retired from IUP in 2010 and currently holds the position of Emeritus Professor of Physics. He maintains his memberships in the American Physical Society and American Association of Physics Teachers.

Section I
Introduction

Light

Its Nature and History of Study

The ultimate nature of reality is number.

Pythagoras of Samos (570–480 BC)

Because photons have wave and particle characteristics, perhaps all forms of matter have wave as well as particle properties.

De Broglie (1892–1987)

1.1 INTRODUCTION

Light, as we know today, is a form of energy whose behavior, for the most part, remained a mystery for most of humankind's interaction with it. The documented history of the development of our understanding of light dates back as far as 3000 years ago when the Greeks initiated tentative ideas on the reflection of light from metal surfaces. With little done until AD 1000, quantitative observations on reflections were reported by the tenth-century scientist Alhazen, who tested some of those earlier attempts and documented an analysis on one of the laws of reflection. Sporadic but serious endeavors to understand the nature of light continued with simple ideas about vision and the use of lenses. This long period of modest progress may be labeled as the first phase of optics, which perceived light as small particles emitted by their sources and traveling in straight lines, in apparent agreement with straight line motions of solid objects.

Although other theories presented light as waves, notably Huygens (~1668), the particle model prevailed about three more centuries before being questioned seriously. This marked the beginning, at the dawn of the nineteenth century, of a new phase in the study of light in which its wave properties were experimentally investigated and verified by Young (1801), Fresnel (~1821), Fraunhofer (1823), and others. In addition, around 1873, Maxwell developed a comprehensive physical–mathematical wave theory of electromagnetism, which supported nearly all previous experimental data concerning the behavior of light as a wave. This property was quantitatively verified by Michelson and Morley in a series of experiments from 1881 to 1886 and Hertz in 1888. It was the electromagnetic theory that defined light as wave form oscillations of electric and magnetic fields, perpendicular to each other and both perpendicular to their direction of propagation. The theory also stated that the speed c with which light travels in free space (vacuum) is constant and equals $\sqrt{1/\varepsilon_o \, \mu_o} = 3.00 \times 10^8 \, \text{m/s}$, where ε_o and μ_o are the permittivity and permeability, respectively, of free space. The value $c = 3.00 \times 10^8 \, \text{m/s}$ was experimentally confirmed by several experimentalists; most important among them was Michelson (1881). His experiment with Morley in 1886 dismissed the ether theory and established a unique property of light that it propagates with no need for a material medium. The remaining decades of the nineteenth century witnessed Planck's radiation law in which he introduced the quantization of radiations upon absorption or emission of energy by atomic or molecular oscillators, a law that had its impact on the explanation of the photoelectric effect in 1905 by Einstein, embracing the particle behavior of light. The following two decades in the twentieth century witnessed several more striking discoveries, where the wave property of light passed all tests with flying colors. It was the third decade of the twentieth century when de Broglie demonstrated a distinct connection between constructive and destructive interference of light on one hand and light emitted in atomic transitions as described by Bohr in 1913 on the other. In these two models, the appearance of integers was common, featuring both effects, leading de Broglie to state that electrons, which had been viewed as physical particles up to this point, should be associated with waves in a manner similar to light. But the electron wave is not an electromagnetic (EM) wave.

It is a matter wave. This then led him to conclude that all physical entities must be dual in nature in that they exhibit both wave and particle properties. By this conclusion, de Broglie established a new principle, the duality of nature! As surprising as it sounds, especially as early as 1923–1924, it took just a few more years until 1927 when this principle was experimentally observed for electrons, with many similar affirmations exhibited by other fundamental particles. In a way, light seems to be pointing toward the end of the long tunnel of classical physics, signaling an era of a new physics, which today we call "modern physics." This representation developed from Planck's theory of the black-body radiation in 1900, where light previously analyzed as a wave was instead treated as a localized packet that is now called a photon, which exhibits particle properties but propagates like a wave. This established the dual nature of light. All of these developments helped to initiate a third phase for the understanding of light that is called "quantum optics" where light is treated as quanta of photons of zero rest mass and definite energy and momentum. By the end of the third decade of the twentieth century, this new understanding of light and matter led to a solid foundation for the establishment of a new discipline in physics called "quantum mechanics" to demonstrate a third phase for the understanding of light. And with the birth of well-formulated quantum mechanics near the end of the third decade of the twentieth century, quantum optics emerged, bringing our understanding of the third phase of light closer toward a more complete perception. With the invention of the laser and the development of newer optical materials, our study and understanding of light continue to expand.

1.2 LIGHT: THE CORE OF OPTICS

The layout of this rather broad introduction about light calls for defining a host of quantities that together comprise the basis for the theory of numerous phenomena embraced by the first two phases of study, geometric and physical optics. Whenever possible, an attempt will be made to spell out the connection between each quantity and the part of optics to which it is linked.

The formulation of the electromagnetic theory of light by Maxwell offers an appropriate starting point for this discussion. The theory assures one that light consists of electric and magnetic fields oscillating perpendicular to each other, and both are perpendicular to the direction of propagation. Light emitted from an arbitrary source propagates in three dimensions. Therefore, a mathematical representation of the wave is a three-dimensional function that characterizes its direction of propagation, direction of the oscillation of the electric and magnetic fields in the wave, and the wave's periodic change in time and space. In the following sections, features of a light wave are presented through the simplest type of waves known as a plane wave propagating in one dimension.

1.3 PLANE WAVES

The mathematical form of a one-dimensional arbitrary plane wave propagating in the x direction can be represented by a sinusoidal function:

$$\Psi(x,t) = A \sin\left[(kx - \omega t) + \varphi\right], \tag{1.1}$$

where

    A represents the amplitude of the wave

    x is the position of a point in space relative to some convenient reference origin, through which the wave passes

    k is called the wave number and is given by $2\pi/\lambda$, with λ being the wavelength of the wave

    ω is the angular or radian frequency

    t is the time

    φ is the reference phase, also known as the initial phase of the wave or the wave's phase constant

A wave identical to that expressed in (1.1), propagating in the negative direction, takes the form

$$\Psi(x, t) = A \sin\left[(kx + \omega t) + \varphi\right]. \tag{1.2}$$

Equations 1.1 and 1.2 can also be written in terms of the wave linear velocity v. The two forms combined in one give

$$\Psi(x, t) = A \sin\left[k\left(x \mp v t\right) + \varphi\right].$$

Note that the argument of the sine function is dimensionless. As will be demonstrated, Equation 1.1 proves to be extremely successful in elucidating the fundamental features of a light wave.

1.3.1 PHASE

Among the infinitely possible zero values of $\Psi(x, t)$, the values where x, t, and φ are all zero is the trivial case, which is of no interest unless it designates the wave's initial conditions. As $\Psi(x, t)$ is a function of only the variables x and t, the constant φ, called the phase angle, usually has some definite value.

Normally, it is the value at which the wave starts its sinusoidal form; more specifically, it is the value of φ when x and t are zero in a defined reference frame. Figure 1.1a shows a sine wave with 90° phase angle. This corresponds to being one fourth of a cycle ahead of another sine wave of zero phase angle (see Figure 1.1b). Such a sine wave is identical to a cosine wave of the form

$$\Psi(x, t) = A \cos\left[(kx - \omega t) + \varphi\right] \tag{1.3}$$

whose initial phase angle φ = 0.

As a consequence, the phase angle is one of the features that characterizes a particular wave throughout its propagation at any position x and time t. If φ = 0 radians, Equation 1.1 becomes

$$\Psi(x, t) = A \sin\left(kx - \omega t\right). \tag{1.4}$$

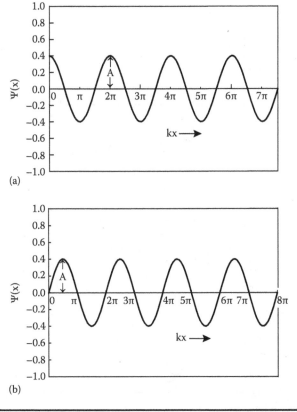

(a)

(b)

FIGURE 1.1 (a) shows a sine wave with 90° phase angle. This corresponds to being one fourth of a cycle ahead of another sine wave having zero phase angle as shown in (b).

The part (kx – ωt) in Equation 1.4 is called the phase of the wave; also note that two other quantities appear here: k and ω; and since the argument of the sine function is dimensionless, k and ω, in the SI system, are in m⁻¹ and s⁻¹, respectively.

Many of the features of the wave described by Equation 1.4 can be explained in Figures 1.2a, b and 1.3a, b. Figure 1.2a demonstrates a snapshot at a certain instant of time that may be considered the initial or t = 0 s configuration and shows the variation of the wave with position x or equivalently kx as shown in Figure 1.2b.

Figure 1.3a shows a similar plot for a fixed position and describes the progression of the wave in time t. Equivalently, in Figure 1.3b, the progression of the wave is shown with ωt. The two cases are for a wave of zero initial phase angle.

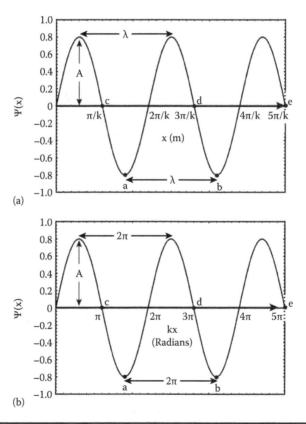

FIGURE 1.2   Snapshot of a wave at t = 0 as a function of (a) position x and (b) phase angle θ = kx.

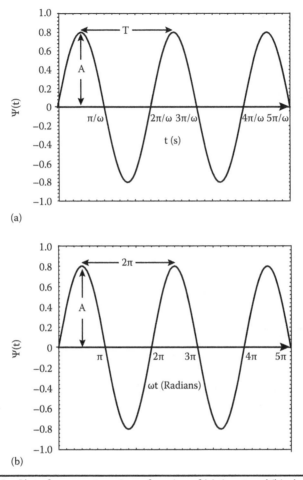

FIGURE 1.3   Plot of a wave at x = 0 as a function of (a) time t and (b) phase angle θ = ωt.

1.3.2 AMPLITUDE

The constant A in Equation 1.4 is called the amplitude. As the maximum absolute value of a sine is unity, the amplitude represents the maximum displacement of the wave above or below its zero level that can be considered the wave reference line. In Figure 1.2b, the maximum displacement occurs when the $kx = \pi/2$ or $3\pi/2,....$ At these values, $\sin(kx) = \pm1$, respectively, and

$$\Psi = \pm A \qquad (1.5)$$

and

$$|\Psi|^2 = |A|^2. \qquad (1.6)$$

1.3.3 WAVELENGTH

λ is called the wavelength. This is one of the fundamental properties of a wave and is best described in association with Figure 1.2a. It is defined as the distance between two successive points along the wave that have a phase difference of 2π between them. Examples of that are the distances between two successive troughs (a, b) or the two successive points (c, d). The units of wavelength commonly used in the optical region, ultraviolet, visible, and infrared are the angstrom Å ($=10^{-10}$ m) and the nanometer (nm = 10^{-9} m). This makes

$$1.0\,\text{Å} = 0.10\,\text{nm}.$$

1.3.4 PERIOD

The period T of a wave is the time elapsed between the two nearest points along the wave that have a phase difference of 2π between them. This is equivalent to describing it as the time needed by one point to cover a distance of one full wavelength. The period is the time of each cycle.

1.3.5 LINEAR FREQUENCY

Defining the repetitive distance covered by one wavelength as a cycle, the frequency f, also known as the linear or cyclic frequency, is the number of cycles executed every second. From its definition, it corresponds to the inverse of the period T of the wave and hence has units of s^{-1}; this unit is also known as Hz. Thus

$$f = \frac{1}{T}. \tag{1.7}$$

1.3.6 ENERGY OF A LIGHT WAVE

The concept of quantization of radiation introduced by Planck states that the energy of a light wave is bundled in quanta, each of which is a quantum that was called later a photon of an energy given by

$$E_{ph} = hf, \tag{1.8}$$

where h is Planck's constant that has a value of 6.626×10^{-34} J·s ($=4.14 \times 10^{-15}$ eV·s). The electron volt is an appropriate energy unit used more commonly to describe energies of fundamental particles when atomic and nuclear dimensions are encountered. It is related to the mechanical unit joule, J, as follows:

$$1.00\,\text{eV} = 1.60 \times 10^{-19}\,\text{C V} = 1.60 \times 10^{-19}\,\text{J}. \tag{1.9}$$

1.3.7 INTENSITY

The intensity of a wave is defined as the power it carries per unit area directed normal to the direction of propagation of the wave. This is equal to the energy carried by the wave per unit area crossed normally every second. Thus

$$\text{Intensity} = \frac{\text{Power }(W)}{\text{Area }(m^2)},$$

or

$$\text{Intensity} = \frac{\left[\text{Energy per unit time}(J/s)\right]}{\text{Area }(m^2)}. \tag{1.10}$$

From another equivalent perspective, the intensity of a light wave emitted by a source may be defined as the number of photons per unit time per unit area present in the wave. From a quantum mechanical perspective, the intensity of a wave, represented by any of the wave equations (Equations 1.1 through 1.3), is proportional to the absolute square of the amplitude A. This will be discussed in a later chapter.

1.3.8 WAVE NUMBER

The wave number is denoted by k. If one takes a snapshot of the wave configuration and calls this the configuration at t = 0, then Equation 1.4 implies that A sin k(x + λ) = A sin(kx + 2π). Therefore, the association kλ = 2π is valid and

$$k = \frac{2\pi}{\lambda}; \tag{1.11}$$

k is also called the propagation constant, and notice that its dimensions are in radians per meter, abbreviated as rad/m or m^{-1}.

1.3.9 ANGULAR FREQUENCY

Consider the progression of the wave in time at the point x = 0, then Equation 1.4 leads to

$$A \sin\left[-\omega t\right] = -A \sin\left[\omega t\right] = -A \sin\left[\omega t + 2\pi\right].$$

Therefore

$$A \sin\left[\omega(t + T)\right] = A \sin\left[\omega t + 2\pi\right].$$

So

$$\omega T = 2\pi.$$

That is,

$$\omega = \frac{2\pi}{T}; \qquad (1.12)$$

ω is called the angular or radian frequency, and note that its units are radians per seconds, abbreviated as rad/s or s^{-1}. Using Equation 1.7, one can see that ω and f are related as

$$\omega = 2\pi f. \qquad (1.13)$$

1.3.10 VELOCITY

From the definitions of the wavelength λ and the period T of a wave, its velocity v is

$$v = \frac{\left(\text{Distance in one cycle}\right)}{\left(\text{Duration of one cycle}\right)}.$$

Thus

$$v = \frac{\lambda}{T}. \qquad (1.14)$$

This is called the phase velocity. From Equation 1.7, v also is

$$v = \lambda f. \qquad (1.15)$$

As for the speed of a light wave in vacuum, v is usually denoted by c, where

$$c = \lambda f. \qquad (1.16)$$

Air is a medium of very low density and is treated as vacuum. That is, the speed of light in vacuum (or air), as stated earlier, has the value of 3.00×10^8 m/s. *Light traveling in any medium other than air slows down to a speed v of a value given by*

$$v = \frac{c}{n} \qquad (1.17)$$

TABLE 1.1
Indices of Refraction of Several Selected Substances for the 589 nm
Wavelength of Sodium Light

Material	Index of Refraction	Material	Index of Refraction
Ice (H_2O)	1.31	Water (H_2O)	1.33
Glass crown	1.52	Methanol	1.33
Quartz	1.54	Benzene	1.50
Diamond	2.42	Glycerin	1.47

where n is known as the index of refraction of the medium and is always larger than unity.

For any material including liquids, the index of refraction is a frequency-dependent quantity. Therefore, its values are defined for a particular wavelength propagating through the material. At standard temperature and pressure, the variations are very small and usually ignored. Disregarding its variation with temperature of the material, a few substances and their indices of refraction for sodium light of 589 nm are listed in Table 1.1.

1.4 ENERGY AND MOMENTUM OF ELECTROMAGNETIC WAVES

If one associates a photon energy, hf, to a particle of mass m, then it is observed that light must also carry momentum just as a particle does. Therefore, Einstein's mass–energy relation may be used to obtain the effective momentum of a photon of light, p_{ph}. Consider the following:

$$E = mc^2; \tag{1.18}$$

m is the mass of the particle. From Equations 1.16 and 1.18

$$mc^2 = hf$$

from which

$$p_{ph} = \frac{hf}{c}; \quad p_{ph} = mc.$$

That is,

$$p_{ph} c = hf \tag{1.19}$$

or

$$P_{ph}\, c = \frac{hc}{\lambda_{ph}}. \tag{1.20}$$

That is,

$$P_{ph} = \frac{h}{\lambda_{ph}}. \tag{1.21}$$

Again, the wavelength λ_{ph} is in meters (m). Equations 1.19 and 1.20 are, in essence, the same. Therefore, the energy of a photon is

$$E = P_{ph}\, c. \tag{1.22}$$

This then provides a new relation for the momentum of the photon:

$$P_{ph} = \frac{E}{c}. \tag{1.23}$$

It is clear from Maxwell and others that EM radiation encompasses a broad range of frequencies and wavelengths, which include visible light, and that this radiation has energy and momentum and exhibits the duality properties predicted by quantum physics (Figure 1.4).

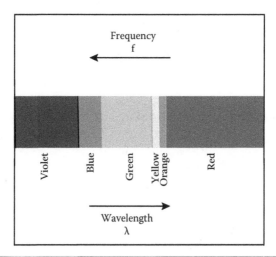

FIGURE 1.4   The visible part of EM radiation. Saleh, B.E.A., and Teich, M.C., *Fundamentals of Photonics*, 2nd edn., John Wiley & Sons, Hoboken, NJ, 2007, p 39.

EXAMPLE 1.1

Given that the intensity of radiation of 540 nm wavelength emitted by a light bulb is 0.024 W/m²

(a) Determine the angular frequency of this radiation.
(b) Determine the wave number of this wave.
(c) Express the intensity u of the bulb in eV per second per meter squared.

Solution

From the relation

$$\lambda = \frac{c}{f},$$

the linear frequency is

$$f = \frac{c}{\lambda} = \left(3.00 \times 10^8 \text{ m/s}\right) / \left(540 \text{ nm} \left(\frac{1.00 \text{ m}}{10^9 \text{ nm}}\right)\right)$$

or

$$f = 5.5 \times 10^{14} \text{Hz}.$$

(a) The angular frequency is, simply, $\omega = 2\pi f$. Thus

$$\omega = 2\pi \left(5.5 \times 10^{14} \text{ Hz}\right) = 3.5 \times 10^{15} \text{ rad/s}.$$

(b) The wave number of the emitted waves is $k = 2\pi/\lambda$. That is

$$k = 2\pi/(540 \text{ nm} \times 10^{-9} \text{ m/nm}) = 1.2 \times 10^7 \text{ m}^{-1}.$$

(c) The intensity denoted by u of 0.024 W/m² is equal to an amount of energy per unit area per sec. This intensity in eV per second per meter squared is

$$u = 0.024 \text{ J/s} \cdot \text{m}^2$$

$$= 0.024 \left(\text{J} \frac{1.0 \text{ eV}}{1.60 \times 10^{-19} \text{ J}}\right) /\text{s} \cdot \text{m}^2$$

$$= 1.5 \times 10^{17} \text{ eV/s} \cdot \text{m}^2.$$

EXAMPLE 1.2

Reconsider Example 1.1 to determine how many photons of the 540 nm wavelength per second are emitted by the bulb.

Solution

The energy that each photon of the 540 nm wavelength carries is

$$E_{ph} = hf = hc/\lambda$$

$$= \left(6.626 \times 10^{-34} \text{ J} \cdot \text{s}\right)\left(3.00 \times 10^8 \text{ m/s}\right)/\left(5.40 \times 10^{-7} \text{ m}\right)$$

$$= 3.7 \times 10^{19} \text{ J}.$$

That is

$$E_{ph} = 3.7 \times 10^{-19} \text{ J} = 3.7 \times 10^{-19} \text{ J}\left(\frac{1.0 \text{ eV}}{1.6 \times 10^{-19} \text{ J}}\right) = 2.0 \text{ eV}.$$

From part (c) of Example 1.1, an intensity of 0.024 W/m² is equal to an amount of energy/s·m²:

$$E = 0.024 \text{ J/s} \cdot \text{m}^2 = 1.5 \times 10^{17} \text{ eV/s} \cdot \text{m}^2.$$

Therefore, the number of photons N_{ph} emitted by the bulb each second is

$$N_{ph} = E/E_{ph} = 1.5 \times 10^{17} \text{ eV}/2.0 \text{ eV} = 7.5 \times 10^{16} \text{ photons}.$$

EXAMPLE 1.3

Considering the EM wave

$$\psi = 8 \sin\left[\left(1.05 \times 10^{-6}\right)x + 314\,t + \pi/4\right],$$

for x in meters and t in seconds, find its (a) wavelength and wave number, (b) linear and angular frequency, (c) velocity, and (d) initial phase angle.

Solution

(a) Comparing the given wave form with the general equation, we see that the wave number

$$k = 1.05 \times 10^{-6} \, m^{-1} = 2\pi/\lambda$$

$$\lambda = 6.00 \times 10^{6} \, m.$$

(b) Also, as

$$\omega = 314 \, rad/s = 2\pi f,$$

$$f = 50.0 \, Hz.$$

(c) As this is an EM wave, its velocity is 3.00×10^8 m/s, which may be checked from $v = \lambda f$.
That is,

$$v = \left(6.00 \times 10^6 \, m\right)\left(50.0 \, s^{-1}\right) = 3.00 \times 10^8 \, m/s \text{ as it should.}$$

(d) Lastly, the comparison shows that the phase angle is

$$\varphi = \pi/4 \, rad/s.$$

EXAMPLE 1.4

Reconsider Example 1.3 to determine

(a) The energy in eV and joules and momentum of the photon described by that wave
(b) In what spectrum band does this photon lie

Solution

(a) From Equation 1.6, the energy of the photon is

$$E = hf = \left(6.626 \times 10^{-34} \, J\right)\left(50.0 \, s^{-1}\right) = 3.32 \times 10^{-32} \, J.$$

Converting the above energy to eV, it becomes

$$E = \left(3.32 \times 10^{-32} \text{ J} \cdot \text{s}\right) / \left(1.60 \times 10^{-19} \text{ ev/J}\right) = 2.08 \times 10^{-13} \text{ eV}.$$

(b) Looking into Table 1.2, the value of E is very low and the wavelength is at the short end of the listed values. The wave is below the radio waves in frequency. In fact, it is in the audio frequency range but cannot be heard by us because the wave is not a mechanical wave but an EM wave, and humans cannot detect EM waves, except as visible light with our eyes and our skin as heat.

TABLE 1.2
Spectral Range of Frequency, Energy, and Wavelength for Several Bands of the Electromagnetic Spectrum

		Electromagnetic Spectrum		
The Electromagnetic Band	**Order of Magnitude of Wavelength Range (m)**	**Order of Magnitude of Wavelength Range (nm)**	**Approximate Frequency Order of Magnitude (Hz)**	**Energy (eV)**
Gamma rays	10^{-14}–10^{-12}	10^{-5}–10^{-2}	10^{22}–10^{24}	10^{8}–10^{6}
X rays	10^{-12}–10^{-9}	10^{-3}–1.0	10^{19}	10^{6}–10^{3}
Ultraviolet (UV)	10^{-9}–10^{-7}	1–10^{2}	10^{16}	1.24×10^{3}–3.1
Visible (Vis)	$4 \times 10^{-7} \times 10^{-7}$	4×10^{2}–7×10^{2}	$\sim 3 \times 10^{14}$	3.1–1.77
Near infrared (NIR)	10^{-6}–10^{-3}	10^{3}–10^{6}	10^{13}	1.77–1.24
FM radio	10–100	10^{10}–10^{11}	10^{7}	1.24×10^{-7}–1.24×10^{-8}
AM radio	10^{2}–10^{3}	10^{11}–10^{12}	6×10^{5}	1.24×10^{-8}–1.24×10^{-9}
Low-frequency AM radio	10^{4}–10^{5}	10^{13}–10^{14}	10^{3}	1.24×10^{-6}–1.24×10^{-11}

Source: Riza, N.A., *Photonic Signals and Systems: An Introduction*, McGraw Hill, Inc., New York, 2013, p. 30.

Note: Among all, the visible band occupies the narrowest range.

PROBLEMS

1.1   Verify the value of 3.00×10^8 m/s for the speed of light in vacuum by substituting for $\varepsilon_o = 8.85 \times 10^{-12}$ C²/Nm² and $\mu_o = 4\pi \times 10^{-7}$ Tm/A in the relation $c = \sqrt{\dfrac{1}{\varepsilon_o \mu_o}}$.

1.2   The wave sketched below is a snapshot of an EM wave at a fixed instant, t_o. Use the plot to determine
   (a)   The wavelength of this wave
   (b)   The period and frequency of this wave
   (c)   The energy E (in joules and eV) of a photon that has the frequency of this wave

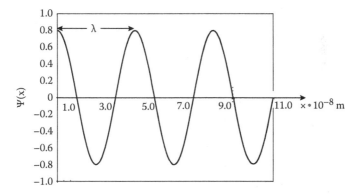

1.3   Consider a golf ball of 55 g moving with a speed of 22 m/s. Using de Broglie's relation
   (a)   Calculate the wavelength of the wave associating this ball.
   (b)   Calculate the ball's kinetic energy.
   (c)   Compare the wavelength calculated in part (a) with the wavelength of a gamma ray photon.

1.4   The wave equations

$$\Psi_1(z) = 3.0 \cos[kz - \omega t], \ \Psi_2(z) = 6.0 \cos\left[(kz - \omega t) + \pi/4\right]$$

are two waves emitted by two identically coherent sources; one wave is ahead of the other by one eighth of a wavelength (45° phase angle). The waves are propagating along the z direction; z is in meters and t is in seconds. For $f = 2.5 \times 10^{14}$ Hz and in two separate sketches of the two waves, display a snapshot plot at t = 0 s of $\Psi_1(z)$ and $\Psi_2(z)$, each against a properly scaled z axis and draw a sketch to determine the difference in phase between the two waves.

1.5   The wave depicted below illustrates the wave form of the electric field of an EM wave. Use the plot below (x is in meters) to extract enough information to write its wave form on which the amplitude and the wave number are explicitly represented.

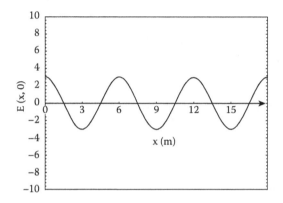

1.6   The wave sketched below is a snapshot of an EM wave at a fixed position. Use the plot to determine
(a)   The frequency of this wave
(b)   The wavelength of this wave
(c)   The energy in eV of this wave

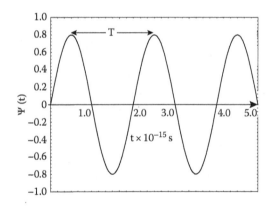

1.7   The two waves shown in the figures below are emitted by two identically coherent sources with a time lag of $t = 1.2 \times 10^{-15}$ s between them.
(a)   Determine the linear and angular frequency of each wave.
(b)   Determine the wavelength and wave number of each wave.
(c)   Write an equation for each of these waves such that the amplitude, the angular frequency, and the phase constants are numerically included.

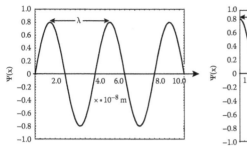 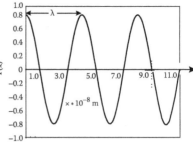

1.8   For a photon of energy E = 1.55 eV
(a)   What is its energy in joules?
(b)   Calculate the wavelength of such a photon.
(c)   In what region of wavelength does this fit?

1.9   Consider two waves, one in the gamma and the other in the UV range, having the wavelengths 5.00×10^{-13} m and 5.00×10^{-8} m, respectively. For each of these waves, calculate
(a)   The wave number
(b)   The frequency
(c)   The energy

1.10   Consider two waves, one in the visible and another in the radio range, having the wavelengths 5.50×10^{-7} m and 5.50×10^{2} m, respectively. For each of these, calculate
(a)   The wave number
(b)   The frequency
(c)   The energy

1.11   The plot below represents a snapshot of the wave function ψ as a function of position x, where x is in meters. Assume that the plot represents an EM wave in free space and calculate the following:
(a)   The period of the wave
(b)   The amplitude
(c)   The angular frequency

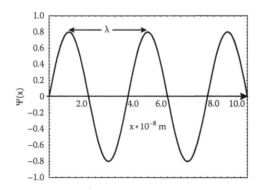

1.12   The plot below represents a snapshot of the wave function ψ as a function of time t, where t is in seconds. Assume that the plot represents an EM wave in free space and calculate the following:
(a)   The period and the angular frequency of the wave
(b)   The amplitude and wavelength of the wave

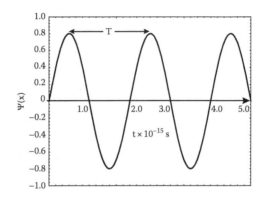

1.13   Consider a source with an intensity, I, given by $I = 5.00 \times 10^{-3}$ W/m². If the source is emitting 1.20×10^{9} photons/m² s, determine
(a)   The energy of each photon
(b)   The frequency of the wave

1.14   Consider a lamp, with power rating of 75 W, emitting radiation with an average wavelength of $\lambda = 540.0$ nm. Determine the number of photons it emits per second. Hint: 1.0 W = 1.0 J/s.

1.15   For the intensity $I = 5.00 \times 10^{-3}$ W/m², given in Problem 1.13, expressed by $I = \alpha |A|^2$, $A = 0.080$ nm$^{-1/2}$, calculate α and state its units.

Section II
Geometrical
Optics of Light

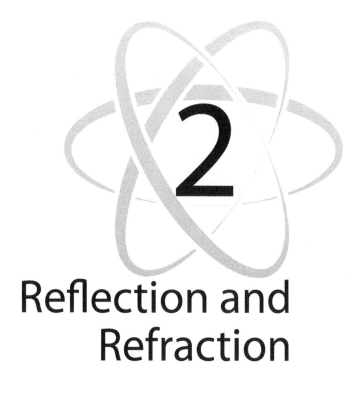

Reflection and Refraction

The repetitive occurrence of an incident does not guarantee that the same thing will happen again.

Francis Bacon (1561–1626)

2.1 INTRODUCTION

The phenomena of reflection and refraction are among the oldest properties of light that had attracted attention. The earliest attempts date back to the Greeks reporting on reflection a few hundred years BC. Some of the reported quantitative observations on reflection of light were known to have taken place in the eleventh century, and empirical relations were attributed to Alhazen. With the current perception of the particle and wave-like models of light that underlie geometric and wave optics, we can say that both reflection and refraction of light are well understood. However, geometric optics, which adopts straight line propagation of light, offers a simpler, clearer, and more concise description of both phenomena. This chapter starts with defining these phenomena and presents some rules that govern a sign convention that facilitates the understanding of the various features that surround the formation of images from these effects. Of interest to us are the position, size, and nature of the formed images.

2.2 REFLECTION

We can classify reflection as diffuse or specular (Figure 2.1). Diffuse reflection is similar to what happens when light is incident on a rather rough or irregular surface, similar to the surface of a screen or bed of sand. Light gets reflected from each spot of the screen; thus, light rays incident on an overwhelming number of nearby spots reflect off in all directions, crossing each other and illuminating the surrounding space. So, a narrow beam of parallel light rays incident on the screen hits its irregularities with different angles of incidence and reflect off the screen surface with no correlation between their angles of reflection (Figure 2.1a).

The reflected intensity, then, is the result of superimposed waves reflected in random directions producing uniform illumination with the intensity spread over the whole viewing area. However, specular reflection is what happens when the surface is smooth and possibly shiny, mirror like, as in a silvered glass plate (Figure 2.1b). In such a case, any light ray incident on the surface with a certain angle gets reflected along a specific direction, obeying the following condition: Angle of incidence equals the angle of reflection, which is known as the *law of reflection*; this law will be discussed and used later in a few examples. Geometrically speaking, it is much easier to explain certain phenomena like reflection and refraction through the straight line propagation of light rays.

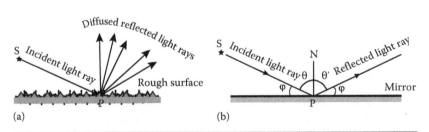

(a)   (b)

FIGURE 2.1   (a) Diffused versus (b) specular reflection.

2.2.1 BASIC DEFINITIONS

For a more detailed discussion of specular reflection, the following definitions are needed:

1. Light ray: The surface of any advancing wave is called wave front. In case the source is far away from a viewer, the wave front at the observation site becomes a plane. The line normal to the wave front,

spherical or planar, defines a light ray in that direction. The ray is indicated by the straight line drawn from a point on the source to the point of interest. The straight line propagation of light is used here to describe this phenomenon.

2. The angle of incidence of a light ray on a surface is defined as the angle between the incident ray and the normal to the surface at the point of incidence. The angle of reflection of a light ray off a surface is defined as the angle between the reflected ray and the normal to the surface at the point of incidence.

3. The plane of incidence is the plane containing the incident ray and the normal to the surface at the point of incidence. The plane of reflection is the plane that contains the reflected ray and the normal to the surface. For specular surfaces, the incident and reflected rays along with the normal line all form a single plane.

2.2.2 Laws of Reflection

The above definitions play an important role in the statements of the following two laws of reflection:

1. A light ray incident on a surface with an angle of incidence, θ, gets reflected with an angle of reflection, θ', such that the angle of incidence equals the angle of reflection. That is,

$$\theta = \theta' \tag{2.1}$$

2. In any reflection, the incident ray, the normal to the surface at the point of incidence, and the reflected ray all lie in one plane. This is equivalent to saying that the plane of incidence coincides with the plane of reflection. Figure 2.2 demonstrates the above laws of reflection.

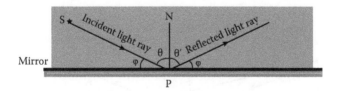

FIGURE 2.2 Demonstration of the first law of reflection.

2.3 IMAGE FORMATION VIA REFLECTION

Light rays emanating from a point source and directed toward a mirror (plane or spherical) get reflected into the same side from which the rays originate. The reflected rays may or may not converge to a point. If they converge in the same space as the source, then the intersection is formed from real reflected rays that have bounced off the mirror. This should mean that such a point is the position of a real image of the source point that emitted the rays. If the reflected rays do not converge but their extensions on the other side of the mirror do, then the image would not be real because the intersection happens behind the mirror. The energy carried by the rays has been blocked by the mirror in this case, that is, real reflected rays are not reaching the point of intersection. Such a point is the presumed image, and hence the image is said to be virtual. This distinction between a real image and a virtual one applies to all kinds of mirrors, planar or curved. In addition to this image labeling, additional ways are presented in the next few sections to describe all features of images: their nature, position, and size.

2.3.1 PLANE MIRRORS

Consider the figure below, where a source at a distance o from a flat horizontal mirror emits two rays, 1 and 2, that are incident on the mirror with angles of incidence θ_1 and θ_2, respectively, and let $\theta_2 > \theta_1$. The two rays reflect off the surface of the mirror with two angles of reflection, $\theta_1' (= \theta_1)$ and $\theta_2' (= \theta_2)$, respectively. Obviously, the angle $\theta_2' > \theta_1'$. Thus, the angle $\beta (= 90° - \theta_2')$ is less than $\phi (= 90° - \theta_1')$. Therefore, the two reflected rays 1' and 2' diverge from each other, and hence they never meet in front of the mirror. However, the extensions of 1' and 2' meet at S' behind the mirror (Figure 2.3).

As the angles of incidence and reflection, θ_1 and θ_1', are equal, the two other angles, one between the incident ray 1 and the surface and the other between the reflected ray 1' and the surface, are equal; each of these is labeled as ϕ. Geometry shows that the angle PP_1S' is also ϕ. Thus, the two triangles SP_1P and $S'P_1P$ are similar; they have P_1P in common, and also each has the two angles, ϕ and $\pi/2$, on the two sides of P_1P. Folding the two triangles, SP_1P

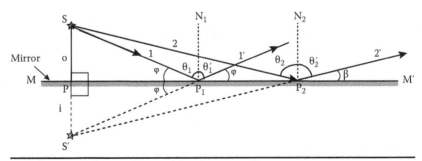

FIGURE 2.3   Reflection from a plain mirror - ray diagram.

and $S'P_1P$, onto each other along line P_1P results in $SP = S'P$. As these represent the distances o and i of the object and image, respectively, from the mirror in magnitude, i = o. That is

$$|i| = |o|. \tag{2.2}$$

Further elaboration on the signs of the distances i and o will be deferred until a comprehensive sign convention for mirrors is developed in full. However, at this point of the discussion, we still can elude to the point that the image is not formed from (real) reflected rays but from the intersection of their extensions. For now, this is good enough of a criterion to describe the image as *virtual*.

EXAMPLE 2.1

A small object is placed 12.0 cm from a plane, horizontal mirror. Among the rays emitted by the object, consider the two rays, 1 and 2, that are incident on the mirror with 60° and 45° angles of incidence, respectively. Following the laws of reflection and by ray tracing only, locate the object's image.

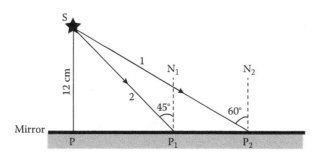

Solution

A careful and accurate depiction of the reflected rays 1′ and 2′ with angles of reflection of 60° and 45°, respectively, shows that their extensions behind the mirror intersect at the point S′, which is the image position. Measuring the distance PS′, we find it equal to SP. Thus, the image is 12.0 cm away, behind the mirror (see scale in the following figure), and is virtual because it is resulting from the intersection of the extensions of the reflected rays 1′ and 2′, rather than the rays themselves.

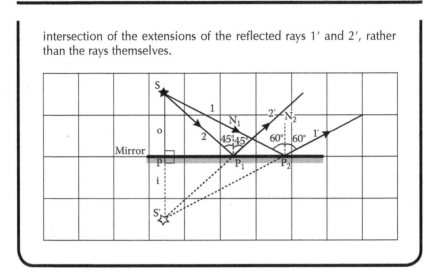

2.3.2 Spherical Mirrors

2.3.2.1 Definitions

Spherical mirrors in general, convex or concave, have several features in common. In the following, several definitions are introduced that apply to all spherical mirrors. Figure 2.4 illustrates these quantities:

1.   *Center of curvature*: As every spherical surface is part of a spherical shell, each has a center of curvature, which is the center of the shell, of whose surface, the mirror is a segment.
2.   *Optic axis or principal axis*: It is the symmetry line passing through the center of curvature that bisects the spherical surface segment forming the mirror. The optic axis in a simple mirror (or refracting) surface acts as a reference axis connecting object and image positions.
3.   *The vertex*: This is the point that results from the intersection of the principal axis and the mirror's surface. This is usually denoted by V.

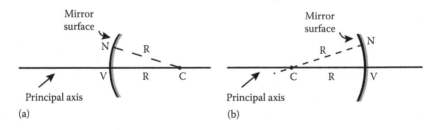

FIGURE 2.4   Spherical mirrors (a) convex and (b) concave.

4. *Paraxial rays*: Rays emanated from the object and those reflected by a mirror (or refracted by a refracting surface) that make very small angles with the optic axis are called paraxial rays.

2.3.2.2 Sign Convention

From design and geometry perspectives, spherical mirrors are more involved than flat ones. They can be concave or convex, and that is in reference to how light rays propagating toward the mirror "see" the mirror's surface. Several quantities like the radius of curvature, object and image distances from a mirror's surface, and heights of the object and image are in one way or another involved in image formation, necessitating a systematic convention of signs for these quantities. Therefore, prior to laying out an analytic treatment of mirrors, consider Figure 2.5 in which a convex mirror (a) and a concave mirror (b) are displayed. In the following is a set of rules that describe a sign convention that can be followed in handling object- and image-related quantities:

1. In all cases, the object is positioned to the left of the mirror (object side). A virtual object is not plausible without special arrangement. Also, since light rays emitted by the object and incident on the mirror get reflected, a real image formed from the intersection of real rays is expected to be on the left side of the mirror. That is why in the present convention, the space on the left side of the mirror is considered the object side and is the "real-image side" (R-side) as well.

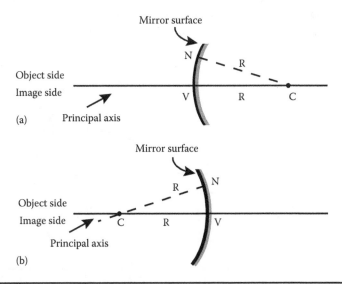

FIGURE 2.5   Spherical mirrors: (a) convex and (b) concave; the vertex V, the center of curvature C, and the line of symmetry connecting V and C are shown.

2. The object distance o on the left of the mirror is positive. A parallel argument goes for the image distance i; it is positive if the image is on the left side of the mirror (real image, R-side) and negative if it happens to be to the right of the *vertex* "behind" the mirror.

3. As a consequence, a positive sign for i means the image is real, and a negative sign for i means it is virtual. Object distance o is always positive for real objects.

4. The sign of the radius of curvature R for a concave mirror as seen by the incident rays is negative, while for a convex mirror, it is positive. As will be demonstrated later, this convention was introduced so that the focal length becomes positive for a concave mirror and negative for a convex mirror. This is equivalent to saying that a spherical surface for which the center of curvature is to the right of the surface has a positive radius of curvature, while that whose center of curvature is to the left of the surface has a negative radius of curvature.

5. Heights measured above the optic (principal) axis are positive, while those below it are negative.

2.3.2.3 Mirror Equation

Consider the concave mirror in Figure 2.6 that has a radius of curvature R and center of curvature C. Let us have an object at O emitting two rays: one ray, ON, incident with an angle of incidence θ that gets reflected along NI with angle of reflection θ, and another ray, OV, along the principal axis that reflects onto itself ($0.0°$ angle of incidence = $0.0°$ angle of reflection). The intersection at I of the reflected rays NI and VI is the image position.

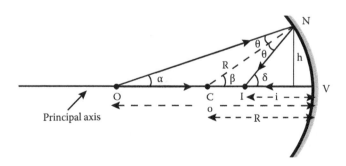

FIGURE 2.6 The diagram shows a concave mirror on which a ray emanating from a point object, O, is reflected, forming a point image, I.

As we are working with paraxial rays, the angles α, β, and δ are all small, and thus the tangent of each angle equals the angle itself.

From geometry, with respect to the triangle ONI

$$\delta = 2\theta + \alpha,$$

or

$$\delta - \alpha = 2\theta, \tag{2.3}$$

and considering the triangle ONC

$$\beta = \theta + \alpha$$

or

$$\beta - \alpha = \theta. \tag{2.4}$$

Combining Equations 2.3 and 2.4 gives

$$\delta - \alpha = 2\beta - 2\alpha$$

or

$$\delta + \alpha = 2\beta.$$

Replacing each angle in the above equation by its tangent, that is, labeling

$$\tan \delta = \delta, \quad \tan \alpha = \alpha, \quad \tan \beta = \beta,$$

we get from the geometry

$$\frac{h}{i} + \frac{h}{o} = \frac{2h}{R},$$

which reduces to

$$\frac{1}{i} + \frac{1}{o} = \frac{2}{R}. \tag{2.5}$$

It is worth noting that the left-hand side of the above equation is positive. However, since the radius of curvature was adopted within the sign convention to be negative, this apparent inconsistency could be remedied by entering a negative sign in front of the right side of Equation 2.5 so that upon substitution for R with a negative value, the above equation would retain its integrity. Modifying the above equation to conform to the sign convention, we write

$$\frac{1}{o} + \frac{1}{i} = -\frac{2}{R}. \tag{2.6}$$

Introducing another quantity, called the focal length f, such that

$$\frac{1}{f} = -\frac{2}{R},\qquad(2.7)$$

Equation 2.6 could then be written as

$$\frac{1}{o} + \frac{1}{i} = \frac{1}{f}.\qquad(2.8)$$

Note that the definition of f in Equation 2.7 though seems optional, it was introduced in that specific way so that concave or converging mirrors always have positive focal length. A concave mirror converges parallel rays into a point on the R-side for the image.

This makes the focal length of a concave mirror positive. The position at a distance, $f(=|R/2|)$, from the vertex of the mirror is called the mirror's focal point F.

With a similar but parallel argument, which we could follow for a convex mirror, Equations 2.6 through 2.8 turn out to be of the same form. However, since R for a convex mirror is positive, f in Equation 2.7 must be negative. Equation 2.8 then can be applied for both kinds of mirrors, concave and convex, as long as the proper sign for f is used in Equation 2.8, which is called the *mirror equation.*

Another quantity of interest is the *magnification* M of an image produced by a spherical mirror. This is defined as

$$M = \frac{h_i}{h_o}.$$

With proper use of the sign convention for h_i, geometry shows that $\frac{h_i}{h_o} = -\frac{i}{o}$.
Therefore

$$M = -\frac{i}{o}.\qquad(2.9)$$

Conforming to the sign convention, the sign of M has the following relevance:

1. M being positive means the image is erect or upright.
2. M being negative means the image is inverted.

COMMENTS

1. Notice that switching i with o and o with i in Equation 2.8 keeps its form unchanged. That is, the object and image positions may exchange roles such that rays traced from the object O toward the image I can be traced backward. This demonstrates the property of reversibility of paths of light rays. A virtual image consequently allows us to conceive a virtual object.

2. Equation 2.7 shows that the focal point F of a spherical mirror lies halfway between the vertex and center of curvature and is on the left side of a concave mirror, while it is on the right side of the convex mirror. Every mirror has one focal point.

3. As R for a plane mirror is ∞, the focal point of a plane mirror is at infinity, and hence f = ∞. Therefore, from the mirror equation

$$\frac{1}{o} + \frac{1}{i} = 0.$$

Hence

$$i = -o.$$

That is

$$|i| = |o|.$$

This is a relation that was established earlier. The negative sign of the image distance i implies that the image is virtual, o being positive.

2.3.2.4 Special Cases

1. For an object at infinity, o = ∞, Equation 2.8 results in i = f. Since f for a concave mirror is positive, i is positive, that is, the image formed by a concave mirror is real; however, since f for a convex mirror is negative, i is negative, that is, the image formed by a convex mirror is virtual. Since rays coming from infinity are parallel, such rays incident on a concave mirror parallel to the principal axis reflect back, passing through its focal point F (Figure 2.7a). But if the mirror is convex, the rays reflect off the surface, diverging in a manner that their extensions pass through the focal point F. This situation is demonstrated in Figure 2.7b.

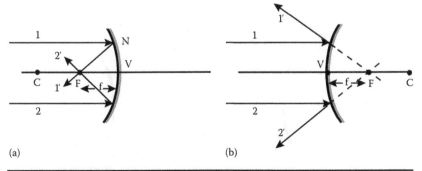

(a)   (b)

FIGURE 2.7   Parallel rays incident on the surface of (a) concave and (b) convex mirrors. In (a) the reflected rays intersect at the focal point, while in (b) the extensions of the reflected rays intersect at the focal point.

2. For an object at the focal point F, that is, o = f, Equation 2.8 results in i = ∞. Therefore, light rays emitted by an object at the focal point of a concave mirror reflect off the surface parallel to the principal axis (Figure 2.8a). However, for a convex mirror, incident rays that appear to pass through the focal point, representing a virtual object at F, reflect back parallel to the optic axis (Figure 2.8b). Notice that Figure 2.8a can be produced by a reverse tracing of rays in Figure 2.7a. By the same token, Figure 2.8b can be produced by a reverse tracing of rays in Figure 2.7b. These cases also demonstrate reversibility of light paths.

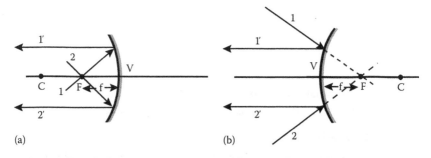

FIGURE 2.8   Rays originating from the focal point in (a) reflect parallel to the principal axis, while in (b) the incident rays whose extensions appear to be passing through the focal point get reflected parallel to the principal axis.

3. A special case of interest is when an object in front of a spherical mirror emits a light ray whose path or its extension passes through the mirror's center of curvature (Figure 2.9). As the angle of incidence of the ray is zero, the angle of reflection is zero. Therefore, in both cases, the ray reflects back along the same path.

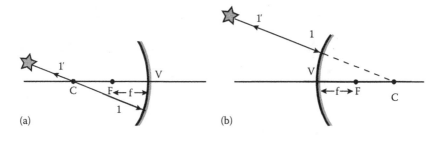

FIGURE 2.9   A ray incident on (a), a concave mirror, passing through its center, and on (b), a convex mirror, where its extension is passing through the center; both do not suffer any shift in their paths.

EXAMPLE 2.2

Consider the following figure where a concave mirror with a focal length f = 8.0 cm. Find the location and magnification of the image of an object (an arrow), 8.0 mm tall, standing vertically at 12 cm from the mirror.

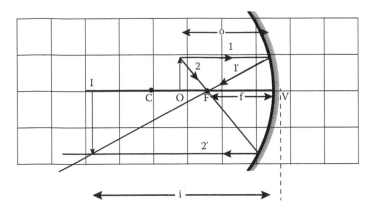

Solution

Following the properties stated earlier for a concave mirror, two rays, 1 parallel to the optic axis and another, 2, passing through the focal point, reflect off the mirror along 1' and 2', respectively, as demonstrated in the figure below. Their intersection at I defines the image position. By measuring the object and image heights, an estimate of the magnification, $\frac{h_i}{h_o}$, is found to be negative 2; the negative sign means the image is inverted. The preceding diagram, sketched to scale, demonstrates the situation rather well.

EXAMPLE 2.3

In the previous example, keeping the same mirror, use the mirror equation to find the location and magnification of the image of the 8.0 mm tall arrow placed at

(a) 12 cm from the mirror
(b) 4.0 cm from the mirror

Solution

(a) From Equation 2.8

$$\frac{1}{o} + \frac{1}{i} = \frac{1}{f}.$$

Substituting for o and f, Equation 2.8 becomes

$$\frac{1}{12\,(\text{cm})} + \frac{1}{i} = \frac{1}{8.0\,(\text{cm})}.$$

This results in

$$\frac{1}{i} = \frac{1}{8.0\,(\text{cm})} - \frac{1}{12\,(\text{cm})},$$

or

$$\frac{1}{i} = 0.12 \text{ cm}^{-1} - 0.083 \text{ cm}^{-1}.$$

Thus

$$i = 24\,\text{cm}.$$

As the sign of i is positive, the image is real.
The magnification is

$$M = -\frac{i}{o} = -\frac{24}{12} = -2.0.$$

As the sign of M is negative, the image is inverted and is twice the size of the object.

(b) Again, from Equation 2.8

$$\frac{1}{o} + \frac{1}{i} = \frac{1}{f}.$$

Substituting for o and f, it becomes

$$\frac{1}{4.0\,(\text{cm})} + \frac{1}{i} = \frac{1}{8.0\,(\text{cm})}.$$

This results in

$$\frac{1}{i} = \frac{1}{8.0\,(\text{cm})} - \frac{1}{4.0\,(\text{cm})},$$

or

$$\frac{1}{i} = 0.13 \text{ cm}^{-1} - 0.25 \text{ cm}^{-1} = -0.13 \text{ cm}^{-1}.$$

Thus

$$i = -8.0 \text{ cm}.$$

As the sign of i is negative, the image is virtual.
The magnification is

$$M = -\frac{i}{o} = -\frac{-8.0 \text{ cm}}{4.0 \text{ cm}} = 2.0.$$

As the sign of M is positive, the image is upright and is twice the size of the object.

EXAMPLE 2.4

Consider a convex mirror with a focal length f = −6.00 cm. Find the location and magnification of the image of an object (an arrow) 2.0 cm tall standing vertically at 8.00 cm from the mirror.

Solution

From Equation 2.8

$$\frac{1}{o} + \frac{1}{i} = \frac{1}{f}.$$

Substituting for o and f, Equation 2.8 becomes

$$\frac{1}{8.00 \, (\text{cm})} + \frac{1}{i} = \frac{1}{-6.00 \, (\text{cm})}.$$

This results in

$$\frac{1}{i} = \frac{1}{-8.00 \, (\text{cm})} - \frac{1}{6.00 \, (\text{cm})},$$

from which

$$\frac{1}{i} = -0.125 \text{ cm}^{-1} - 0.0833 \text{ cm}^{-1}.$$

Thus

$$i = -3.8 \text{ cm}.$$

As the sign of i is negative, the image is virtual, lying on the virtual side of the mirror, behind it.
   The magnification is

$$M = -\frac{i}{o} = -\frac{-3.8}{8.0} = 0.50.$$

As the sign of M is positive, the image is upright and is about half the size of the object. The diagram below, sketched to scale (each square on the grid is 2 cm in length), illustrates the situation well.

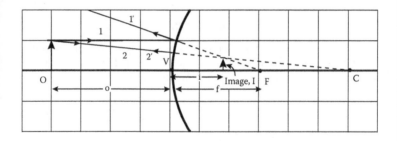

2.4 REFRACTION

Consider a medium made up of two transparent materials, each characterized by a dimensionless parameter called the index of refraction, symbolized by n_1 and n_2 for the respective materials. Suppose that in the two materials, light travels at speeds $v_1(=c/n_1)$ and $v_2(=c/n_2)$. The boundary between the two materials is distinct and $n_1 < n_2$. When a light ray traveling in the material of index n_1 is incident upon the boundary with the material of index n_2, it changes its original path to one closer to the line normal to the boundary surface. This phenomenon, called refraction, is due to the difference in the speed of light in the two media. This process is governed by the law of refraction as developed in the next section.

2.4.1 DEFINITIONS AND LAWS OF REFRACTION

1. The ray's angle of refraction in the second medium is defined as the angle between the refracted ray and the normal to the boundary at the point of refraction (Figure 2.10a).
2. The plane of refraction in the refracting medium is the plane that combines the refracted ray and the normal to the surface at the point of refraction.

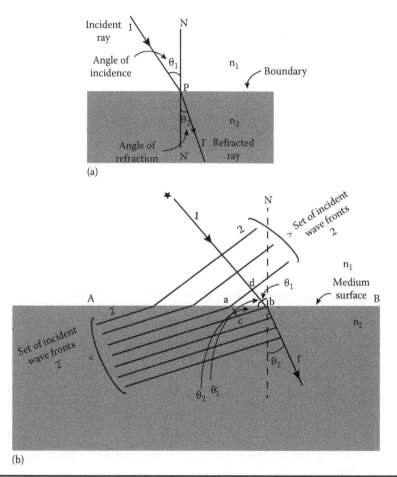

FIGURE 2.10   (a) A typical sketch of refraction with refraction parameters displayed on the sketch; (b) the wave fronts of an incident wave (ray 1) and those of the refracted wave (ray 1′) are displayed.

The basic laws of refraction may be deduced from experiment but may also be developed from the fundamental theory of waves by considering a plane wave train of frequency f and wavelength λ impinging upon a transparent material and traveling into it:

1. When a light ray, 1, from a material of low index of refraction, n_1, enters a material of high index, n_2, the ray path $1'$ is shifted closer to the normal. This is illustrated in the diagram, Figure 2.10a and b, where the light rays 1 and $1'$ and wave fronts 2 and $2'$ (normal to the light rays) in the two media are shown.

    Letting the wave fronts be spaced by one wavelength apart, λ_1 and λ_2 in media n_1 and n_2, respectively,

$$n_1 = c/v_1, \quad n_2 = c/v_2,$$

from which

$$n_2/n_1 = v_1/v_2. \tag{2.10}$$

As the frequency in the two media is the same, we have

$$f = v_1/\lambda_1 = v_2/\lambda_2$$

or

$$v_1/v_2 = \lambda_1/\lambda_2. \tag{2.11}$$

From the geometry in Figure 2.10b, $\theta_1' = 90 - \theta_1$; thus, $\cos\theta_1' = \sin\theta_1$. Now

$$\cos\theta_1' = \frac{bd}{ab};$$

that is,

$$\sin\theta_1 = \frac{bd}{ab}.$$

Also

$$\sin\theta_2 = \frac{ac}{ab}.$$

Dividing the above two equations gives

$$\frac{\sin\theta_1}{\sin\theta_2} = \frac{v_1}{v_2} = \frac{\lambda_1}{\lambda_2}.\tag{2.12}$$

Combining Equations 2.10 through 2.12 results in

$$\frac{\sin\theta_1}{\sin\theta_2} = \frac{n_2}{n_1}.$$

Thus

$$n_1\sin\theta_1 = n_2\sin\theta_2.\tag{2.13}$$

This is known as Snell's law, where n_1 and n_2 are the indices of refraction of the first and second media, respectively.

2. The second law states that the incident ray, the normal to the surface at the point of incidence, and the refracted ray are all contained in one plane.

2.5 IMAGE FORMATION VIA REFRACTION

In refraction, two light rays incident on a boundary between two different media with two different angles of incidence get refracted with two different angles of refraction. Upon following the paths of the refracted rays, one can locate the image formed from their point of intersection. Since the second medium is where the real refracted rays appear, a real image is expected if they intersect in that medium. However, if these rays do not intersect but their backward extensions do, then the image is virtual. This rule applies to all refracting media, regardless of the nature of the surface, plain or spherical. In the following sections, several cases of refraction are discussed in detail.

2.5.1 REFRACTION OFF A PLANE BOUNDARY

Consider Figure 2.11, where a source of light S in one medium, of index of refraction n_1, is at a distance o from a flat horizontal boundary, AB, that represents the top surface of another medium of an index of refraction n_2; let $n_2 > n_1$. Take two rays, 1 and 2, that are incident on the surface with angles of incidence 0.0° and θ_1, respectively. According to the law of refraction, the two rays refract into the second medium along paths 1' and

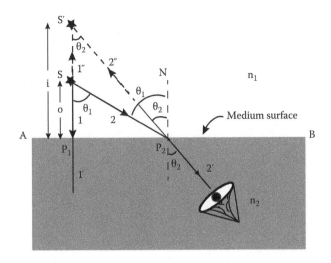

FIGURE 2.11 Refraction through a plane boundary, object in medium of lower refractive index.

2′ with angles of refraction, 0.0° and θ_2, respectively. The extensions of these rays along 1′ and 2′, respectively, intersect at the position labeled as S′ that represents the image of the source S as seen by an observer below and located within medium 2.

Looking into this diagram, we can make two observations: First, from geometry, the image distance i of S′ from the surface is larger than the object distance o; second, the image S′ is in the same side as the object, that is, both lie in the medium of incidence. Since this image is formed from the extension of the refracted rays, the image is virtual.

Also from the geometry and considering small angles only, $\tan \theta_1 = \sin \theta_1$; $\tan \theta_2 = \sin \theta_2$. Thus

$$\tan \theta_1 = \sin \theta_1 = \left(P_1\, P_2 / o \right),$$

where $P_1\, P_2$ is a segment of the boundary AB;

$$\tan \theta_2 = \sin \theta_2 = \left(P_1\, P_2 / i \right).$$

Dividing the two relations gives $(\sin \theta_1 / \sin \theta_2) = [(P_1\, P_2 / o)/(P_1\, P_2 / i)]$.
That is

$$\sin \theta_1 / \sin \theta_2 = \left(i/o \right). \tag{2.14}$$

But from the law of refraction for ray 2, $n_1 \sin \theta_1 = n_2 \sin \theta_2$.

That is

$$\sin\theta_1/\sin\theta_2 = (n_2/n_1).\qquad(2.15)$$

Combining Equations 2.14 and 2.15 gives

$$(i/o) = (n_2/n_1)$$

or

$$i = (n_2/n_1)o.\qquad(2.16)$$

Since

$$(n_2/n_1) > 1,\quad i > o.$$

That is, for a viewer in the optically more dense medium n_2, the image S' formed in the less dense medium n_1 seems farther from the surface than the object position at S.

One can reverse the scenario by taking a case of an object in medium n_2 and graphically determine the position of the image as seen by an observer in medium n_1. Following the argument we used for the previous case, a careful look at Figure 2.12 shows that i < o, which more specifically is

$$i = (n_1/n_2)o.\qquad(2.17)$$

Therefore, for a viewer in medium n_1, the image S' in medium n_2 seems closer to the surface than the object S.

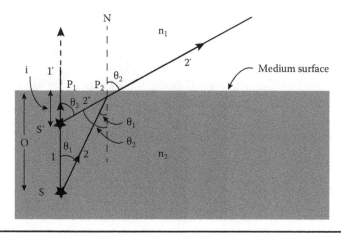

FIGURE 2.12 Refraction through plane boundary, object in medium of higher refractive index.

EXAMPLE 2.5

A bulb is attached to the side wall of a swimming pool, 2.0 m below the water surface. For an electrician, checking the bulb and standing on the edge of the pool, looking straight down into the water (n_w = 1.33), how far beneath the surface would he see the formed image?

Solution

From Equation 2.17,

$$i = (n_{air}/n_w)o.$$

Therefore

$$i = (1.00/1.33)(2.0 \text{ m}) = 1.5 \text{ m.}$$

2.5.2 REFRACTION AND TOTAL INTERNAL REFLECTION

From the property of reversibility of light paths, the law of refraction expressed in Equation 2.13 can be applied to any ray going in either direction from a medium of low index of refraction to one of a higher or from higher index of refraction to one of lower index and vice versa. Figure 2.13 shows two media of indices of refraction, n_1 and n_2, where $n_2 > n_1$. Following a set of rays, 1, 2, and 3, emitted in medium n_2 and incident on the boundary, they all get refracted along paths 1′, 2′, and 3′, respectively. For an angle of incidence, θ_2, for ray 1 in the medium n_2, the angle of refraction is θ_1 in medium n_1, and as $n_2 > n_1$, $\theta_2 < \theta_1$. Following the remaining rays, 2, 3,... emitted in medium n_2 with continuously increasing angles of incidence, the angles of refraction for the refracted rays 2′, 3′,... get bigger and bigger. At some critical value of the angle of incidence of a ray (labeled in the figure as ray 4) in medium n_2, the angle of refraction of the ray (labeled in the figure as 4′) in medium n_1 happens to be 90°.

Applying the law of refraction

$$n_2 \sin \theta_2 = n_1 \sin \theta_1.$$

For this ray, the above reduces to

$$n_2 \sin \theta_c = n_1 \sin 90°; \quad \sin 90° = 1.$$

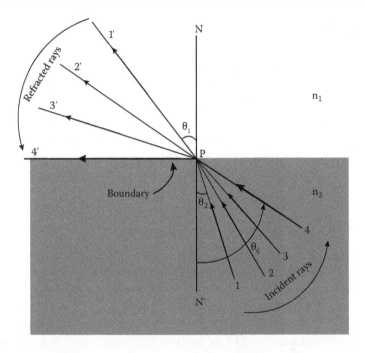

FIGURE 2.13 Phenomenon of total internal reflection that is observed when a light ray originating in medium n_2 with an angle larger than the critical angle gets reflected within the same medium.

Thus

$$n_2 \sin \theta_c = n_1.$$

That is

$$\sin \theta_c = n_1/n_2. \tag{2.18}$$

The angle θ_c is called the critical angle, and as noticed, it depends on the indices of refraction of the two media through which light is propagating.

Any ray in the second medium incident on the boundary with an angle larger than the critical angle gets reflected back into the same medium without a chance for it to emerge into the first medium. This phenomenon is called total internal reflection. The boundary on the side of the higher index of refraction medium acts as a mirror reflecting all light rays of angles of incidence $\theta_2 > \theta_c$.

COMMENTS

1. From the above expression of the critical angle, one can see that the total internal reflection may occur only when light is traveling from the medium with the higher index of refraction toward the medium with the lower index of refraction, not the other way around. This results from the fact that $\sin \theta_c$ cannot be larger than unity.

2. As θ_c depends only on the ratio (n_1/n_2), the smaller the ratio, the bigger the chance for total internal reflection to occur for a given angle of incidence. That is especially interesting for the case of diamond that has a high index of refraction, ~2.6.

2.5.3 SPHERICAL REFRACTING SURFACES

Light rays emanating from a source in one medium and directed toward a spherical surface serving as a boundary to another medium get refracted into that other medium. Since a point can be defined by the intersection of two lines, the refracted rays may or may not intersect. If they do at a position within the medium, then the intersection is formed from real refracted rays that have been transmitted by the surface. This should mean that such point is the position of a real image of the source point that emitted the rays. If the refracted rays do not intersect but their extensions into the first medium do, then the image would not be real; the intersection occurs on the side of the source. Such point is the presumed image, and hence the image is said to be virtual. This distinction between a real image and a virtual one applies to all kinds of surfaces, planar or spherical, concave or convex. In addition to this image labeling, additional ways in the next few sections are presented to describe all features of images: nature, position, and size.

2.5.3.1 Sign Convention

From the perspective of design and geometry, spherical refracting surfaces are as involved as spherical mirrors. The main difference between the two types resides in the function of each; mirrors reflect rays, so real images are formed on the same side of the object, while for refracting surfaces, real images form on the opposite side of the surface (Figure 2.14a and b).

The sign convention applied for refracting surfaces addresses object and image distances and radius of curvature as well as angles of incidence and refraction.

Here are the main elements of the sign convention:

1. In all cases, the object is positioned by convention on the left side of the refracting surface (object side), and since light rays emitted by the object and incident on the surface get refracted, a real image

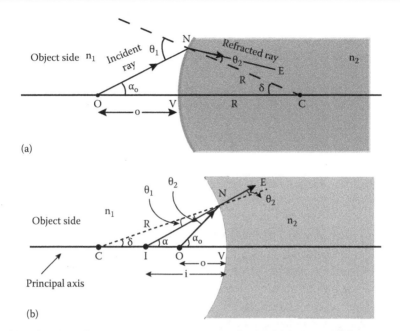

FIGURE 2.14   Refraction through (a) a convex surface and (b) a concave surface.

formed from the intersection of real rays is expected to be on the *right* side of the surface. Thus for the present convention, the space on the right side of the refracting surface is considered the *real-image side*, R-side.

2. The object distance o on the left side of the surface is positive. However, the image distance i is positive if it is on the right side and negative if it happens to be on the left side of the surface.

3. A positive sign for i means the image is real image, and a negative sign for i means it is virtual.

4. The sign of the radius of curvature R for a concave surface is negative, while for a convex surface it is positive. This is equivalent to saying that a spherical surface for which the center of curvature is to the right of the surface (R-side) has a positive radius of curvature, while a surface whose center of curvature is to the left of the surface has a negative radius of curvature. *The notion concave or convex is as seen by the incident light coming from the object on the left side of the surface.*

5. Heights measured above the optic axis are positive, while those below it are negative.

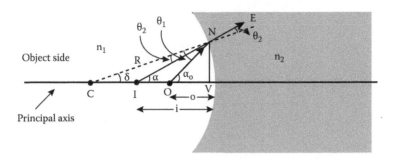

FIGURE 2.15   Image formation I of a point object O via refraction through a concave surface.

2.5.3.2 Equation of Refraction for a Spherical Surface

Consider a cylindrical glass rod of index of refraction n, its left end being concave with a radius of curvature R (Figure 2.15). The ray ON incident on the surface with an angle of incidence, θ_1, gets refracted into the glass along NE with an angle of refraction, θ_2. Here, we assume paraxial approximations or small angles in radians, not degrees.

Considering the triangle CNO, from geometry

$$\alpha_o = \theta_1 + \delta,$$

giving

$$\theta_1 = \alpha_o - \delta. \tag{2.19}$$

In the triangle CNI

$$\alpha = \theta_2 + \delta,$$

from which

$$\theta_2 = \alpha - \delta. \tag{2.20}$$

From the law of refraction

$$n_1 \sin \theta_1 = n_2 \sin \theta_2.$$

Using the small angle approximation the above equation becomes

$$n_1 \theta_1 = n_2 \theta_2. \tag{2.21}$$

Upon substitutions from Equations 2.19 and 2.20, Equation 2.21 becomes

$$n_1(\alpha_o - \delta) = n_2(\alpha - \delta).$$

Thus

$$n_1\alpha_o - n_1\delta = n_2\,\alpha - n_2\,\delta$$

or

$$n_1\alpha_o - n_2\,\alpha = (n_1 - n_2)\delta.$$

Again, as for small angles, the angle and its tangent are equal:

$$n_1(h/o) - n_2(h/i) = (n_1 - n_2)(h/R)$$

The above reduces to

$$n_1/o - n_2/i = (n_1 - n_2)/R. \qquad (2.22)$$

The above argument can be applied to a convex surface as well (Figure 2.16), where from geometry in triangle CNI,

$$\delta = \theta_2 + \alpha,$$

giving

$$\theta_2 = \delta - \alpha. \qquad (2.23)$$

Similarly for triangle CNO

$$\theta_1 = \alpha_o + \delta. \qquad (2.24)$$

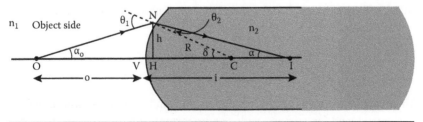

FIGURE 2.16 Image formation I of a point object, O, via refraction through a convex surface.

From the law of refraction

$$n_1 \sin \theta_1 = n_2 \sin \theta_2,$$

which reduces for small angle approximation to

$$n_1 \theta_1 = n_2 \theta_2.$$

Using Equations 2.23 and 2.24, the above becomes

$$n_1(\alpha_o + \delta) = n_2(\delta - \alpha).$$

Again, as all angles involved are small, each of them can be replaced by its sine or tangent. Therefore, the above equation reduces to,

$$n_1\left[(h/o) + (h/R)\right] = n_2\left[(h/R) - (h/i)\right].$$

Thus

$$\left(n_1/o\right) + \left(n_1/R\right) = \left[\left(n_2/R\right) - \left(n_2/i\right)\right]$$

$$\left[\left(n_1/o\right) + \left(n_2/i\right)\right] = \left(n_2 - n_1\right)/R. \tag{2.25}$$

As observed in Equations 2.22 and 2.25, these may be unified into one general form that would apply to both concave and convex surfaces as follows. Neither of the individual surface derivations considered the defined sign conventions on i, o, and R. They were intuitively all considered as numbers of positive magnitude based upon geometry alone. One observes that the sign convention injected into Equation 2.25 for a concave surface requires a negative sign for the image distance i and a negative sign for the radius R. This, in turn, makes Equation 2.25 identical to Equation 2.22. Thus, Equation 2.25 may be used as the general form for both concave and convex surfaces if we observe the proper sign conventions for the respective surface. Thus, Equation 2.22 for the concave refracting surface can also be rewritten as

$$\left[\left(n_1/o\right) + \left(n_2/i\right)\right] = \left(n_2 - n_1\right)/R. \tag{2.26}$$

The magnification of a refracting surface is given by an expression similar to that introduced for mirrors, with one important difference. In mirrors, the medium of incidence and medium of reflection are the same. However, in refraction, the medium of incidence (n_1) and medium of refraction (n_2) are obviously different. This is taken care of through the following generalized definition of magnification:

$$M = -\left(n_1/n_2 \right)\left(i/o \right). \tag{2.27}$$

2.5.3.3 Special Cases

1. For an object whose image is at infinity, Equation 2.26 holds:

$$\left[\left(n_1/o\right) + \left(n_2/i\right)\right] = \left(n_2 - n_1\right)/R.$$

After substitution for $i = \infty$, the above becomes

$$\left[\left(n_1/o\right) + 0\right)\right] = \left(n_2 - n_1\right)/R.$$

This reduces to

$$\frac{n_1}{o} = \frac{n_2 - n_1}{R},$$

giving

$$o = \frac{n_1}{\left(n_2 - n_1\right)}R. \tag{2.28}$$

Since R for a convex surface is positive, o is also positive, and according to the sign convention, that means it lies to the left of the surface. This defines what is called the primary focal point F for a convex surface. However, for a concave surface, R is negative; thus, o would also be negative, and that means the object O ought to be to the right of the surface. This can be understood in the following manner. Light rays that are incident on the concave surface, converging toward O, get refracted parallel to the principal axis. This defines the *primary focal point* F of the concave surface. Figure 2.17a and b demonstrates the two cases.

Thus, denoting o by f, the above equation becomes

$$f = \frac{n_1}{n_2 - n_1} R. \qquad (2.29)$$

2.   For an object at infinity, o = ∞, Equation 2.26 results in

$$\frac{n_2}{i} = \frac{n_2 - n_1}{R},$$

giving

$$i = \frac{n_2}{\left(n_2 - n_1\right)} R.$$

Using the argument, we just applied for the convex surface with R being positive, i would then be positive, that is, it is on its proper side. The image is on the right side. The distance i is known as the secondary focal length, and the point at which light rays converge defines the *secondary focal point* F′ for a convex surface. Figure 2.18 demonstrates the two cases.

However, for a concave surface, R is negative; the image then is on the left of the surface defining what is called the *secondary focal point* F′ for the concave surface.

Thus, denoting i by f′, the above equation becomes

$$f' = \frac{n_2}{n_2 - n_1} R. \qquad (2.30)$$

3.   A special case of interest is when an object in front of a spherical surface emits a light ray whose path or its extension passes through the mirror's center of curvature. As the angle of incidence of the ray is zero, the angle of refraction ought to be zero; the ray in both cases refracts back along the same path.

COMMENT

Special cases listed above represent a useful set of rules for locating images via ray tracing. In cases like this, meticulous drawing of dimensions to scale is critical (Figures 2.17–2.19).

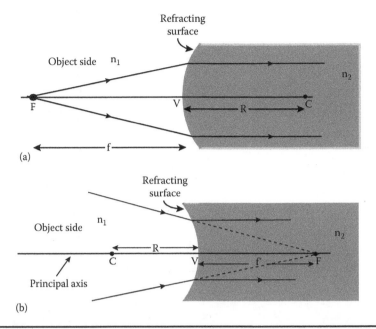

FIGURE 2.17   Special cases that are used to define the primary focal points F for (a) convex and (b) concave spherical surfaces. (Distances of R and F are not to scale.)

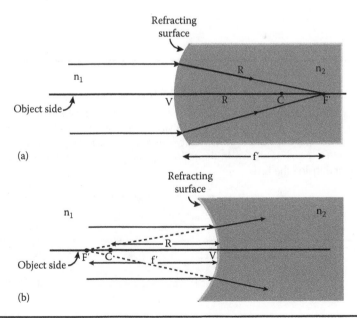

FIGURE 2.18   Special cases that are used to define the secondary focal points F′ for (a) convex and (b) concave spherical surfaces. (Distances of R and F′ are not to scale.)

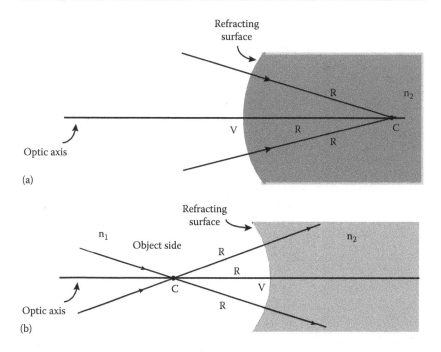

FIGURE 2.19   Rays incident on (a) a convex or (b) a concave surface passing through the center do not shift their paths.

EXAMPLE 2.6

Consider an object in the form of a 2.00 cm arrow placed vertically at a distance of 40.0 cm from the left side of a long glass rod (n = 1.40) whose two ends are symmetrically curved with a radius of curvature R = 10.0 cm, as shown in the following diagram. Find the image distance inside the rod.

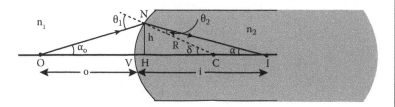

Solution

Substituting for the given information in

$$\left[(n_1/o) + (n_2/i)\right] = (n_2 - n_1)/R,$$

we then have

$$\left[(1.00/40.0 \text{ cm}) + (1.40/i)\right] = (1.40 - 1.00)/10.0 \text{ cm}.$$

Thus

$$(1.40/i) = 0.0400 - 0.0250.$$

That is

$$(1.40/i) = 0.0150,$$

giving

$$i = 93.0 \text{ cm}.$$

PROBLEMS

2.1   A boy is standing in front of a long plane mirror. Find (i) the position and (ii) nature of image for the following distances of his position from the mirror:
(a)   1.0 m
(b)   2.0 m

2.2   Locate the images formed by the two mirrors in the following diagram for the solid square-shaped object placed at point O between them.

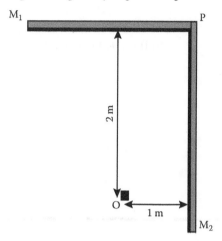

2.3   Consider the two mirrors set as shown in the following diagram. A light ray incident along path 1 on mirror M_1 with a 30° angle of incidence gets reflected along path 2, hitting mirror M_2. Find the angle of incidence that ray 2 makes with mirror M_2.

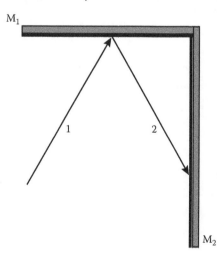

2.4   Consider a concave mirror of 8.0 cm focal length. Find the image of an object placed in front of the mirror at the following distances from it:
(a)   16.0 cm.
(b)   12.0 cm.
(c)   Determine the magnification of the image in both cases.

2.5   Redo Problem 2.4 for the same mirror to find the location of an object placed in front of the mirror if the image is to form at the following distances from the mirror:
(a)   16.0 cm
(b)   24.0 cm

2.6   For Problems 2.4 and 2.5
(a)   Determine the nature of the image for each of the stated cases.
(b)   Correlate between the given object distances and their image positions.

2.7   For the same mirror in Problem 2.4 (f = 8.0 cm), determine (a) the location and (b) nature of the image of an object placed 4.0 cm away from the mirror's front surface. Comment on your answer.

2.8   Consider a convex mirror of 8.0 cm focal length. Find the image of an object placed in front of the mirror at the following distances from it:
(a)   16.0 cm
(b)   12.0 cm
(c)   4.00 cm

2.9   Consider an object on the left of a concave mirror of a focal length f = 8.0 cm. A screen placed on the same side of the mirror at a distance i displayed an image twice as big as the object:
(a)   Determine the object and image distances from the lens.
(b)   What would be the position of the object so that the image is half the size of the object? Comment on what property of the paths of light rays is relevant to your answer.

2.10   Apply the equation of a refracting surface (Equation 2.26) to a plane boundary (R = ∞) to determine the image distance in terms of the object distance, showing that the result is consistent with the discussion of Section 2.5.1, Equation 2.16.

2.11   A light ray originating in air is incident on the top surface of water in a glass container with an angle of incidence of 42.0°, figure below. Ignoring the effect of the glass container, determine
(a)   The angle with which the light ray emerges from the lower surface of the water (n = 1.33). How does the direction of the light emerging from the block relate to the incident light?

(b)   If the water is 22 cm deep, find the distance d along the bottom of the container that marks the deviation of the emerging light ray from the point of incidence on the top surface.

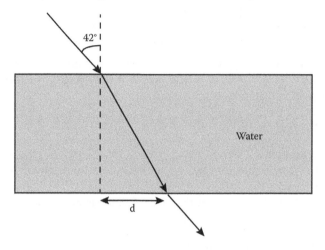

2.12   Consider a small light source near the bottom of a pond shielded by a spherical dome with an adjustable slit. The source emits a narrow beam of light that hits the water oil boundary with an angle of incidence θ_1. If there is a thick layer of oil (n_o = 1.40) covering the surface of water (n_w = 1.33), what is the minimum angle with which the light ray emerges from the source, so that no light escapes from the oil?

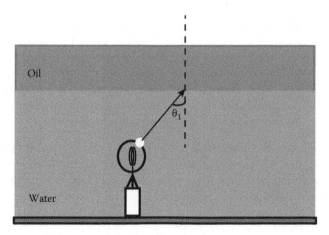

2.13   Consider a 60.0 cm long glass block (n = 1.40) whose left side is convex, having a 12.0 cm radius of curvature. An object in the form of

a 1.00 cm arrow is placed left of this side. Determine the (i) position and (ii) magnification of the image of the arrow in the following
cases:

(a)   o = 120 cm

(b)   o = 12.0 cm

Correlate your answers with the nature of these images.

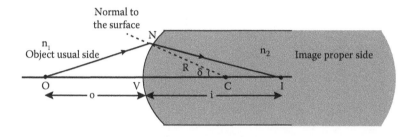

2.14   Consider a glass rod (n = 1.40) whose left side is convex, having 12.0
cm radius of curvature, while its right surface is flat. An object in the
form of a 1.00 cm arrow is placed left of the rod. Assuming that the
rod is 60.0 cm long, determine the (a) position and (b) magnification
of the final image of the arrow for an object distance o = 12 cm.

2.15   Redo the previous problem for a concave left surface of a rod that has
all remaining features described in the previous problem unchanged.
Determine the (i) position and (ii) magnification of the image of the
arrow formed by the left surface in the following cases:

(a)   o = 120 cm

(b)   o = 12.0 cm

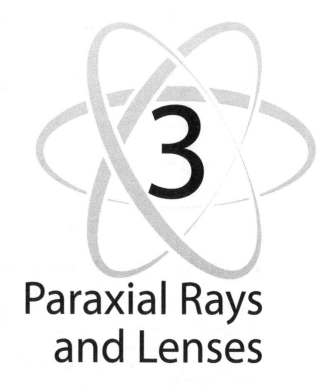

Paraxial Rays and Lenses

Without the rule of law, the life of man would be 'solitary, poor, nasty, brutish, and short.'

Thomas Hobbes (1588–1679)

3.1 INTRODUCTION

A lens consists of a finite width of a transparent medium bound by two surfaces, both spherically shaped, or one plane with the other spherical. They form images based on the property of refraction (see Chapter 2). Lenses are classified as thick or thin depending on the width of the refracting medium between the two surfaces relative to the diameter of the lens. For a thick lens, analyzing the optics of light rays incident on it requires that the laws of refraction be applied for the two surfaces independently. However, thin lenses as we will show are a special case of thick lenses, and formulation of a thin lens reduces to a simple one residing in treating it as one entity. For this reason, it is very insightful and deserves special attention making it worthy of the major part of this chapter.

Since this chapter is the second in the domain of geometric optics, except for limited cases that need clarity, light waves are dealt with as rays that follow straight line propagation. As will be shown, in any optics problem where lenses are used, light rays emitted by an object and incident on a lens get refracted, converging or diverging, forming an image.

3.2 THIN LENS: KINDS AND SHAPES

The surfaces of a lens may be spherical, both convex; spherical, both concave; spherical, one convex and the other concave; plano-convex; and plano-concave. Figure 3.1 shows several typical shapes of thin lenses.

3.3 IMAGE FORMATION IN THIN LENSES

In any thin lens, each surface acts as a refracting boundary. Thus whether the surface is convex, or concave, or plane, the laws of refraction apply to light rays incident on it. From the discussion of a single refracting surface (Section 2.4), we would expect the geometry of the two surfaces in a lens to be rather involved, necessitating the adoption of a sign convention that is inclusive of all quantities involved in image formation. This includes object and image distances from the lens, radii of curvature, focal points, and heights of object and image above or below the principal line (also called the optic axis). The following two subsections present a few common definitions, which in one way or another are involved in optics problems. We have also included a sign convention that pertains to many of these quantities.

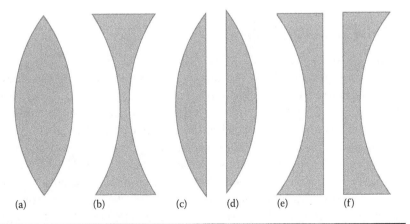

(a)   (b)   (c)   (d)   (e)   (f)

FIGURE 3.1   Selection of typical lens shapes: (a) double convex, (b) double concave, (c)–(d) plano-convex, and (e)–(f) plano-concave.

3.3.1 DEFINITIONS

Figure 3.2 illustrates essential quantities that relate to the lens properties. Alongside, we present definitions of the most prominent ones.

1. *Center of curvature*: For each surface of the lens, the center of curvature is the center of the sphere of which the lens surface is cut.
2. *Optic axis*: It is the line of symmetry that passes through the centers of curvature of the two (lens) surfaces.
3. The center of the lens is defined at the middle of its width halfway between V_1 and V_2.

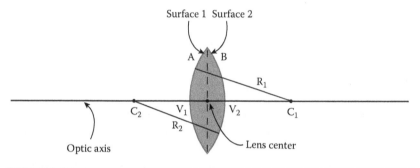

FIGURE 3.2 Lens basic features.

3.3.2 SIGN CONVENTION

Note that some of these rules are similar to those introduced in Chapter 2 for single surface refraction. The repetition is worthwhile due to the special role they play for lenses:

1. In all cases, the object is positioned to the left of the lens (object side), and light rays emitted by the object and incident on the lens get doubly refracted, due to the presence of two surfaces. Real refracted rays continue on the right side of the lens; hence, a real image, if formed from the intersection of real rays, is expected to be on the *right* side of the lens (image side). Thus, in the present convention, the space on the left side of the lens is to be the object side, while the space on the right side of the lens is considered the image proper, that is, the real image side.
2. The object distance o on the left side of the surface is positive. However, the image distance i is positive if it lies on the right side of the lens (real image side) and negative if it happens to form on the left side of the lens (object side); in such a case, the image is not on its proper side, and the image is virtual.

3. An image is real when refracted rays pass through it and the image can be captured on a screen. An image is virtual if rays coming from the object only appear to the observer on the image side to emanate from the image. Such an image cannot be projected onto a screen.

4. As for the surfaces of the lens, viewed from the left, the sign of the radius of curvature R for a concave surface is negative, while for a convex surface, it is positive. This is equivalent to saying that a spherical surface for which the center of curvature is to the right of the surface has a positive radius of curvature, while a surface whose center of curvature is to the left of the surface has a negative radius of curvature. For example, in Figure 3.3, R_1 is positive, and R_2 is negative.

5. Heights measured above the optic axis are positive, while those below it are negative.

3.4 LENS EQUATION

The starting point of detailing the optics of an object image formation in a thin lens is to resort to refraction of spherical surfaces that was presented in Section 2.5.3. In Figure 3.3, let the lens be of an index of refraction n_2, placed in a medium of index of refraction n_1, and let the lens spherical surfaces 1 and 2 be of radii of curvature R_1 and R_2, respectively.

For an object O at a distance o from the lens, that is, a distance o_1 from the first surface, Equation 2.26 becomes

$$\left[(n_1/o_1) + (n_2/i_1) \right] = (n_2 - n_1)/R_1. \tag{3.1}$$

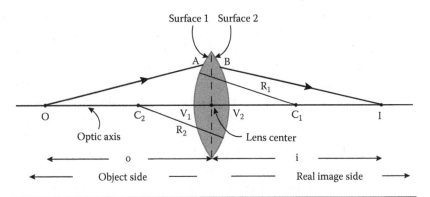

FIGURE 3.3   Display of lens features relevant to sign convention.

The image formed by the first surface will act as an object for the second surface, which forms a final image at some distance i_2 from it. Thus, applying Equation 2.26 again for the second surface, we have

$$\left[(n_2/o_2)+(n_1/i_2)\right]=(n_1-n_2)/R_2. \tag{3.2}$$

Adding Equations 3.1 and 3.2 gives

$$\left[(n_1/o_1)+(n_2/i_1)\right]+\left[(n_2/o_2)+(n_1/i_2)\right]=(n_2-n_1)/R_1+(n_1-n_2)/R_2.$$

As $o_2 = -i_1$, the above equation becomes

$$\left[(n_1/o_1)+(n_1/i_2)\right]=(n_2-n_1)/R_1+(n_1-n_2)/R_2.$$

Factoring out $(n_2 - n_1)$ in the right-hand side, the above equation reduces to

$$\left[(n_1/o_1)+(n_1/i_2)\right]=(n_2-n_1)[1/R_1-1/R_2].$$

If the thickness of the lens is assumed negligible (*thin lens approximation*) and referring the object and image distances to the center of the lens, we can replace o_1 by o and i_2 by i. Thus, the above equation becomes

$$\frac{1}{o}+\frac{1}{i}=\frac{(n_2-n_1)}{n_1}\left(\frac{1}{R_1}-\frac{1}{R_2}\right). \tag{3.3}$$

Introducing a new quantity called the focal length such that

$$\frac{1}{f}=\frac{(n_2-n_1)}{n_1}\left(\frac{1}{R_1}-\frac{1}{R_2}\right), \tag{3.4}$$

Equation 3.3 becomes

$$\frac{1}{o}+\frac{1}{i}=\frac{1}{f}. \tag{3.5}$$

Equation 3.5 is called the *lens equation* and is also known as the *Gaussian formula*. The distance f from the lens defines the position of the lens focal point F.

A close look into Equation 3.4 and following the sign convention described earlier, one can make specific observations about the sign of the focal length for any lens of known radii of curvature. For example, with regard to the following two lenses shown in Figure 3.4, the signs of the radii in lens (a) are positive for R_1 and negative for R_2. Thus for a lens in air, the right-hand side in Equation 3.4 is positive, and f then is positive. However, the signs of the radii in lens (b) are negative for R_1 and positive for R_2. Thus for such a lens, the right-hand side of the equation is negative, and f then is negative. That is why the lens (a) is labeled as a positive lens, while lens (b) is labeled as a negative lens (Figure 3.4).

As noted, Equations 3.3 and 3.4 were derived for a convex lens. It can be shown that these equations apply to all kinds of thin lenses. Although establishing these equations for other kinds of thin lenses where the two surfaces are concave would be a worthwhile exercise, one can assert more confidence about these equations through using them in a variety of examples. Of course, the sign conventions for o, i, and radii of surfaces all have to be strictly observed.

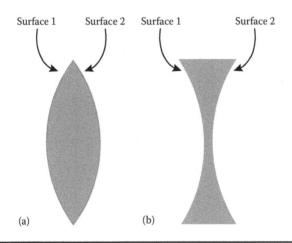

FIGURE 3.4   (a) Positive and (b) negative lens.

3.4.1 FOCAL POINTS AND FOCAL LENGTHS OF A THIN LENS DERIVED FROM THE LENS EQUATION

Consider a positive lens for the purpose of analyzing two special cases of object image formation:

1. For an object whose image is at infinity, Equation 3.5, after substitution for i = ∞, becomes

$$\frac{1}{o} + \frac{1}{\infty} = \frac{1}{f}.$$

That is,

$$\frac{1}{o} = \frac{1}{f},$$

giving

$$o = f. \tag{3.6}$$

Note that Equation 3.6 tells one that a point source object placed at $o = f$ on the left side of the lens will refract all rays from the source such that they are parallel to the optic axis after passing through the lens as shown in Figure 3.5. This point is called the *primary focal point*. In other words, *the primary focal point F is on the left of a positive lens* (Figure 3.5).

2. For an object far away from the lens, light rays incident on the lens are parallel. Substituting for $o = \infty$, Equation 3.5 becomes

$$\frac{1}{\infty} + \frac{1}{i} = \frac{1}{f}.$$

This reduces to

$$\frac{1}{i} = \frac{1}{f},$$

giving

$$i = f. \tag{3.7}$$

For a positive lens, f is positive; hence i is positive, implying that the image I is on the right side of the lens. That is, all rays incident on the

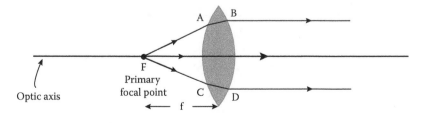

FIGURE 3.5 Primary focal point for a positive lens.

lens parallel to the optic axis (source at infinity) get refracted such that all converge at a point on the right side of the lens at a distance equal to the focal length. However, it is denoted by f′ so that it gets distinguished from f on the left side of the lens. This defines a point on the right side called the *secondary focal point,* F′. Figure 3.6 demonstrates case 2 above.

3. In a similar way, the focal lengths for a negative lens can be analyzed. The *primary focal point* F for a negative lens is on the right side of the lens, while the secondary focal point F′ is on the left side of the lens. These cases are demonstrated in Figure 3.7a and b, respectively.

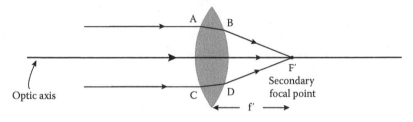

FIGURE 3.6   Secondary focal point for a positive lens.

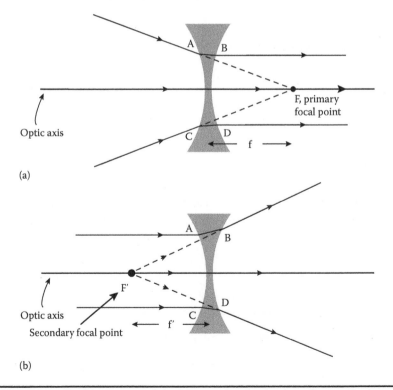

FIGURE 3.7   The (a) primary and the (b) secondary focal points of a negative lens.

3.4.1.1 Comments and Analyses

1. The primary and secondary focal points of a lens can equivalently be defined as follows: *The primary focal point* is a point on the optic axis where all rays emerging from it, passing through it, or traveling toward it get refracted by the lens along a direction parallel to the optic axis. However, *the secondary focal point* is a point on the optic axis where all rays incident along a direction parallel to the optic axis get refracted by the lens along directions that converge to intersect or appear to intersect at it.

2. The primary and secondary focal lengths for a thin lens are equal, that is, $f = f'$.

3. The center of the lens defined in Section 3.3.1 has the property that any light ray incident on the lens and passing through its center does not get refracted.

3.4.2 IMAGE MAGNIFICATION

The size of an image height being larger or smaller than the object height is called lateral magnification. Its value is given by the ratio between the image size to the object size. Thus, the magnification of an image formed by a thin lens for a given object depends on the object position with respect to the lens; for a two-dimensional object, the height and width are both affected. To derive a formal relation between the magnification and the object image distances, we consider Figure 3.8 where a thin positive lens is shown. The object is represented by an arrow at location O. Through the ray tracing method, we need to follow only two rays, 1 and 2, emitted from the tip of the object and traveling toward the lens. For convenience, these two rays are selected such that ray 1 is traveling parallel to the optic axis and ray 2 is passing through the primary focal point. The refracted rays 1' and 3' intersect at H' below the optic axis, producing the image at I.

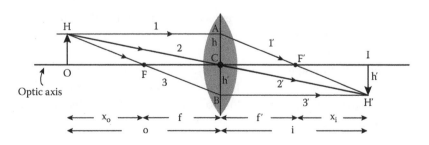

FIGURE 3.8   Primary and secondary focal points of a positive lens and center of the lens assist in ray tracing.

For calculating the magnification, consider the similar triangles H′ B C and HAC, where

$$\frac{H'B}{BC} = \frac{HA}{AC}.$$

That is,

$$\frac{i}{|h'|} = \frac{o}{|h|},$$

or

$$\frac{i}{o} = \frac{|h'|}{|h|}. \tag{3.8}$$

Since according to the sign convention h′ is negative, we have $\frac{|h'|}{|h|} = -\frac{h'}{h}$. As i and o in this object–image configuration are both positive (see sign convention), the magnification in this case is negative and can be expressed as

$$M \equiv \frac{h'}{h} = -\frac{|h'|}{|h|}, \tag{3.9}$$

or

$$M = -\frac{i}{o}. \tag{3.10}$$

A physical relevance can now be associated with the negative sign in the equation for M, which means the image is inverted. A positive value of M means it is upright.

Note that although this derivation is derived for a positive lens, the expression takes the same form for any image produced by any thin lens, provided that the sign convention for i and o is strictly observed and followed.

EXAMPLE 3.1

Consider an arrow of 2.00 cm height placed at a distance of 40.0 cm, on the left side of a thin positive lens (n = 1.40); the radii of curvature R_1 and R_2 of the lens are 10 cm and –10.0 cm, respectively. Determine

(a) The focal length of the lens
(b) The image position from the lens
(c) The magnification of the image

Solution

(a) From Equation 3.4

$$\frac{1}{f} = \frac{(n_2 - n_1)}{n_1}\left(\frac{1}{R_1} - \frac{1}{R_2}\right).$$

Substituting for the given information, the equation becomes

$$\frac{1}{f} = \frac{(1.40 - 1.00)}{1.00}\left(\frac{1}{10.0\text{ cm}} - \frac{1}{-10.0\text{ cm}}\right).$$

Thus,

$$\frac{1}{f} = 0.08\text{ cm}^{-1}.$$

Hence,

$$f = 12.6\text{ cm}.$$

(b) From Equation 3.5

$$\frac{1}{o} + \frac{1}{i} = \frac{1}{f}.$$

Substituting for o and f, we get

$$\frac{1}{40.0\text{ cm}} + \frac{1}{i} = \frac{1}{6.25\text{ cm}}.$$

That is,

$$\frac{1}{6.25\,\text{cm}} - \frac{1}{40.0\,\text{cm}} = \frac{1}{i}$$

$$\frac{1}{i} = 0.135\,\text{cm}^{-1}.$$

This makes

$$i = 7.41\,\text{cm}.$$

(c) The magnification of the image is given by the equation $M = -(i/o)$. Substituting for $i = 7.41$ cm and $o = 40.0$ cm, the above gives

$$M = -(7.40/40.0) = -0.185.$$

As described earlier, the negative sign for M means that the image is inverted. Its value being less than 1.00 means the image size is smaller than the object size.

3.5 RAY TRACING

Images can be located by ray tracing of any two rays coming from the object. The properties of the primary and secondary focal points (Section 3.4.1) prove to be helpful in choosing the rays for this purpose. In cases like this, careful drawing of light rays and properly scaling lens geometric properties are critical. All it takes to locate the point image of any point source is to follow two rays emanating from the object. For a positive lens (Figure 3.9), one of these rays, parallel to the optic axis, emerges passing through the secondary focal point. The second ray passing through the primary focal point emerges parallel to the optic axis. The intersection of the refracted rays would define the image position. Similar analysis can be done for a thin negative lens. The figures below show four cases, two for a positive lens and two for a negative lens, where ray tracing demonstrates the type of images: real or virtual, inverted or erect, and image size compared to the object size.

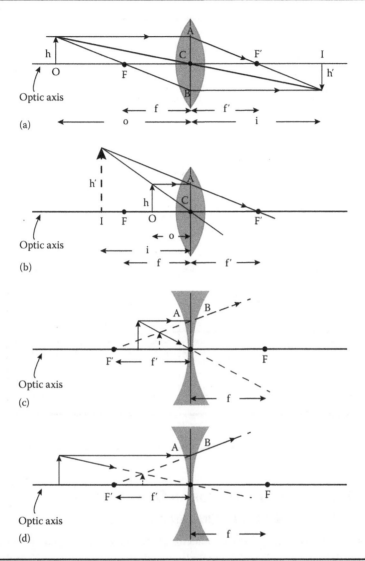

FIGURE 3.9 Image types: (a) real and inverted for o > f (positive lens) and (b)–(d) virtual and erect for o < f (positive lens) and all negative lenses.

COMMENTS

As observed in the diagrams depicted in Figure 3.9, one can make the following observations:

1. Images formed by a converging (positive) lens can be erect and virtual or inverted and real.
2. Images formed by a diverging, that is, negative, lens are always erect (upright) and virtual.

EXAMPLE 3.2

Let us reconsider the lens in the previous example. From the value of f = 6.25 cm as a given, draw all dimensions to scale and use the ray tracing method to locate the position of the image and determine its size.

Solution

Following the argument described in the ray tracing discussion, the two rays 1 and 2 coming from the tip of the arrow at O after refraction intersect, forming the tip of the image at I. As described, the image is inverted and real (see the following diagram).

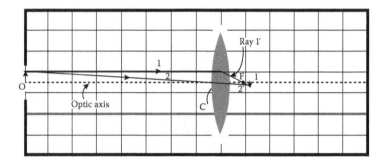

3.5.1 NEWTONIAN FORM FOR AN OBJECT–IMAGE RELATIONSHIP IN THIN LENSES

Consider a positive lens with an object O placed at a distance o on its left. Let us have three rays, 1, 2, and 3, emanating from the object and 1′, 2′, and 3′ as their refracted rays, respectively, forming the image at I (Figure 3.10). The two triangles IH′ F′ = ACF′ are similar. Thus

$$\frac{|h'|}{h} = \frac{IF'}{CF'}.$$

That is,

$$\frac{|h'|}{h} = \frac{IF'}{f'},$$

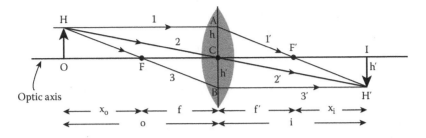

FIGURE 3.10   Object and image distances from the focal points that appear in the Newtonian formula.

or

$$\frac{|h'|}{h} = \frac{i-f'}{f'}.$$ (3.11)

Similarly, the two triangles OHF and BCF are similar. Thus

$$\frac{h}{|h'|} = \frac{OF}{CF}.$$

That is,

$$\frac{h}{|h'|} = \frac{o-f}{CF}$$

$$\frac{|h'|}{h} = \frac{f}{o-f}.$$ (3.12)

In Equations 3.11 and 3.12, as the two left-hand sides are the same, the two right-hand sides are equal, and since $f = f'$,

$$\frac{f}{o-f} = \frac{i-f}{f}$$

or, equivalently,

$$(I-f)(o-f) = f^2.$$

Letting

$$(i-f) = x_i, \quad (o-f) = x_o,$$

the above relation can be rewritten as

$$x_o\, x_i = f^2,\tag{3.13}$$

where x_o denotes the distance of the object from the *primary focal point* F and x_i denotes the distance of the image from the focal point F′, *the secondary focal point*. The above form is known as the *Newtonian equation*, or *Newtonian formula*, and shows that the product of the object and image distances from the focal points of a thin lens is equal to the square of the lens focal length.

EXAMPLE 3.3

Use the data described in Example 3.2 to verify Newton's formula, $x_o\, x_i = f$.

Solution

Since

$$f = 6.25, \quad x_o = o - f, \quad 40.0 - 6.25\ \text{cm} = 33.7\ \text{cm}.$$

Also,

$$i = 7.41 - 6.25 = 1.16\ \text{cm}.$$

The right-hand side of Newton's formula is equal to $x_o\, x_i = (33.75\ \text{cm})$ $(1.16\ \text{cm}) = 39.2\ \text{cm}^2$, and the left side of Newton's formula is equal to $f^2 = 39.1\ \text{cm}^2$. Rounding off the final answers to three significant figures, the two sides are almost exact.

3.5.2 POWER AND VERGENCE OF A THIN LENS

In the thin lens equation (3.5), we have the following quantities: $\frac{1}{o}$, $\frac{1}{i}$, and $\frac{1}{f}$. Any surface of radius r has its curvature defined as $1/r$; the value of $1/r$ is indicative of the sharpness of the surface curvature or lack thereof. For a flat surface, that is, a plane, the radius of curvature is infinite, and hence its curvature is zero. Thus, $1/o$ is a measure of the curvature of a spherical surface whose center is at O. This can be looked at from the perspective of light being a spherical wave emanating from the object at O and traveling toward the lens. The distance o from the object O to vertex point V is then the radius of the wave that has traveled toward the lens, touching its surface at V, and gets refracted by the lens first surface. Accordingly, $1/o$ is a measure of the curvature of the incident wave. A similar description can be laid out for $1/i$, where i then represents the radius of the refracted wave that has traveled a distance i before reaching the image at point I. The wave emitted by the object in this case, like all waves emitted by a source on the left side of the lens, is a diverging wave, while the refracted wave is a converging one.

The power P of a thin lens placed in air is directly related to its focal length through the following definition:

$$P = 1/f, \tag{3.14}$$

which, as suggested by Equation 3.4, for a thin lens of an index of refraction n_1 placed in a medium of an index of refraction n_2, becomes

$$P = \frac{(n_2 - n_1)}{n_1}\left(\frac{1}{R_1} - \frac{1}{R_2}\right). \tag{3.15}$$

For any thin lens, the sign of the power P follows the sign of its focal length f. Thus for a positive lens, P is positive, and for a negative lens, P is negative. For a focal length in meters, the units of P are in m^{-1}. In optics, this unit is called the diopter, symbolized by D.

Another notion that is well known and frequently used in optics is called vergence. The vergence, denoted by V, is a criterion for the sharpness of a curve or surface. Its value could be positive, demonstrating that the surface is expanding away from the center, that is, diverging, and could be negative, demonstrating that the surface is converging so that it collapses at the center. Use of this new notion can be made by rewriting the lens equation

$$\frac{1}{o} + \frac{1}{i} = \frac{1}{f}$$

as

$$V_o + V_i = P, \tag{3.16}$$

where

$$V_o = 1/o; \quad V_i = 1/i; \quad P = 1/f.$$

To bring Equation 3.16 in harmony with the sign convention introduced earlier, we see that for an object on the left side of the lens, the light wave emitted by the object at O is diverging from it and hence that V_o is positive. A parallel description applies to V_i. For an image on the right side of the lens, the refracted light wave received at the image site I is converging toward it, and hence V_i is positive.

EXAMPLE 3.4

A thin positive lens of index of refraction 1.5 has two identical convex surfaces of radii $R_1 = 14$ cm and $R_2 = -14$ cm. An object of 2.0 cm height is placed to the left of the lens 14.0 cm away from it. Determine

(a) The focal length and power of the lens
(b) The position of the image
(c) The nature, magnification, and size of the image

Solution

(a) From Equation 3.4

$$\frac{1}{f} = \frac{(n_2 - n_1)}{n_1}\left(\frac{1}{R_1} - \frac{1}{R_2}\right).$$

Substituting for the given information, the equation becomes

$$\frac{1}{f} = \frac{(1.40 - 1.00)}{1.00}\left(\frac{1}{14.0\,\text{cm}} - \frac{1}{-14.0\,\text{cm}}\right) = 0.0571\,\text{cm}^{-1}.$$

Thus, the power of the lens is

$$P = \frac{1}{f} = 0.0571\text{D},$$

and the focal length is

$$f = 17.5\,\text{cm}.$$

(b) The image position can be found from Equation 3.5:

$$\frac{1}{o} + \frac{1}{i} = \frac{1}{f}.$$

Substituting for o and f, we get

$$\frac{1}{14.0\,cm} + \frac{1}{i} = \frac{1}{17.5\,cm}$$

$$\frac{1}{i} = 0.0571 - 0.0714 = -0.0143.$$

This makes

$$i = -69.9\,cm.$$

(c) As the sign of the image distance i is negative, which means it lies to the left of the lens, that is, not in its proper space, all indicating that the image then is virtual.

The magnification is given by

$$M = -(i/o).$$

That is,

$$M = -(-69.9\,cm/14.0\,cm) = +4.93.$$

As described earlier, the positive sign of M means the image is upright. Its magnification is almost five times the object size, making it about 9.90 cm in height.

EXERCISE

Make a comparison between the object and image distances of Examples 3.2 and 3.4, and correlate these to the nature of the image obtained in each case. You may need to consider the lens focal lengths.

3.6 COMBINATION OF LENSES

In the discussion of thin lenses, an optical system consisting of two or more thin lenses is of interest. The case of two lenses placed next to each other such that the separation distance is very small compared to the focal length of each lens is described here in detail.

Consider two positive lenses shown below, which from left to right have focal lengths f_1 and f_2, respectively. Let an object O be placed in front of the first lens. Analysis of the image formation via the two-lens system can be detailed as follows (Figure 3.11). The first lens forms an image I_1 at a distance i_1 from the first or second lens. The lenses being thin, their thicknesses can be ignored. The image I_1 would be seen by the second lens as an object for which it forms an image; call it I_2. Thus, i_1 would represent for the second lens the object distance, o_2, with $o_2 = -i_1$, as this object lies to the right of this lens. The image I_2 is the final image formed by this double-lens system.

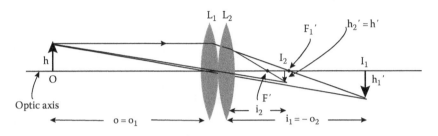

FIGURE 3.11   Image formation by a combination of lenses.

Applying the lens equation for the first lens, we have

$$\frac{1}{o_1} + \frac{1}{i_1} = \frac{1}{f_1}. \tag{3.17}$$

For the second lens, since $o_2 = -i_1$, the equation becomes

$$\frac{1}{-i_1} + \frac{1}{i_2} = \frac{1}{f_2}. \tag{3.18}$$

Adding the above two equations gives

$$\frac{1}{o_1} + \frac{1}{i_2} = \frac{1}{f_1} + \frac{1}{f_2}.$$

As o_1 is the object distance from the system and as i_2 is the final image formed by the system, better denoted as i, the above can be rewritten as

$$\frac{1}{o} + \frac{1}{i} = \frac{1}{f_1} + \frac{1}{f_2},$$

or

$$\frac{1}{o} + \frac{1}{i} = \frac{1}{f_{\text{eff}}},$$

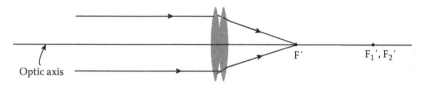

Optic axis

FIGURE 3.12   Two thin positive lenses.

where f_{eff} is the effective focal length of the double-lens system that is equal to

$$\frac{1}{f_{eff}} = \frac{1}{f_1} + \frac{1}{f_2}. \tag{3.19}$$

In a simplified diagram, considering a pair of parallel light rays incident on the first lens, the image would form at the focal point F of the double-lens system whose focal length is f_{eff} derived in Equation 3.19. For $f_1 = f_2 = f$, the effective focal length $f_{eff} = f/2$. This is illustrated in Figure 3.12.

EXAMPLE 3.5

A thin lens L_1 of 5.00 cm focal length is placed in front of another lens L_2 of 8.00 cm focal length. The two lenses are barely touching each other. A small object is positioned at a distance of 10.0 cm on the left of the first lens. Using the lens equation for each

(a) Determine the final image position.
(b) Find the magnification of the final image.
(c) Calculate the effective focal length of the combination.
(d) Treat the combination as one effective lens with effective focal length as found in part (c), and determine the final position of the image. How does this answer compare to that found in part (a)?

Solution

(a) As for the first lens, from Equation 3.5

$$\frac{1}{o} + \frac{1}{i} = \frac{1}{f}.$$

Substituting for o and f, we get

$$\frac{1}{10.0\,cm} + \frac{1}{i} = \frac{1}{5.00\,cm},$$

or

$$\frac{1}{i} = 0.100 \, \text{cm}^{-1}.$$

This makes i = 10.0 cm.

As the sign of i is positive, this image is to the right of the lens, that is, to the right of the second lens as well. This image will act as an object for the second lens.

(b) As for the second lens, o = −10.0 cm, f = 8.00 cm. Thus, the lens equation

$$\frac{1}{o} + \frac{1}{i} = \frac{1}{f},$$

after substitution, becomes

$$\frac{1}{-10.0 \, \text{cm}} + \frac{1}{i} = \frac{1}{8.00 \, \text{cm}}.$$

This gives

$$i = 4.44 \, \text{cm.}$$

Again, as the sign of i is positive, this image is to the right of the second lens. This image will act as an object for the second lens. The magnification of the image by the first lens is

$$M_1 = -\left(10.0 \, \text{cm}/10.0 \, \text{cm}\right) = -1.00, \text{inverted.}$$

The magnification of the image by the second lens is

$$M_2 = -\left(i_2/o_2\right) = -\left(i_2/o_2\right) = -\left(4.44/-10.0\right).$$

That is,

$$M_2 = 0.444, \text{upright with respect to the second lens.}$$

The total magnification of the two lenses combined is

$$M = M_1 M_2 = \left(-1.00\right)\left(0.444\right) = \left(-0.444\right).$$

Therefore, the final image relative to the object is inverted.

(c) From Equation 3.19

$$\frac{1}{f_{eff}} = \left(\frac{1}{f_1} + \frac{1}{f_2}\right).$$

That is,

$$= \left(\frac{1}{5.00\,cm} + \frac{1}{8.00\,cm}\right)$$

$$= (0.200 + 0.125)$$

or

$$\frac{1}{f_{eff}} = (0.325\,cm^{-1}).$$

Thus

$$f_{eff} = 3.08\,cm.$$

(d) For the combination treated as one equivalent thin lens of $f = 3.08$, the lens equation

$$\frac{1}{o} + \frac{1}{i} = \frac{1}{f}$$

becomes

$$\frac{1}{10.0\,cm} + \frac{1}{i} = \frac{1}{3.08\,cm}.$$

That gives $i = 4.45$ cm; image is real and to the right of the combination.

This answer is in full agreement with the answer found in part (a). A slight difference is due to rounding of some of the answers to three significant figures.

Analysis

The magnification obtained by this combination treated as one effective lens is given by

$$M = -(i/o); \quad o = 10.0\,cm, \quad and \quad i = 4.45\,cm.$$

That is,

$$M = -(4.45\,cm/10.0\,cm) = -0.445, \text{ inverted.}$$

PROBLEMS

3.1   Consider the lens shapes shown in the following figure where the radii of curvature of all curved surfaces of the lenses are equal. Use the sign convention to determine whether the focal length for each lens is positive or negative.

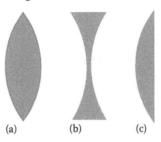

(a)   (b)   (c)

3.2   Consider the lenses described below in (a), (b), and (c). If each is made of glass of an index of refraction n = 1.5, follow the sign convention to determine the focal length of each:
  (a)   A thin lens made from glass whose index of refraction n = 1.5 and its radii of curvature R_1 = 7.0 cm and R_2 = −7.0 cm
  (b)   A thin lens made from glass whose index of refraction n = 1.5 and having its radii of curvature R_1 = −7.0 cm and R_2 = 7.0 cm
  (c)   A plano-convex lens whose back surface has a radius of curvature R_2 = −7.0 cm. Is the lens positive or negative?

3.3   Consider a thin crescent-shaped lens (n = 1.4) that consists of two concave surfaces of radii of curvature R_1 = −24.0 cm and R_2 = −12.0 cm, respectively, as shown in the following diagram. If an object is placed 12.0 cm in front of the lens, find
  (a)   The focal length of the lens
  (b)   The position and nature of the image

3.4   Redo the previous problem for a mirror image of the described crescent, so that the crescent in this problem is facing in the opposite direction. That is, radii of curvature R_1 = 24.0 cm and R_2 = 12.0 cm,

respectively, as shown in the following diagram. If an object is placed 12.0 cm in front of the lens, determine
(a)   The lens focal length
(b)   The position and nature of the image

3.5   Consider an object located at a distance of 8.00 cm on the left side of a positive thin lens (n = 1.40) whose two radii of curvature R_1 and R_2 are 12.0 cm and −12.0 cm, respectively.
(a)   Use the lens equation to calculate the image distance from the lens
(b)   Determine the magnification of the image

3.6   Using the data given in the previous problem
(a)   Draw a sketch with properly scaled dimensions that enables you to determine the image distance position from the lens.
(b)   Compare your answer to the value obtained in the previous problem.

3.7   Consider an object located at a distance of 8.00 cm on the left side of a negative thin lens (n = 1.40) whose two radii of curvature R_1 and R_2 are −12.0 cm and 12 cm, respectively, as shown in the following diagram.
(a)   Use the lens equation to calculate the image distance from the lens.
(b)   Determine the magnification of the image.

3.8   Consider an object placed on the left at a distance o from a positive lens of a focal length f = 8.00 cm. A screen placed on the other

side of the lens at a distance i showed an image twice as big as the object.
(a)   Determine the object and image distances from the lens.
(b)   What would be the position of the object, so that the image is half the size of the object?

3.9   Consider an object placed vertically at a distance o from the left side of a positive thin lens (n = 1.40) of a focal length f = 20.0 cm. Determine the position of the image in the following cases:
(a)   o = 120 cm
(b)   o = 24.0 cm
(c)   o = 12.0 cm
(d)   o = 8.00 cm

3.10   Look into the answers obtained for each of the cases in the previous problem, and comment on the obtained images, being real or virtual, inverted or upright, and their magnification.

3.11   Redo the previous problem for a negative thin lens (n = 1.40) with an object placed vertically at a distance o from the left side of the lens of a focal length f = −20.0 cm. Determine the position of the image in the following cases:
(a)   o = 120 cm
(b)   o = 24 cm
(c)   o = 12 cm
(d)   o = 8.0 cm

3.12   Look into the answers obtained for each of the cases in the previous problem, and comment on the obtained images, being real or virtual, inverted or upright.

3.13   Consider two thin lenses of focal lengths 4.0 cm and −4.0 cm, respectively. The lenses are placed next to each other; the positive lens is on the left. A 2.0 cm high object is placed vertically 20.0 cm away, to the left of the positive lens. Determine
(a)   The image position from the lens
(b)   The magnification and size of the image

3.14   Two identical thin positive lenses of 14.0 cm focal length each are placed next to each other side by side. For an object placed 21.0 cm to the left of the first lens
(a)   Determine the focal length of the combination.
(b)   Determine the position of the final image.

3.15   Two identical thin positive lenses, each of 8.00 cm focal length, are placed on the same axis with no separation between them. For an object placed 12.0 cm to the left of the first lens,

(a)   Determine the effective focal length of the combination.
(b)   Determine the position of the final image by two separate ways, applying the lens equation for each.
(c)   Redo part (b) using the effective lens formula.

3.16   Two identical thin positive lenses, each of focal length 8.00 cm, are placed on the same axis with a separation of 10.0 cm between them. For an object placed 12.0 cm to the left of the first lens, determine the position of the final image.

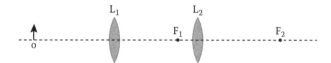

3.17   Find the focal length f_{123} equivalent to three thin positive lenses of focal lengths f_1, f_2, and f_3 placed next to each other. (Hint: Do that in two steps, one through which you find the focal length f_{12} equivalent to f_1 and f_2, and in the second step you find f_{123} equivalent to f_1, f_2, and f_3.)

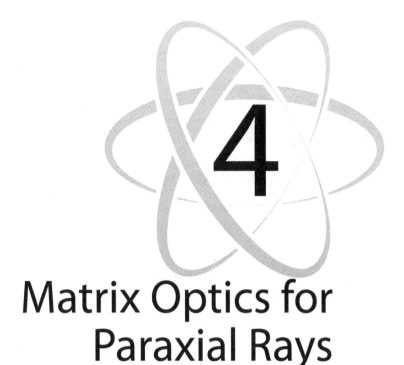

Matrix Optics for Paraxial Rays

The mind at birth is like a blank slate, waiting to be written on by the world of experience.

John Locke (1632–1704)

4.1 INTRODUCTION

In this chapter, a new method of handling paraxial geometric optics is introduced. All phenomena of reflection from planar or spherical surfaces and refraction through single or multiple surfaces can be treated via matrix optics. The treatment extends to the propagation of light through thick and thin lenses as well. This is possible because the equations that describe any of these processes are linear. As will be demonstrated in the sections that follow, a ray, at every point, is represented by two parameters, height and inclination, of a 2 × 1 column matrix. The matrices for the input and output rays are called input and output vectors. An optical operator that modifies a ray is represented by a 2 × 2 matrix. The optical operator can be as simple as a mere translation of light between two points, or as complex as multiple surface refractions which can be represented by a 2 × 2 matrix. Several cases are treated: (a) simple translation, (b) single-surface refraction, (c) thick and thin lens two-surface refraction, and (d) compound systems comprised of two or more lenses and translation. The treatments will demonstrate the relative ease in calculations and convenience that matrix optics offers over the use of lens equations for complex optical systems.

4.2 TRANSLATION MATRIX

Consider a light ray along the line AB that makes an angle α with the horizontal reference line taken as the x axis. Let A and B be points of heights h_o and h, respectively, above the x axis, and let the distance between their projections on the x axis be d. The ray angle at h_o would be labeled with the same subscript for easy association (Figure 4.1).

From the geometry,

$$h = h_o + \Delta h,$$

or

$$h = h_o + d \tan \alpha_0; \quad \Delta h = d \tan \alpha_0.$$

For small angles, $\tan \alpha_0 = \alpha_0$, and the above equation becomes

$$h = h_o + d \times \alpha_o \qquad (4.1)$$

$$\alpha = \alpha_o. \qquad (4.2)$$

Equations 4.1 and 4.2 can be rewritten in a form where each of the ray parameters (h, α) on the left side is expressed in terms of the ray parameters (h_o, α_o) on the right side as follows:

$$h = h_o + d \times \alpha_o;$$

$$\alpha = 0 \cdot h_0 + \alpha_o$$

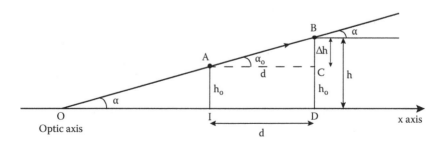

FIGURE 4.1 Translation of a light ray through one medium.

The above expressed in a matrix form becomes

$$\begin{bmatrix} h \\ \alpha \end{bmatrix} = \begin{bmatrix} 1 & d \\ 0 & 1 \end{bmatrix} \cdot \begin{bmatrix} h_o \\ \alpha_o \end{bmatrix}. \tag{4.3}$$

In condensed symbolic terms, the above is written as

$$O = M_T\, I, \tag{4.4}$$

where

$$O = \begin{bmatrix} h \\ \alpha \end{bmatrix} \tag{4.5}$$

is the light ray output and

$$I = \begin{bmatrix} h_o \\ \alpha_o \end{bmatrix} \tag{4.6}$$

is the light ray input. The matrix

$$M_T = \begin{bmatrix} 1 & d \\ 0 & 1 \end{bmatrix} \tag{4.7}$$

is called the translation matrix.

COMMENT

Notice that the determinant of the matrix M_T denoted by $|M_T|$ is unity, and that relates to the energy conservation principle. Thus, in a translation operation of a light ray from one position to another through one medium, air, for example, the determinant

$$|M_T| = 1. \tag{4.8}$$

EXAMPLE 4.1

Referring to the following diagram, consider a light ray originating at A and propagating toward B along a 22 cm path that makes an angle $\alpha_o = 18°$ (=0.10 π rads) with the optic axis. For $h_o = 7.0$ cm

(a) What is the translation matrix?
(b) Determine the output parameters of the light ray at B and its column matrix.

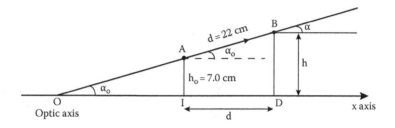

Solution

(a) The horizontal distance d between A and B is

$$d = AB \cos\alpha_0 = (22.0\,\text{cm})\cos(0.10\times\pi) = 22.0(0.95) = 21\,\text{cm}.$$

Therefore, the translation matrix becomes

$$M_T = \begin{bmatrix} 1 & d \\ 0 & 1 \end{bmatrix} = \begin{bmatrix} 1 & 21 \\ 0 & 1 \end{bmatrix}.$$

(b) This operation in a compact form is

$$\begin{bmatrix} h \\ \alpha \end{bmatrix} = \begin{bmatrix} 1 & d \\ 0 & 1 \end{bmatrix} \cdot \begin{bmatrix} h_o \\ \alpha_o \end{bmatrix}$$

or

$$\begin{bmatrix} h \\ \alpha \end{bmatrix} = \begin{bmatrix} 1 & 21 \\ 0 & 1 \end{bmatrix} \cdot \begin{bmatrix} 7.00 \\ 0.10\,\pi \end{bmatrix} = \begin{bmatrix} 14 \\ 0.31 \end{bmatrix}.$$

That is,

$$\begin{bmatrix} h \\ \alpha \end{bmatrix} = \begin{bmatrix} 14 \\ 0.31 \end{bmatrix}.$$

Thus,

$$h = 14\,\text{cm}, \quad \text{and} \quad \alpha = 0.31\,\text{rad}.$$

4.3 REFRACTION MATRIX

The phenomenon of refraction occurs as light crosses a boundary between two different media, regardless of whether the boundary is flat or curved. Refraction at the boundary of spherically curved surfaces is of special interest. However, before analyzing refraction through spherical surfaces, a much simpler case, refraction by a flat, that is, plane boundary, will be considered. Upon extending the treatment of matrix optics to spherical surfaces, one will see that a flat boundary is just a special case.

4.3.1 REFRACTION MATRIX OF A PLANE BOUNDARY

Consider Figure 4.2 where a ray ON, originating in a uniform medium of index of refraction n_1, is incident on a flat boundary at point N, refracts, and proceeds through another uniform medium of index of refraction n_2 $(>n_1)$ to the right of the boundary.

From the geometry, the input parameters of the light ray ON are h_o and α_o, and its parameters at N′ on the other side of the boundary are h and α_r; at N, the heights h_o and h from the x axis are equal. That is,

$$h = h_o. \tag{4.9}$$

However, from Snell's law,

$$n_1 \sin \alpha_o = n_2 \sin \alpha_r,$$

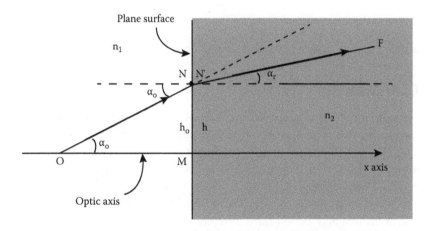

FIGURE 4.2 Refraction of light at the boundary of two different media.

which, for small angles, becomes

$$n_1 \, \alpha_o = n_2 \, \alpha_r.$$

That is,

$$\alpha_r = \left(\frac{n_1}{n_2} \right) \alpha_o. \tag{4.10}$$

Equations 4.9 and 4.10 expressed in a matrix form become

$$\begin{bmatrix} h \\ \alpha_r \end{bmatrix} = \begin{bmatrix} 1 & 0 \\ 0 & \dfrac{n_1}{n_2} \end{bmatrix} \cdot \begin{bmatrix} h_o \\ \alpha_o \end{bmatrix}. \tag{4.11}$$

The above operation expressed in symbols can be written as

$$\mathbf{O} = \mathbf{M_{RP}} \, \mathbf{I}, \tag{4.12}$$

where **I** and **O** are the input and output ray column matrices, respectively, and

$$\mathbf{M_{RP}} = \begin{bmatrix} 1 & 0 \\ 0 & \dfrac{n_1}{n_2} \end{bmatrix} \tag{4.13}$$

is the refraction matrix for a plane boundary denoted by the subscript RP.

COMMENT

Notice that the determinant of the matrix $\mathbf{M_{RP}}$ is (n_1/n_2). As will be seen, this is a general rule for any single optical operation resulting from refraction of light by a boundary separating two different media of indices n_1 and n_2; n_1 is for the medium of incidence and n_2 is for the medium of emergence. As will be seen later, this applies to thin lenses as well, where the thickness of the lens is assumed to be zero, reducing the two individual operations of refraction at the first and second surfaces to one effective operation. Note that the determinant $\mathbf{M_{RP}} \neq 1$.

EXAMPLE 4.2

Referring to following diagram, consider a light ray AB, originating at A, 7.0 cm above the optic axis, and incident on a glass block of a flat front CD. If the index of refraction of the block is 1.40, the ray path AB is 22 cm long, and its angle of inclination equals 18° (0.10 π rads), determine

(a) The refraction matrix of the glass block.
(b) The overall matrix of this optical system. (Hint: The overall matrix involving two operations is the product of the two matrices representing the two operations.)
(c) The output parameters of the ray at point B′ inside the incident boundary of the block.

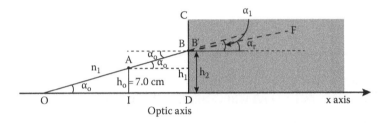

Solution

(a) From Example 4.1, we found that the translation matrix is

$$M_T = \begin{bmatrix} 1 & 21 \\ 0 & 1 \end{bmatrix}.$$

(b) However, the refraction matrix of the block surface is

$$M_{RP} = \begin{bmatrix} 1 & 0 \\ 0 & \dfrac{n_1}{n_2} \end{bmatrix}. \tag{4.14}$$

The system matrix M_S is

$$M_S = M_{RP} \cdot M_T = \begin{bmatrix} 1 & 0 \\ 0 & \dfrac{n_1}{n_2} \end{bmatrix} \begin{bmatrix} 1 & 21 \\ 0 & 1 \end{bmatrix}.$$

That is,

$$M_S = \begin{bmatrix} 1 & 21 \\ 0 & \dfrac{n_1}{n_2} \end{bmatrix}.$$

(c) Given the input ray vector $I = \begin{bmatrix} h_o \\ \alpha_o \end{bmatrix} = \begin{bmatrix} 7.0 \\ 0.10\,\pi \end{bmatrix}$, the system combined operation becomes

$$O = M_{RP} \cdot I = \begin{bmatrix} 1 & 0 \\ 0 & \dfrac{n_1}{n_2} \end{bmatrix} \begin{bmatrix} 7.0 \\ 0.10\,\pi \end{bmatrix}.$$

That is,

$$\begin{bmatrix} h \\ \alpha \end{bmatrix} = \begin{bmatrix} 7.0 \\ 0.071\,\pi \end{bmatrix}.$$

Thus, the ray output parameters are

$$h_2 = 7.0\,\text{cm} \quad \text{and} \quad \alpha_2 = 0.07\,\pi\,\text{rad}.$$

4.3.2 Refraction Matrix: Spherical Surfaces

Consider a cylindrical glass rod of index of refraction n_2, its left end being convex with a radius of curvature R (Figure 4.3). For the ray ON incident on the surface, the refracted ray within the glass is NF. The input ray parameters are h_o and α_o, while the output ray parameters are h and α.

From the geometry,

$$\alpha = \theta_r - \delta,$$

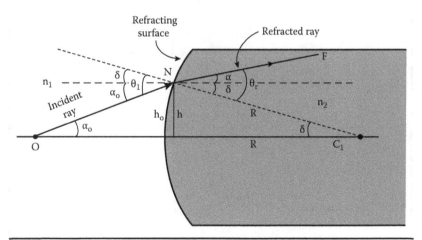

FIGURE 4.3 Illustration of refraction of light at a spherical surface.

or for paraxial rays,

$$\alpha = \theta_r - (h/R). \tag{4.15}$$

From Snell's law, for small angle approximation,

$$n_1 \theta_1 = n_2 \theta_r.$$

Therefore,

$$\theta_r = (n_1/n_2)\theta_1.$$

Thus, (4.15) becomes

$$\alpha = (n_1/n_2)\theta_1 - (h/R). \tag{4.16}$$

Also from geometry,

$$\theta_1 = \alpha_o + \delta$$
$$= \alpha_o + (h/R). \tag{4.17}$$

Substituting from Equation 4.17 in Equation 4.16 gives

$$\alpha = (n_1/n_2)\left[\alpha_o + (h_o/R)\right] - (h_o/R).$$

As for the heights of the input and output light rays, they are h_o and h, respectively, and they are equal. That is,

$$h = h_o.$$

With conveniently adding zero and some rearrangement, the above two equations may be rewritten as

$$h = h_o + 0(\alpha_o)$$

$$\alpha = \left[\frac{1}{R}\left(\frac{n_1}{n_2} - 1\right)\right]h_o + \left(\frac{n_1}{n_2}\right)\alpha_o.$$

In matrix form, the above pair of equations becomes

$$\begin{bmatrix} h \\ \alpha \end{bmatrix} = \begin{bmatrix} 1 & 0 \\ \dfrac{1}{R}\left(\dfrac{n_1}{n_2} - 1\right) & \dfrac{n_1}{n_2} \end{bmatrix} \cdot \begin{bmatrix} h_o \\ \alpha_o \end{bmatrix}, \tag{4.18}$$

which in a compact form becomes

$$O = M_{RS}\, I, \tag{4.19}$$

where

$$
\mathbf{M_{RS}} =
\begin{bmatrix}
1 & 0 \\
\dfrac{1}{R}\left(\dfrac{n_1}{n_2}-1\right) & \dfrac{n_1}{n_2}
\end{bmatrix}
\tag{4.20}
$$

is the matrix for a spherical refracting surface denoted by **RS** as labeled in the subscript for avoiding confusion with refraction from a plane **RP**; **O** and **I** are the ray output and input column matrices, respectively, as defined earlier in Equations 4.5 and 4.6.

COMMENTS

1. The above matrix applies to all refracting surfaces, convex, concave, or plane ($R = \infty$). As for the sign of R, we follow the same convention defined earlier for spherical surfaces (Chapter 3).
2. As noted, the determinant of matrix in (4.20) is not unity, simply because it describes an optical system with two different media on its two sides.

EXAMPLE 4.3

In the following diagram, light ray ON is incident on a cylindrical glass block of index of refraction 1.45. The block's left end is convex with a 28.0 cm radius of curvature. If ray ON makes an angle of 12.0° (=0.209 rad) with the optic axis and is incident on the surface at a height of 7.00 cm from the optic axis, determine

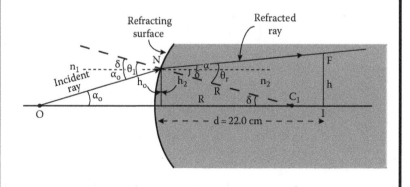

(a) The refraction matrix of the spherical surface
(b) The height of the refracted ray at a point F inside the block 22.0 cm away from the surface

Solution

In this problem, we have two operations: the first is refraction at point N, and the second is the translation of 22.0 cm through the body of the block. Matrix representations of these operations are M_{RS} and M_T, respectively, and are of the forms

$$
M_{RS} = \begin{bmatrix} 1 & 0 \\ \dfrac{1}{R}\left(\dfrac{n_1}{n_2}-1\right) & \dfrac{n_1}{n_2} \end{bmatrix} = \begin{bmatrix} 1 & 0 \\ \dfrac{1}{28.0}\left(\dfrac{1}{1.45}-1\right) & \dfrac{1}{1.45} \end{bmatrix} = \begin{bmatrix} 1 & 0 \\ -0.0111 & 0.610 \end{bmatrix},
$$

$$
M_T = \begin{bmatrix} 1 & d \\ 0 & 1 \end{bmatrix} = \begin{bmatrix} 1 & 0.220 \\ 0 & 1 \end{bmatrix}.
$$

The product of the matrices represents the matrix of the system. That is,

$$
M_S = M_T M_{RS} = \begin{bmatrix} 1 & 0.220 \\ 0 & 1 \end{bmatrix}\begin{bmatrix} 1 & 0 \\ -0.0111 & 0.610 \end{bmatrix} = \begin{bmatrix} 0.998 & 0.134 \\ -0.0111 & 0.610 \end{bmatrix}.
$$

$$(4.21)$$

Thus, the overall operation in matrix form becomes

$$
\begin{bmatrix} h \\ \alpha_r \end{bmatrix} = \begin{bmatrix} 0.998 & 0.134 \\ -0.0111 & 0.610 \end{bmatrix}\begin{bmatrix} h_o \\ \alpha_o \end{bmatrix}.
$$

Substituting for $h_o = 7.00$ cm and $\theta_o = 12.0°$ ($=0.209$ rad), the above becomes

$$
\begin{bmatrix} h \\ \alpha_r \end{bmatrix} = \begin{bmatrix} 0.998 & 0.134 \\ -0.0111 & 0.610 \end{bmatrix}\begin{bmatrix} 7.00 \\ 0.209 \end{bmatrix} = \begin{bmatrix} 7.01 \\ 0.0498 \end{bmatrix}.
$$

$$(4.22)$$

Thus, the ray output parameters at F are

$$h = 7.05\,\text{cm} \quad \text{and} \quad \alpha = 0.0498\,\text{rad}.$$

4.4 MULTI OPERATION MATRIX: LENSES

Consider an optical system of a thick spherical lens of thickness L and index of refraction n_2 (Figure 4.4). For a light ray originating at O, propagating along path ON_1, it encounters three operations before it emerges at N_2: first, refraction at N_1 by the first surface; second, translation through the system along path $N_1 N_2$; and third, refraction by the second surface at N_2 beyond which light travels along path N_2 P.

The three operations presented in compact matrix forms are as follows:

i. A refraction operation $\mathbf{M_{RS\text{-}1}}$ by the first surface, which for an input ray $\mathbf{I_o}$ produces the first output O_1, that is,

$$\mathbf{O_1} = \mathbf{M_{RS-1}}\,\mathbf{I_o},$$

where

$$\mathbf{O_1} = \begin{bmatrix} h \\ \alpha \end{bmatrix}; \quad \mathbf{M_{RS\text{-}1}} = \begin{bmatrix} 1 & 0 \\ \dfrac{1}{R_1}\left(\dfrac{n_1}{n_2}-1\right) & \dfrac{n_1}{n_2} \end{bmatrix}; \quad \mathbf{I_o} = \begin{bmatrix} h_o \\ \alpha_o \end{bmatrix}. \qquad (4.23)$$

ii. A translation operation $\mathbf{M_T}$ through the body of the object (see Equation 4.4) in which $\mathbf{O_1}$ becomes the input $\mathbf{I_1}$ for the translation

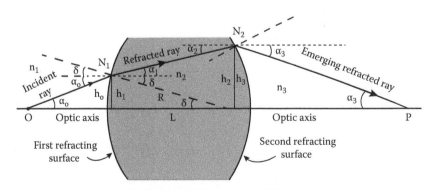

FIGURE 4.4   Illustration of a compound process involving successive translation and refraction operations due to the two surfaces having radii R_1 and R_2.

operation, resulting in an output light ray O_2 at point N_2 just prior to emerging from surface 2:

$$O_2 = M_T\, O_1,$$

where

$$O_2 = \begin{bmatrix} h_2 \\ \alpha_2 \end{bmatrix}; \quad M_T = \begin{bmatrix} 1 & L \\ 0 & 1 \end{bmatrix}; \quad I_1 = O_1 = \begin{bmatrix} h \\ \alpha \end{bmatrix}. \tag{4.24}$$

iii. A refraction by the second surface $M_{RS\text{-}2}$, resulting in the third output O_3, through the operation

$$O_3 = M_{RS\text{-}2}O_2$$

where

$$O_3 = \begin{bmatrix} h_3 \\ \alpha_3 \end{bmatrix}; \quad M_{RS-2} = \begin{bmatrix} 1 & 0 \\ \dfrac{1}{R_2}\left(\dfrac{n_2}{n_3}-1\right) & \dfrac{n_2}{n_3} \end{bmatrix}; \quad I_2 = O_2 = \begin{bmatrix} h_2 \\ \alpha_2 \end{bmatrix}. \tag{4.25}$$

The overall operation is

$$O_3 = M_{RS\text{-}2}\, O_2 = M_{RS\text{-}2}\, M_T\, O_1 = M_{RS\text{-}2}\, M_T\, M_{RS\text{-}1}\, I_0.$$

That is,

$$O_3 = M_{RS\text{-}2}\, M_T\, M_{RS\text{-}1}\, I_0. \tag{4.26}$$

The product of the system three composite matrices $M_{RS\text{-}2}$, M_T, $M_{RS\text{-}1}$ results in one matrix that represents the system effective matrix. Denoting that by M_S, one can write

$$M_S = M_{RS\text{-}2}\, M_T\, M_{RS\text{-}1}. \tag{4.27}$$

That is,

$$
\mathbf{M_S} =
\begin{bmatrix}
1 & 0 \\
\dfrac{1}{R_2}\left(\dfrac{n_2}{n_3}-1\right) & \dfrac{n_2}{n_3}
\end{bmatrix}
\begin{bmatrix}
1 & L \\
0 & 1
\end{bmatrix}
\begin{bmatrix}
1 & 0 \\
\dfrac{1}{R_1}\left(\dfrac{n_1}{n_2}-1\right) & \dfrac{n_1}{n_2}
\end{bmatrix}. \qquad (4.28)
$$

Upon writing the explicit form of the product on the right-hand side of Equation 4.28 and realizing that $n_1 = n_3$, the above becomes the matrix of a thick lens of width L, that is,

$$
\mathbf{M_L} =
\begin{bmatrix}
1 & 0 \\
\dfrac{1}{R_2}\left(\dfrac{n_2}{n_1}-1\right) & \dfrac{n_2}{n_1}
\end{bmatrix}
\begin{bmatrix}
1 & L \\
0 & 1
\end{bmatrix}
\begin{bmatrix}
1 & 0 \\
\dfrac{1}{R_1}\left(\dfrac{n_1}{n_2}-1\right) & \dfrac{n_1}{n_2}
\end{bmatrix}
$$

$$
=
\begin{bmatrix}
1 & 0 \\
\dfrac{1}{R_2}\left(\dfrac{n_2-n_1}{n_1}\right) & \dfrac{n_2}{n_1}
\end{bmatrix}
\begin{bmatrix}
1 & L \\
0 & 1
\end{bmatrix}
\begin{bmatrix}
1 & 0 \\
\dfrac{1}{R_1}\left(\dfrac{n_1-n_2}{n_2}\right) & \dfrac{n_1}{n_2}
\end{bmatrix}. \qquad (4.29)
$$

Dealing with $\mathbf{M_L}$ in the form shown on the right-hand side of Equation 4.29 is less cumbersome than executing the product of the three matrices to find the general form of the combination. However, for the sake of satisfying any nagging sense of curiosity, the product was executed and is reduced to a single matrix of the form

$$
\mathbf{M_L} =
\begin{bmatrix}
1+\dfrac{L}{R_1}\left(\dfrac{n_1-n_2}{n_2}\right) & \dfrac{L\,n_1}{n_2} \\[2mm]
\dfrac{1}{R_2}\left(\dfrac{n_2-n_1}{n_1}\right)\left[1+\dfrac{L}{R_1}\left(\dfrac{n_1-n_2}{n_2}\right)\right]+\dfrac{n_2}{n_1}\dfrac{1}{R_1}\left(\dfrac{n_1-n_2}{n_2}\right) & \dfrac{1}{R_2}\left(\dfrac{n_2-n_1}{n_1}\right)\dfrac{L\,n_1}{n_2}+1
\end{bmatrix}.
$$
$$ (4.30) $$

EXERCISES

1. Verify the above calculation by carrying out the multiplication in Equation 4.29.
2. Check the matrix that would come out of (4.30) if L = 0.

4.5 THIN LENS: REVISITED

The matrix of a compound optical system, comprised of two refracting surfaces and a finite length of the refracting medium, as expressed in Equation 4.29 represents the matrix form of a *thick lens*. For a lens whose width along the optic axis is much smaller than its height and much smaller than other dimensions like the lens radii of curvature and distances of object and image from the lens, L is considered negligible and can be equated to zero. This would reduce the above to a *thin lens* matrix form. Denoting the thin lens matrix by M_ℓ, the matrix becomes

$$M_\ell = \begin{bmatrix} 1 & 0 \\ \dfrac{1}{R_2}\left(\dfrac{n_2}{n_3}-1\right) & \dfrac{n_2}{n_3} \end{bmatrix} \begin{bmatrix} 1 & 0 \\ 0 & 1 \end{bmatrix} \begin{bmatrix} 1 & 0 \\ \dfrac{1}{R_1}\left(\dfrac{n_1}{n_2}-1\right) & \dfrac{n_1}{n_2} \end{bmatrix}. \qquad (4.31)$$

With some algebra of matrix multiplication, the *thin lens equation* in (4.31) takes the following form:

$$M_\ell = \begin{bmatrix} 1 & 0 \\ \left(\dfrac{n_2-n_1}{n_1}\right)\left(\dfrac{1}{R_2}-\dfrac{1}{R_1}\right) & 1 \end{bmatrix}. \qquad (4.32)$$

COMMENTS

1. Notice that the determinant of M_L for a thick lens is rather involved and is not unity.
2. The determinant of M_S for a *thin lens* is unity. Can you draw an analogy between this property and that of a translation operation that the determinant in each is unity?

The element $\left(\dfrac{n_2-n_1}{n_1}\right)\left(\dfrac{1}{R_1}-\dfrac{1}{R_2}\right)$ defines the inverse of the focal length of a thin lens of index of refraction n_2 (see Chapter 3), that is,

$$\frac{1}{f} = \left(\frac{n_2-n_1}{n_1}\right)\left(\frac{1}{R_1}-\frac{1}{R_2}\right).$$

Thus, Equation 4.32 becomes

$$M_\ell = \begin{bmatrix} 1 & 0 \\ -\dfrac{1}{f} & 1 \end{bmatrix}. \qquad (4.33)$$

EXAMPLE 4.4

Consider two thin lenses, each of +20.0 mm focal length and an index of refraction of 1.4 separated by a distance d = 16 mm. Use the matrix method to determine

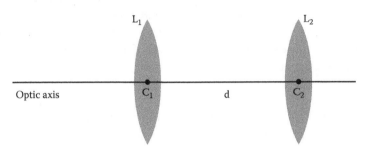

(a) The composite matrix M_S of the combination of this system
(b) The composite matrix M of the combination of the two lenses in the case when d = 0 cm
(c) The focal length of the combination in parts (a) and (b)

Solution

(a) The composite matrix represents the system's matrix $M_S = M_2 M_T M_1$. Thus

$$M_S = \begin{pmatrix} 1 & 0 \\ -\dfrac{1}{20.0} & 1 \end{pmatrix} \begin{pmatrix} 1 & 16 \\ 0 & 1 \end{pmatrix} \begin{pmatrix} 1 & 0 \\ -\dfrac{1}{20.0} & 1 \end{pmatrix} = \begin{pmatrix} 1 & 0 \\ -\dfrac{1}{20.0} & 1 \end{pmatrix} \begin{pmatrix} 1-\dfrac{16}{20.0} & 16 \\ -\dfrac{1}{20.0} & 1 \end{pmatrix}$$

$$= \begin{pmatrix} 1 & 0 \\ -\dfrac{1}{20.0} & 1 \end{pmatrix} \begin{pmatrix} \dfrac{4}{20.0} & 16 \\ -\dfrac{1}{20.0} & 1 \end{pmatrix} = \begin{pmatrix} \dfrac{4}{20.0} & 16 \\ -\dfrac{4}{(20.0)^2} - \dfrac{1}{20.0} & -\dfrac{16.0}{20.0}+1 \end{pmatrix}$$

$$= \begin{pmatrix} \dfrac{1}{5} & 16 \\ -0.060 & \dfrac{1}{5} \end{pmatrix} = \begin{pmatrix} \dfrac{1}{5} & 16 \\ -0.060 & \dfrac{1}{5} \end{pmatrix}.$$

(b) When d = 0, the above system becomes

$$M_S = \begin{pmatrix} 1 & 0 \\ -\dfrac{1}{20.0} & 1 \end{pmatrix} \begin{pmatrix} 1 & 0 \\ -\dfrac{1}{20.0} & 1 \end{pmatrix} = \begin{pmatrix} 1 & 0 \\ -\dfrac{1}{10.0} & 1 \end{pmatrix}.$$

(c) For part (a), $-\dfrac{1}{f} = -0.060$, then $f = 17$ mm, and for part (b),

$-\dfrac{1}{f} = -\dfrac{1}{10.0}$, giving $f = 10$ mm.

In conclusion, Equations 4.30 and 4.31 demonstrate that the matrix description of an optical system, comprised of multiple components, always reduces to the form

$$O = M_s\, I$$

or

$$\begin{bmatrix} h \\ \alpha \end{bmatrix} = \begin{bmatrix} A & B \\ C & D \end{bmatrix}\begin{bmatrix} h_o \\ \alpha_o \end{bmatrix},\qquad (4.34)$$

where, as defined earlier, h and α are the elements of the output ray vector and h_o and α_o the elements of the input ray vector.

4.5.1 EFFECTIVE MATRIX OF AN OPTICAL SYSTEM: FURTHER ANALYSIS

For any case where a compound system is involved and a multimatrix operation is executed, a 2 × 2 matrix of the form

$$M_S = \begin{bmatrix} A & B \\ C & D \end{bmatrix}\qquad (4.35)$$

is obtained. The above is called the ABCD matrix. Each of the elements A, B, C, and D has a special significance, especially in cases when among all a particular element is zero.

An explicit correlation between the parameters of the input light ray and those of the output light ray can be obtained from the matrix Equation 4.34 if we extract from it the following two linear equations:

$$h = A\,h_o + B\,\alpha_o\qquad (4.36)$$

$$\alpha = C\,h_o + D\,\alpha_o\qquad (4.37)$$

Of course, the values of A, B, C, and D are those that would determine the output, especially if the output happened to be where the image is located. In this regard, the following cases may be analyzed:

i. $A = 0$. This means that $h = B\alpha_o$. That is, h is independent of the input height h_o. It is only a function of the initial input angle α_o such that all input rays having the same angle of inclination α_o (incident rays are parallel to each other) will have their output light rays of the same height, that is, all meet at one point, which is a point on the focal plane of the system.

ii.   B = 0. This means that h = Ah$_o$. That is, h is independent of the input angle α_o. It is only a function of the initial input height h$_o$ such that all input rays having the same h$_o$ (input light rays emanating from the object position) will exit with output light rays having the same height h, that is, all output rays are converging to intersect at a point, the point of the image. This is no more than expressing the correspondence between the object and image positions and in a way demonstrates the reversibility property of light that light rays coming from the object and meeting at the image position can be traced backward from the image position toward the object position.

iii.   C = 0. This means that α = D α_o. That is, α is independent of the input height h$_o$. It is only a function of the initial input angle α_o such that all input rays having the same angle α_o (input light rays are parallel) will have their output light rays of the same angle α (output light rays are parallel). Thus, input light rays look as if they were coming from a distant object, and output light rays look as if they are to converge at infinity. More interesting fact is that D becomes

$$D = \alpha/\alpha_o, \tag{4.38}$$

called the system's angular magnification. This relation is of particular relevance to telescopes where the angular magnification is of special significance.

iv.   Finally, D = 0. This means that α = C h$_o$. That is, α is independent of the input angle α_o. It is only a function of the initial input height h$_o$ such that all input rays having the same input height h$_o$ (input light rays emanating from a focal point) will have their output light rays of the same angle α (output light rays are parallel). This property offers a definition of the *focal plane* of a lens, and this case, contrasted against that described by (i), demonstrates the reciprocity behavior of light.

EXAMPLE 4.5

Consider a glass rod (n = 1.50) in the following diagram with its two ends polished and convex as viewed from outside, becoming a thick lens, 22.0 cm long. The two surfaces' radii of curvature are 28.0 cm and 14.0 cm, respectively. For an object placed 20.0 cm to the left of the rod's first surface, find the image position relative to the rod's second surface on the right.

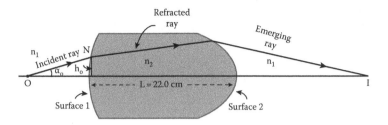

Solution

There will be three individual matrices involved in determining the system matrix for finding the image information. These are given below.

a. Translation matrix through the space between the object and the rod's left surface; denoting this as M_{T1}, then

$$M_{T1} = \begin{bmatrix} 1 & 20.0 \\ 0 & 1 \end{bmatrix}.$$

b. The thick matrix of length 22.0 cm and radii 28.0 cm and –14.0 cm, respectively,

$$\begin{bmatrix} 1 + \dfrac{L}{R_1}\left(\dfrac{n_1-n_2}{n_2}\right) & \dfrac{L\,n_1}{n_2} \\ \dfrac{1}{R_2}\left(\dfrac{n_2-n_1}{n_1}\right)\left[1+\dfrac{L}{R_1}\left(\dfrac{n_1-n_2}{n_2}\right)\right]+\dfrac{n_2}{n_1}\dfrac{1}{R_1}\left(\dfrac{n_1-n_2}{n_2}\right) & \dfrac{1}{R_2}\left(\dfrac{n_2-n_1}{n_1}\right)\dfrac{L\,n_1}{n_2}+1 \end{bmatrix}$$

$$M_\ell = \begin{bmatrix} 1 + \dfrac{22}{28.0}\left(\dfrac{1-1.50}{1.50}\right) & \dfrac{22(1.00)}{1.50} \\ \dfrac{1}{-14.0}\left(\dfrac{1.50-1.00}{1.00}\right)\left[1+\dfrac{22}{28.0}\left(\dfrac{1.00-1.50}{1.50}\right)\right]+\dfrac{1.50}{1.00}\dfrac{1}{R_1}\left(\dfrac{1.00-1.50}{1.50}\right) & \dfrac{1}{-14.0}\left(\dfrac{1.50-1.00}{1.0}\right)\dfrac{22(1.00)}{1.50}+1 \end{bmatrix}.$$

c. Translation matrix through the distance d between the rod's right surface and the image; denoting this as M_{T2}, we get

$$M_{T2} = \begin{bmatrix} 1 & d \\ 0 & 1 \end{bmatrix}.$$

The matrix of the optical system would then become

$$M_S = M_{T2}\, M_L\, M_{T1}$$

$$M_S = \begin{bmatrix} 1 & d \\ 0 & 1 \end{bmatrix} \begin{bmatrix} 1+\dfrac{22}{28}\left(\dfrac{1-1.5}{1.5}\right) \\ \dfrac{1}{-14}\left(\dfrac{1.5-1.0}{1.0}\right)\left[1+\dfrac{22}{28}\left(\dfrac{1.0-1.5}{1.5}\right)\right]+\dfrac{1.5}{1.0}\dfrac{1}{(28)}\left(\dfrac{1.0-1.5}{1.5}\right) \end{bmatrix}$$

$$\left. \begin{array}{c} \dfrac{22(1.0)}{1.5} \\[2mm] \dfrac{1}{-14}\left(\dfrac{1.5-1.0}{1.0}\right)\dfrac{22(1.0)}{1.5}+1 \end{array} \right] \begin{bmatrix} 1 & 20.0 \\ 0 & 1 \end{bmatrix}$$

$$= \begin{bmatrix} 1 & d \\ 0 & 1 \end{bmatrix} \begin{bmatrix} 1.26 & 14.6 \\ -0.0271 & 0.476 \end{bmatrix} \begin{bmatrix} 1 & 20.0 \\ 0 & 1 \end{bmatrix} = \begin{bmatrix} 1 & d \\ 0 & 1 \end{bmatrix} \begin{bmatrix} 1.26 & 39.8 \\ -0.027 & -0.064 \end{bmatrix}$$

$$= \begin{bmatrix} 1.26-0.027d & 39.8-0.064d \\ -0.027 & -0.064 \end{bmatrix} \equiv \begin{bmatrix} A & B \\ C & D \end{bmatrix}.$$

Using the property (ii) by letting the element B = 0 gives

$$39.8 - 0.064d = 0,$$

giving

$$d = 622\,cm.$$

PROBLEMS

4.1   Show that the plane boundary matrix (Equation 4.13) is a special case of the spherical refracting surface matrix given in Equation 4.20 by

$$\mathbf{M_{RS}} = \begin{bmatrix} 1 & 0 \\ \dfrac{1}{R}\left(\dfrac{n_1}{n_2}-1\right) & \dfrac{n_1}{n_2} \end{bmatrix}.$$

4.2   The following diagram shows a side view of a glass block with a large rectangular cross section and an index of refraction $n_2 = 1.50$. The rod, surrounded by air, is 25.0 cm long. The ray OB makes an angle of $\alpha_0 = 12.0°$ (= 0.209 rad) with the optic axis and is incident on the left flat surface at a height $h_1 = 5.00$ cm from the optic axis, labeled in the figure as the x axis. Determine
   (a)   The refraction matrix of the rod consisting of just its left flat surface and a translation along its length
   (b)   The height of the refracted ray at a point F where the ray is about to emerge from the rod

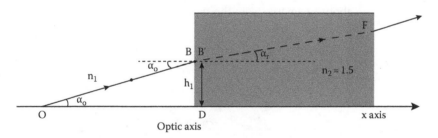

4.3   Redo the previous problem with the beam emerging from the right flat surface of the block.

4.4   In the following diagram is a light ray ON incident on a glass block of index of refraction n = 1.5. The block's left end is convex with 28.0 cm radius of curvature, while its other end is flat. The rod is 22.0 cm long. The ray ON makes an angle of $\alpha_0 = 12.0°$ (= 0.209 rad) with the optic axis and is incident on the surface at a height of $h_0 = 7.00$ cm from the optic axis. Determine
   (a)   The refraction matrix of the block consisting of its left spherical surface, translation along its length, and its plane boundary on the right end
   (b)   The height of the refracted ray at point F where the ray is about to emerge from the block

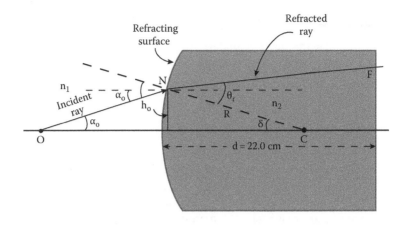

4.5 Consider a block of two ends reversed compared to the case above, the left end flat, while the right end looks concave from outside as shown below. Keeping all entry quantities as stated in the previous problem, determine
   (a) The refraction matrix of the block consisting of its left plane surface, translation along its length, and its spherical convex (as it looks from inside) boundary on the right end
   (b) The height of the refracted ray at a point F where the ray is about to emerge from the spherical end.

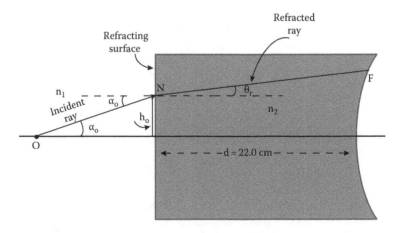

4.6 Let the block in problem 4.4 be doubly convex of radii, $|R_1|=|R_2|=$ 28.0 cm, making it a thick lens 22.0 cm wide. With all input details about height and angle of ray ON_1 unchanged, the ray emerging at point N_2, determine
   (a) The refraction matrix of this thick lens
   (b) The height of the refracted ray at the point N_2

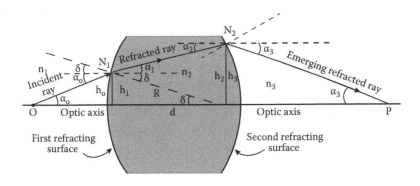

First refracting surface

Second refracting surface

4.7   Consider a thin lens ($d = 0$) with its two surfaces being identical in curvature $|R_1|=|R_2|= 28.0$ cm. Pick a ray, ON, emanating from O and making an angle of α_o with the optic axis. If the ray emerges with an angle α below the horizontal, intersecting the optic axis,

(a)   Prove that an analysis of this process via the matrix optics results in an expression of the focal length of a thin lens given by

$$\frac{1}{o}+\frac{1}{i}=\frac{1}{f},$$

where

$$\frac{1}{f}=\left(\frac{n_2-n_1}{n_1}\right)\left(\frac{1}{R_1}-\frac{1}{R_2}\right).$$

(b)   Find the focal length of the lens.

(c)   Find the image distance from the lens for an object placed at O, 10.0 cm from the lens.

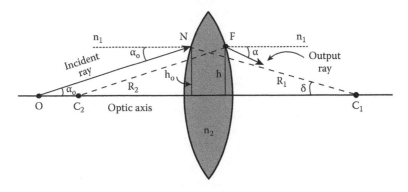

4.8   Letting each of the lenses described below be made of glass of index of refraction n = 1.5, construct the lens matrix for

(a)   A positive thin lens having its radii of curvature R_1 = 7.0 cm and R_2 = −7.0 cm.

(b)   A negative thin lens having its radii of curvature R_1 = −7.0 cm and R_2 = 7.0 cm.

(c)   A plano-convex lens whose back surface has a radius of curvature R_2 = 7.0 cm. Is the lens positive or negative?

(d)   A plano-convex lens whose front surface has a radius of curvature R_1 = − 7.0 cm. Is the lens positive or negative?

4.9   Two identical thin lenses, each of 14.0 cm focal lengths, are placed next to each other side by side. Their substance is of an index of refraction n = 1.40. Employing matrix optics

(a)   Determine the focal length of the combination.

(b)   Check if your answer satisfies the relation $\dfrac{1}{f} = \dfrac{1}{f_1} + \dfrac{1}{f_2}$.

(c)   Comment on the relevance of your result for part (b).

4.10   Consider two thin lenses, L_1 and L_2, of focal lengths 14.0 cm and −7.00 cm, respectively. The two are placed next to each other side by side. Their substance is of an index of refraction n = 1.40. Employing matrix optics, determine the focal length of the combination and check if your answer satisfies the relation $\dfrac{1}{f} = \dfrac{1}{f_1} + \dfrac{1}{f_2}$.

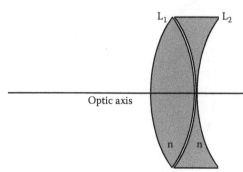

4.11 Redo the previous problem for L_1 and L_2 of focal lengths −7.00 cm and 14.00 cm, respectively, and determine the focal length of the combination.

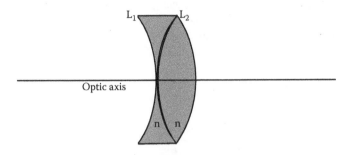

4.12 Consider a thin crescent-shaped lens that consists of two concave surfaces, of radii of curvature $R_1 = -16$ cm and $R_2 = -12$ cm, respectively. For this lens (n = 1.4), determine
(a)   Its matrix
(b)   Its focal length

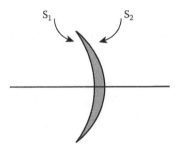

4.13 Redo the previous problem, reversing the two concave surfaces to become of radii of curvature $R_1 = 12.0$ cm and $R_2 = 16.0$ cm, respectively. For this lens (n = 1.4), determine
(a)   Its matrix
(b)   Its focal length

4.14   Consider two thin lenses, one plano-concave and another is plano convex; the two are positioned as shown below. Letting the value of the radius of the spherical surface in each be 28.00 cm and the index of refraction be 1.40 in both, using matrix optics
(a)   Determine the matrix form of the combination.
(b)   Determine the focal length of the combination.
(c)   Comment on the relevance of your result for part (b).

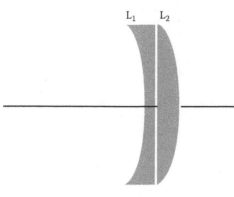

4.15   Verify that the multiplication of the three operations in Equation 4.28 as expressed below

$$
\mathbf{M_S} = \begin{bmatrix} 1 & 0 \\ \dfrac{1}{R_2}\left(\dfrac{n_2}{n_3}-1\right) & \dfrac{n_2}{n_3} \end{bmatrix} \begin{bmatrix} 1 & L \\ 0 & 1 \end{bmatrix} \begin{bmatrix} 1 & 0 \\ \dfrac{1}{R_1}\left(\dfrac{n_1}{n_2}-1\right) & \dfrac{n_1}{n_2} \end{bmatrix}
$$

results in Equation 4.30.

Section III
Wave Optics

5

Light Waves, Properties, and Propagation

The neglected education of my fellow-creature is the grand source of the misery I deplore.

Mary Wollstonecraft (1759–1797)

5.1 INTRODUCTION

This chapter picks up the discussion on light waves and their properties from where the first few sections (Sections 1.1 through 1.4) left off, with special focus here on the treatment of light waves. As will be detailed in the next few sections, light is a wave that consists of electric and magnetic fields oscillating in a sinusoidal form perpendicular to each other and propagating in vacuum (also called free space) with a speed c given by

$$c = \sqrt{\frac{1}{\mu_o \, \varepsilon_o}} = 3.00 \times 10^8 \, \text{m/s,}$$

where ε_o and μ_o are the permittivity and permeability, respectively, of free space. The chapter will address the electromagnetic wave equation derived from Maxwell's equations. Furthermore, specific properties of the electromagnetic wave, its propagation, its electric and magnetic field composites, and relations between them will be detailed.

5.2 MAXWELL'S EQUATIONS

There are two sets of Maxwell's equations. One is a differential form that is also called point form and the other is an integral form. Both can be used to derive the wave equation that is fundamental in describing light's main properties. For a couple of reasons, the derivation presented here is based on the differential form:

1. It is directly related to the early forms of Gauss' laws for electric and magnetic fields, Ampere's law, and Faraday's law of induction.
2. Maxwell's equations in their differential form provide a simpler derivation than that obtained via the integral form.

Considering a nonconducting free space medium that contains no free charge nor conduction current, Maxwell's equations are

$$\nabla \cdot \mathbf{E} = 0 \tag{5.1}$$

$$\nabla \wedge \mathbf{E} = -\frac{\partial \mathbf{B}}{\partial t} \tag{5.2}$$

$$\nabla \cdot \mathbf{B} = 0 \tag{5.3}$$

$$\nabla \wedge \mathbf{B} = \mu_o \varepsilon_o \frac{\partial \mathbf{E}}{\partial t}. \tag{5.4}$$

In the above equations, ε_o and μ_o are the permittivity and permeability, respectively, of free space; $\mathbf{E}$ and $\mathbf{B}$ are the electric and magnetic fields. As for units, the electric field is in newton/coulomb, N/C, and the magnetic field is in tesla, T. The symbol $\mathbf{V}$ is a vector operator known as the del and is defined in Cartesian coordinates as

$$\mathbf{V} = \left(\hat{x}\frac{\partial}{\partial x} + \hat{y}\frac{\partial}{\partial y} + \hat{z}\frac{\partial}{\partial z} \right).$$

The forms in Equations 5.1 and 5.3 are known as Gauss' differential forms of the electric and magnetic fields, respectively. The first, in essence, states that the divergence of the electric field in the absence of a charge is zero everywhere. This is equivalent to stating that the electric field is created by a charge. The latter shows that the divergence of the magnetic field at any point is zero. As the existence of a magnetic monopole has not been established, this equation is always true. Equations 5.2 and 5.4 are two other differential forms of spatial curling in loops of electric and magnetic fields, respectively. The parallel between the two left-hand sides of these two equations highlights a form of mutual dependence of each of the fields on the time rate of the change of the other field, exposing a twin connection between them. The form in (5.2) is Faraday's law of induction, and that expressed in (5.4) is Ampere's law.

From Equation 5.4, the curl of both sides gives

$$\mathbf{V} \wedge \left(\mathbf{V} \wedge \mathbf{B} \right) = \mu_0 \varepsilon_0 \left(\mathbf{V} \wedge \frac{\partial \mathbf{E}}{\partial t} \right),$$

or

$$\mathbf{V} \wedge \left(\mathbf{V} \wedge \mathbf{B} \right) = \mu_0 \varepsilon_0 \left(\frac{\partial \left(\mathbf{V} \wedge \mathbf{E} \right)}{\partial t} \right).$$

The left-hand side, treated as a triple vector product, $\mathbf{V} \wedge (\mathbf{V} \wedge) = \mathbf{V}(\mathbf{V} \cdot) - (\mathbf{V} \cdot \mathbf{V})$, reduces the above equation to

$$\mathbf{V}\left(\mathbf{V} \cdot \mathbf{B} \right) - \left(\mathbf{V} \cdot \mathbf{V} \right)\mathbf{B} = \mu_0 \varepsilon_0 \left(\frac{\partial \left(\mathbf{V} \wedge \mathbf{E} \right)}{\partial t} \right).$$

From Equation 5.3, the first term in the above equation is zero, and substituting for $(\mathbf{V} \wedge \mathbf{E})$ from (5.2), it becomes

$$-\left(\mathbf{V} \cdot \mathbf{V} \right)\mathbf{B} = \mu_0 \varepsilon_0 \left(\frac{\partial \left(-\frac{\partial \mathbf{B}}{\partial t} \right)}{\partial t} \right),$$

which concludes as

$$\nabla^2 \mathbf{B} = \mu_o \varepsilon_o \frac{\partial^2 \mathbf{B}}{\partial t^2}. \tag{5.5}$$

In the above equation, ∇^2 is the Laplacian, which in Cartesian coordinates takes the form

$$\nabla^2 = \frac{\partial^2}{\partial x^2} + \frac{\partial^2}{\partial y^2} + \frac{\partial^2}{\partial z^2},$$

and is frequently encountered in this chapter. A detailed discussion about it is presented in Appendix C. The heavy dot ($\cdot$) and the carrot ($\wedge$) symbols are used to indicate the dot and cross product, respectively.

Equation 5.5 is the differential wave equation for the magnetic field. This derivation can be repeated for $\mathbf{E}$ by taking the curl of Equation 5.2, and using the two equations (5.1) and (5.4). The result is

$$\nabla^2 \mathbf{E} = \mu_o \varepsilon_o \frac{\partial^2 \mathbf{E}}{\partial t^2}. \tag{5.6}$$

5.3 WAVE EQUATION

Equations 5.5 and 5.6 expressed in Cartesian coordinates are

$$\frac{\partial^2 \mathbf{E}}{\partial x^2} + \frac{\partial^2 \mathbf{E}}{\partial y^2} + \frac{\partial^2 \mathbf{E}}{\partial z^2} = \frac{1}{c^2} \frac{\partial^2 \mathbf{E}}{\partial t^2} \tag{5.7}$$

$$\frac{\partial^2 \mathbf{B}}{\partial x^2} + \frac{\partial^2 \mathbf{B}}{\partial y^2} + \frac{\partial^2 \mathbf{B}}{\partial z^2} = \frac{1}{c^2} \frac{\partial^2 \mathbf{B}}{\partial t^2}. \tag{5.8}$$

5.4 TYPES AND PROPERTIES OF ELECTROMAGNETIC WAVE EQUATIONS

The mathematical solution of scalar equations similar to those expressed in Equations 5.5 and 5.6 have long been known. Plane, spherical, or cylindrical wave solutions are among those solutions. The plane wave solution, sin ($\mathbf{k} \cdot \mathbf{r} - \omega t + \varphi$) or cos ($\mathbf{k} \cdot \mathbf{r} - \omega t + \varphi$), which is a special case of the spherical wave, is not only the simplest to work with, but also capable of providing a clear, consistent, and informative understanding of **E** and **B**. Also, the Cartesian components of **E** and **B** satisfy (5.5) and (5.6), and plane waves like $\sin(k_x x - \omega t + \varphi)$ or $\cos(k_x x - \omega t + \varphi)$ are also possible solutions. In these expressions, φ is a constant that describes the phase of either of these functions at $\mathbf{r} = 0$, $t = 0$.

Taking advantage of the properties of the complex function through Euler's identity

$$e^{i(\mathbf{k}\cdot\mathbf{r}-\omega t+\varphi)} = \cos(\mathbf{k}\cdot\mathbf{r}-\omega t+\varphi)+i\sin(\mathbf{k}\cdot\mathbf{r}-\omega t+\varphi), \qquad (5.9)$$

the solution for (5.5) and (5.6) may then be the real or imaginary part of the complex function; also, a linear combination of both parts is a solution. The following complex forms for **E** and **B**,

$$\mathbf{E} = \mathbf{E}_o e^{i(\mathbf{k}\cdot\mathbf{r}-\omega t+\varphi)} \qquad (5.10)$$

and

$$\mathbf{B} = \mathbf{B}_o e^{i(\mathbf{k}\cdot\mathbf{r}-\omega t+\varphi)}, \qquad (5.11)$$

are handy, because the real parts for **E** and **B** are composites of the complex form. Therefore, **E** and **B** in (5.10) and (5.11) both satisfy the wave equations (5.5) and (5.6) provided that $\omega = ck$. The vector fields $\mathbf{E}_o$ and $\mathbf{B}_o$ are called the amplitudes of the oscillating electric and magnetic fields, respectively, and **k** is known as the wave vector or propagation vector directed along the wave direction of propagation. In the case of a single frequency solution, **k** is also in the direction of the phase velocity.

As for spherical waves emanating from a point source, one important aspect about the solution of the three-dimensional wave Equations 5.2 and 5.4 is that the wave fronts in free space are independent of the angular coordinates. Therefore, a wave function $\psi(r, \theta, \varphi)$ reduces to a function of r only, and the Laplacian in r, θ, φ reduces to a differential equation in r only. That is, the Laplacian

$$\nabla^2(r, \theta, \varphi) = \frac{1}{r^2}\frac{\partial}{\partial r}\left(r^2\frac{\partial}{\partial r}\right) + \frac{1}{r^2}\frac{1}{\sin\theta}\frac{\partial}{\partial\theta}\left(\sin\theta\frac{\partial}{\partial\theta}\right) + \frac{1}{r^2\sin^2\theta}\frac{\partial^2}{\partial\phi^2}$$

becomes

$$\nabla^2 = \frac{1}{r^2}\frac{\partial}{\partial r}\left(r^2\frac{\partial}{\partial r}\right).$$

That makes the wave equation for an arbitrary scalar function, $\Psi(r)$,

$$\nabla^2 \Psi(r) = \frac{1}{v^2} \frac{\partial^2 \Psi}{\partial t^2} \tag{5.12}$$

take the form

$$\frac{1}{r^2} \frac{\partial}{\partial r} \left(r^2 \frac{\partial \Psi(r)}{\partial r} \right) = \frac{1}{v^2} \frac{\partial^2 \Psi}{\partial t^2}. \tag{5.13}$$

For a wave propagating away from the source, the solution of a scalar wave equation is one of the following harmonic solutions:

$$\Psi(r, t) = \frac{A}{r} \cos(kr - \omega t + \varphi)$$

or

$$\Psi(r, t) = \frac{A}{r} \sin(kr - \omega t + \varphi),$$

or the complex form

$$\Psi(r, t) = \frac{A}{r} e^{i(kr - \omega t + \varphi)}. \tag{5.14}$$

The above shows a traveling wave of a varying amplitude, $\frac{A}{r}$, implying an intensity varying with $\left(\frac{1}{r^2} \right)$.

5.4.1 Propagation and Transverse Nature of Light Waves

Resorting to Maxwell's equations in free space, we have

$$\nabla \cdot \mathbf{E} = 0 \tag{5.15}$$

$$\nabla \wedge \mathbf{E} = -\frac{\partial \mathbf{B}}{\partial t} \tag{5.16}$$

$$\nabla \cdot \mathbf{B} = 0 \tag{5.17}$$

$$\nabla \wedge \mathbf{B} = \mu_o \varepsilon_o \frac{\partial \mathbf{E}}{\partial t}. \tag{5.18}$$

Using Equation 5.10 in Equation 5.15 gives

$$\nabla \cdot \mathbf{E} = \left(\hat{\mathbf{x}} \frac{\partial}{\partial x} + \hat{\mathbf{y}} \frac{\partial}{\partial y} + \hat{\mathbf{z}} \frac{\partial}{\partial z} \right) \cdot \mathbf{E}_o e^{i(\mathbf{k}\cdot\mathbf{r} - \omega t + \varphi)}$$

$$= \mathbf{k} \cdot \mathbf{E}$$

$$= 0. \tag{5.19}$$

Also, using (5.11) in (5.20) gives

$$\nabla \cdot \mathbf{B} = \left(\hat{\mathbf{x}} \frac{\partial}{\partial x} + \hat{\mathbf{y}} \frac{\partial}{\partial y} + \hat{\mathbf{z}} \frac{\partial}{\partial z} \right) \cdot \mathbf{B}_o e^{i(\mathbf{k}\cdot\mathbf{r} - \omega t + \varphi)}$$

$$= \mathbf{k} \cdot \mathbf{B}$$

$$= 0. \tag{5.20}$$

As the dot products of $\mathbf{k}$ and $\mathbf{E}$ and of $\mathbf{k}$ and $\mathbf{B}$ in Equations 5.19 and 5.20 are zeros, then the propagation vector $\mathbf{k}$ is perpendicular to both $\mathbf{E}$ and $\mathbf{B}$. In other words, the direction of propagation of the light wave is perpendicular to the electric and magnetic fields that comprise the wave. This demonstrates one important property of electromagnetic waves that they are transverse waves. Figure 5.1 shows this property for a plane wave polarized in the y direction; polarization direction for a wave is usually associated with the direction of the electric field. Thus, the electric field is along the y axis, the magnetic field is along the z axis, and the wave vector $\mathbf{k}$ representing the direction of propagation is along the x axis.

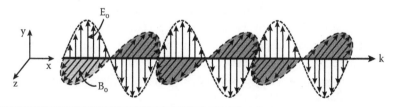

FIGURE 5.1   Electric and magnetic fields associated with a plane polarized electromagnetic wave and the propagation direction, perpendicular to both.

5.4.2 Relation between E and B of a Light Wave

Another important advantage of Maxwell's equations is that they lend them-selves to simple but elegant mathematical derivations that generate more information about light waves. This time, it is the connection between the two composites of the wave, the electric and magnetic fields. Calculating the curl of **E**, Equation 5.10, gives

$$\nabla \wedge \mathbf{E} = -i\mathbf{k} \wedge \mathbf{E}.$$

However, the derivative $\dfrac{\partial \mathbf{B}}{\partial t} = -i\omega \mathbf{B}$.

Therefore, Equation 5.16 implies that $-i\omega \mathbf{B} = -i\,\mathbf{k} \wedge \mathbf{E}$ or

$$\mathbf{B} = \frac{\mathbf{k}}{\omega} \wedge \mathbf{E}.$$

Thus,

$$\mathbf{B} = \hat{\mathbf{k}}\frac{k}{\omega} \wedge \mathbf{E}; \tag{5.21}$$

$\hat{\mathbf{k}} = \mathbf{k}/k$ is a unit vector along the direction of propagation. As for light, $\omega = kc$, then the magnitudes of the electric and magnetic fields in Equation 5.21 in SI units are related as

$$cB = E. \tag{5.22}$$

The vector and scalar relations, Equations 5.21 and 5.22, respectively, demon-strate three fundamental properties:

1. The magnetic field is perpendicular to the electric field.
2. The electric field and magnetic field are perpendicular to the direc-tion of propagation. This was already demonstrated in Equations 5.19 and 5.20. In brief then, a set of three mutually perpendicular directions like **E**, **B**, and **k** in an electromagnetic wave may be cho-sen as fundamentally related, forming a basis of three vectors. Once the direction of propagation **k** is defined, the plane combining **E** and **B** becomes automatically known. In this plane, **E** and **B** can be pointed anywhere perpendicular to **k** as long as they are perpen-dicular to each other.
3. The electric and magnetic fields in an electromagnetic wave are in phase.

EXAMPLE 5.1

Given the electric field for a plane polarized light wave traveling in the x direction described by

$$\mathbf{E}(x,t) = E_o \hat{y} \cos(k_x x - \omega t + \varphi),$$

write down a wave form for the magnetic field **B** of this wave.

Solution

From the given form for **E**, the electric field is directed along y, oscillating in the xy plane. Since in Equation 5.21, the direction of **B** should be along the cross product **k** × **E**, then **B** should be along the z direction, oscillating in the zx plane. Thus

$$\mathbf{B}(x,t) = B_o \hat{z} \cos(k_x x - \omega t + \varphi),$$

or

$$\mathbf{B}(x,t) = \frac{E_o}{c} \hat{z} \cos(k_x x - \omega t + \varphi).$$

A plot is shown in the following figure.

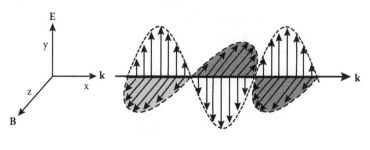

5.4.3 Energy and Power of a Light Wave: Poynting Vector

It has been learned from electrostatics that whenever an electric field exists, it stores energy with a density

$$u_E = \frac{1}{2}\varepsilon_o E^2, \tag{5.23}$$

and from magnetostatics, the energy density stored within a confined space is

$$u_B = \frac{1}{2\mu_o}B^2. \tag{5.24}$$

Also as $E = Bc$, Equations 5.22 and 5.23 show that

$$u_B = \frac{1}{2\mu_o}B^2 = \frac{1}{2\mu_o}\left(\frac{E^2}{c^2}\right) = \frac{1}{2}\left(\varepsilon_o E^2\right) = u_E. \tag{5.25}$$

That is, the energy density in a defined volume of space is equally distributed in the electric and magnetic fields.

Accordingly, the total energy density u stored in a confined space traversed by an electromagnetic wave is twice the density contributed by either field. That is,

$$u = 2u_E = 2u_B$$

$$= \frac{1}{\mu_o}B^2 = \varepsilon_o E^2.$$

The above can equivalently be presented as follows:

$$u = \frac{1}{\mu_o}B^2 = \frac{1}{\mu_o c}EB.$$

The energy density contained in an electromagnetic wave is understood as the energy per unit volume generated by the wave. Thus, selecting a volume in the form of a cylinder of $1.0\ m^2$ cross section (Figure 5.2) indicates that the above represents the energy that has propagated through a $1.0\ m^3$ volume a distance of $1.0\ m$. Multiplying the above by the speed of light c reduces it to

$$c \cdot u = \frac{1}{\mu_o}EB.$$

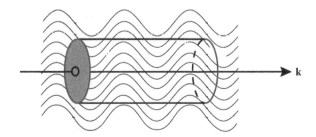

FIGURE 5.2   Unit volume contained in a cylinder of cross-sectional area $1.0\ m^2$ and length $1.0\ m$, used in calculating energy density.

The product c·u represents the energy per 1.0 m³ that has propagated through a distance of 1.0 m. That represents the energy per unit area per second, or power delivered by the wave per m².

This quantity is designated by S, so one may write

$$S = u \cdot c = \frac{1}{\mu_o} E\,B. \qquad (5.26)$$

Realizing that the above represents the energy flow per second per unit area in the direction of propagation, S then may be expressed as a vector **S** called the Poynting vector and written as follows:

$$\mathbf{S} = \frac{\hat{\mathbf{k}}}{\mu_o} E\,B \qquad (5.27)$$

where $\hat{\mathbf{k}}\left(= \dfrac{\mathbf{k}}{k} \right)$ is a unit vector along the direction of propagation. Since **E** and **B** are vectors whose cross product is a vector directed along the direction of propagation, the above may also be expressed as

$$\mathbf{S} = \frac{1}{\mu_o} \mathbf{E} \wedge \mathbf{B}. \qquad (5.28)$$

It has to be noted that **E** and **B** are varying in space and time, and although the representation of each in a wave can be expressed in a complex form (Equations 5.10 and 5.11), a physically real oscillating field is, for clarity, represented by the cosine or sine of these equations. Thus, either of the two representations, the sin or the cosine for **E** and **B**, applies. That is,

(i)   for **E**

$$\mathbf{E} = \mathbf{E}_o \cos\left(\mathbf{k} \cdot \mathbf{r} - \omega t + \varphi \right) \qquad (5.29)$$

or

$$\mathbf{E} = \mathbf{E}_o \sin\left(\mathbf{k} \cdot \mathbf{r} - \omega t + \varphi \right) \qquad (5.30)$$

and
(ii)   for **B**

$$\mathbf{B} = \mathbf{B}_o \cos\left(\mathbf{k} \cdot \mathbf{r} - \omega t + \varphi \right) \qquad (5.31)$$

or

$$B = B_o \sin\left(\mathbf{k.r} - \omega t + \varphi\right) \tag{5.32}$$

Taking the cosine form for **E** and **B** and upon substitution in Equation 5.28, one gets

$$S = \frac{1}{\mu_o} \mathbf{E_o} \wedge \mathbf{B_o} \cos^2\left(\mathbf{k} \cdot \mathbf{r} - \omega t + \varphi\right). \tag{5.33}$$

As the time average of

$$\langle\cos^2\left(\mathbf{k} \cdot \mathbf{r} - wt + \varphi\right)\rangle = \langle\sin^2\left(\mathbf{k} \cdot \mathbf{r} - wt + \varphi\right)\rangle = \frac{1}{2}, \tag{5.34}$$

the power per unit area becomes

$$\langle\mathbf{S}\rangle = \frac{1}{2\mu_o} E_o\, B_o. \tag{5.35}$$

$\langle\mathbf{S}\rangle$ is also called the irradiance or radiant flux density that is in some cases denoted by I, in reference to the radiation intensity. From the relation between E and B, $\langle\mathbf{S}\rangle$ may be expressed in other equivalent forms, two of which are

$$I = \langle\mathbf{S}\rangle = \frac{1}{2} \varepsilon_o c E_o^2 \tag{5.36}$$

$$I = \langle\mathbf{S}\rangle = \frac{1}{2} \frac{c}{\mu_o} B_o^2. \tag{5.37}$$

5.5 ELECTROMAGNETIC WAVE EQUATIONS IN DIELECTRICS

In a homogeneous linear isotropic nonmagnetic dielectric where the conductivity is zero, the wave Equations 5.3 and 5.4 become

$$\nabla^2 \mathbf{B} = \mu\varepsilon \frac{\partial^2 \mathbf{B}}{\partial t^2} \tag{5.38}$$

$$\nabla^2 \mathbf{E} = \mu\varepsilon \frac{\partial^2 \mathbf{E}}{\partial t^2}, \tag{5.39}$$

where μ and ε are the permeability and permittivity of the medium given by

$$\mu = K_M \mu_o, \tag{5.40}$$

$$\varepsilon = K_E \varepsilon_o, \tag{5.41}$$

and K_E and K_M are known as the relative permeability and relative permittivity of the medium. They are dimensionless quantities.

The product $\mu\varepsilon$ in Equations 5.38 and 5.39 is tied to the speed of light in the medium as follows:

$$v = \sqrt{\frac{1}{\mu\varepsilon}}. \tag{5.42}$$

The three-dimensional solutions of Equations 5.38 and 5.39 representing a plane wave for $\mathbf{E}$ and $\mathbf{B}$ take forms similar to those expressed in Equations 5.10 and 5.11. We will let the distinction between the two sets of solutions, those in vacuum and those in the dielectric, be recognized by the phase of the wave, denoted here by θ. Thus

$$\mathbf{E} = \mathbf{E}_o\, e^{i(k\cdot r - \omega t - \theta)} \tag{5.43}$$

and

$$\mathbf{B} = \mathbf{B}_o\, e^{i(k\cdot r - \omega t - \theta)}. \tag{5.44}$$

For a nonmagnetic material, $\mu = \mu_o$, that is, $K_M = 1$, and Equation 5.42 becomes

$$v = \sqrt{\frac{1}{K_E\,\mu_o\,\varepsilon_o}} = c\sqrt{\frac{1}{K_E}},$$

giving

$$\frac{c}{v} = \sqrt{K_E}.$$

Thus,

$$n = \sqrt{K_E}; \quad \frac{c}{v} = n, \tag{5.45}$$

where n is the index of refraction of the material. That is why the relative permittivity K_E is associated with the speed of light in the material. The relations between E and B for an electromagnetic wave propagating in a dielectric medium are similar to those derived for the wave traveling in vacuum. The critical difference in the speed of light in dielectrics is v replacing c in Equations 5.35 through 5.37. Thus, the magnitudes E and B are related as

$$E = B\,v; \quad E_o = B_o\,v \tag{5.46}$$

where again

$$v = c/n. \tag{5.47}$$

The average magnitude of the Poynting vector in a medium keeps the same form, Equations 5.35 through 5.37, but one should remember to use E and B expressed in Equations 5.46 and 5.47.

EXAMPLE 5.2

Consider an electric field for a plane polarized light wave traveling in the x direction described by

$$\mathbf{E}(x,t) = E_o\,\hat{\mathbf{y}}\cos(k_x x - \omega t + \varphi),$$

propagating in a nonmagnetic dielectric of a refractive index n = 1.45. Assuming that the amplitude of the electric field equals 150 V/m,

(a) Find the amplitude of the associating magnetic field and write down a wave form for the magnetic field **B** of this wave.
(b) Calculate the irradiance of this wave.

Solution

(a) $\omega = 2\pi f$. Thus, $\omega = 6.3 \times 10^{15}$ rad/s. The velocity v of the wave in this medium is v = c/n = (3.0 × 10⁸ m/s)/1.45 = 2.07 × 10⁸ m/s. Thus, the amplitude of the magnetic field is Bo = Eo/v = (150 V/m)/2.07 × 10⁸ m/s = 7.2 × 10⁻⁷ T and

$$k_x = \omega/v = \left(6.3 \times 10^{15}\ \text{rad/s}\right)/\left(2.07 \times 10^8\ \text{m/s} = 3.0 \times 10^7\ \text{rad/m}\right).$$

The magnetic field **B** is traveling along the x direction, having the same frequency, same phase, and same wave number that the E wave has. Thus

$$\mathbf{B}(x,t) = B_o\,\hat{\mathbf{z}}\cos(k_x x - \omega t),$$

or more explicitly,

$$\mathbf{B}(x, t) = (7.2 \times 10^{-7}\,\mathrm{T})\hat{\mathbf{z}}\cos\left[(3.0 \times 10^{7}\,\mathrm{rad/m})x - (6.3 \times 10^{15}\,\mathrm{Hz})t\right].$$

(b) The irradiance is given by

$$I = \frac{1}{2}(8.85 \times 10^{-12}\,\mathrm{Nm^2/C^2})^{-1}(2.07 \times 10^{8}\,\mathrm{m/s})(150\,\mathrm{V/m})^2,$$

$$I = 21\,\mathrm{W/m^2}.$$

5.6 PHOTON FLUX DENSITY

As electromagnetic waves are basically quanta of photons (see Chapter 1), other interesting optical quantities in the wave are the photon flux and photon flux density. Assuming that a wave has a flux density of photons defines the number density of photons n_{ph} per unit time per unit area; the photon flux density, however, is the photon flux per unit area that is directed perpendicular to the flux. Therefore, assuming a monochromatic light beam of frequency f, the number of photons per unit area per second is

$$n_{ph} = \frac{I}{hf} = \frac{\langle \mathbf{S} \rangle}{hf}. \tag{5.48}$$

Multiplying the above by the area A crossed by the electromagnetic wave, we obtain what is called the photon flux Φ as

$$\Phi = \frac{A\,I}{hf}. \tag{5.49}$$

COMMENT

Sorting out the units on the right-hand side of Equation 5.49, it can be noticed that the quantity Φ refers to the number of photons N_{ph} per second crossing a particular area A that could be the surface area of a radiation measuring sensor. From Equation 5.48, N_{ph} is numerically equal to Φ:

$$N_{ph} = n_{ph}\,A \tag{5.50}$$

EXAMPLE 5.3

Assuming that earth is spherical and is receiving energy from the sun at a rate of P = 1300 J/s per meter square normal to the sun's radiations, determine the total energy received by the earth in one day. Let the radius of the earth R = 6.38 × 10⁶ m.

Solution

The hemisphere of the earth facing the sun has a circular cross section (sketch below) of area

$$A = \pi\ R_E^2 = 3.59 \times 10^{13}\ m^2.$$

The energy per second received by this cross section, that is, by the earth's hemisphere, is

$$E = P\left(\pi\ R_E^2\right) = \left(1300\ J/s\right)\left(3.59 \times 10^{13}\ m^2\right) = 4.67 \times 10^{16}\ J/s.$$

The total energy received by the earth in one day is then

$$E_{one\ day} = E(24\ h/day) \times (60\ min/\ h \times 60\ s/min)$$

$$= \left(4.67 \times 10^{16}\ J/s\right)\left(8.6 \times 10^4 s\right)$$

$$= 4.03 \times 10^{21}\ J.$$

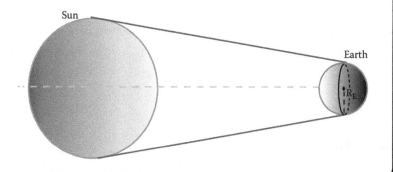

PROBLEMS

5.1 An EM wave, plane polarized along the y direction, propagating in the −x direction with frequency of 1.00×10^{15} s^{-1} has its electric field expressed by

$$E\left(x, t = 5.00 \times 10^{-15}\ s\right) = \left(44.0\ V/m\right)\cos(kx),$$

where x is in meters and t in seconds.
(a) Determine the angular frequency, wavelength, and wave number of the wave.
(b) Write an expression describing the electric field as a function of x and t at another instant of time t.

5.2 An EM wave, plane polarized along the y direction, propagating in the +x direction has its displacement at $x = \lambda/4$ expressed by the equation

$$E\left(x = \lambda/4, t\right) = \left(44.0\ V/m\right)\cos\left(2.00 \times 10^{-15}\ \pi t\right)$$

where λ and x are measured in meters, and t in seconds.
(a) Determine the frequency, wavelength, and wave number of the wave.
(b) Write an expression for the time variation of E at x = 0.

5.3 Consider the following electromagnetic wave in free space:

$$E\left(x, t\right) = E_o \hat{y}\ e^{i\left(k_x x - \omega t + \varphi\right)},$$

where k_x, ω, and φ are constants.
(a) Show that it satisfies Equation 5.4.
(b) Write an appropriate wave form for the **B** field in this wave.

5.4 If the angular frequency of the wave described in the previous problem is 1.00×10^{15} rad/s and $E_o = 0.22$ V/m, use the given form of **E** to determine the value of the irradiance of this wave.

5.5 Given an electromagnetic wave propagating along the x axis, and is plane polarized along the y axis, as in the following figure. The wave's electric field is described by

$$E\left(x, t\right) = E_o \hat{y}\ \cos(kx - \omega t + \varphi),$$

where x is in meters, t in seconds, $E_o = 600.0$ V/m, $\lambda = 3.0 \times 10^{-7}$ m, and $\varphi = 0.0$.
(a) Determine the amplitude of the magnetic field **B** of this wave.
(b) Find positions at which **E** and **B** have their maximum values.

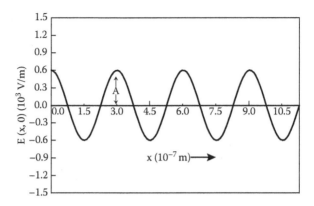

5.6   Consider the electromagnetic wave described in the previous problem again to find
(a)   The wave form of the **B** field for this wave
(b)   The values of **E** and **B** at ($x = 6.0 \times 10^{-7}$ m, $t = 2.00 \times 10^{-15}$ s)

5.7   Consider the plot below as one representing the **E** field of a plane electromagnetic wave polarized in the z direction and propagating along the x axis. Let the wave's amplitude be $E_o = 1000$ V/m. Using the information in the plot,
(a)   Write two equivalent sinusoidal expressions for **E**. State the phase constant φ for each of the expressions.
(b)   Find the period, frequency, wavelength, and magnitude of the propagation vector **k**.

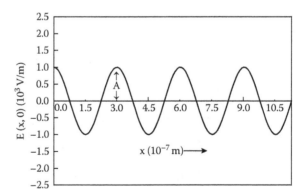

5.8   Given is an electric field of a plane electromagnetic wave polarized in the y direction and propagating along the x axis. Let the amplitude of the wave be 150 V/m, $\varphi = 0.0$ rads, and of frequency $f = 1.0 \times 10^{15}$ Hz. Determine (a) the form of the electric field, (b) the form of the magnetic field, and (c) the irradiance of the wave.

5.9   Consider a plane EM wave whose electric field is described by $E(x,t) = (E_o)\hat{y} \cos(kx - \omega t)$. For this wave,

(a)   Write the appropriate form of the **B** field.

(b)   Find the vector product $E \wedge B$, and comment on the relevance of your answer.

5.10   Consider the plot below as one representing a hypothetical plane EM wave whose electric field **E** is polarized along the z axis and propagating along the x axis.

(a)   Write down the wave form of the **B** field.

(b)   Find the period, frequency, and wavelength.

(c)   Define the range of the EM spectrum to which this wave belongs.

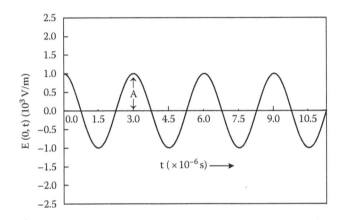

5.11   Equation 5.9 states that $B = \hat{k}\dfrac{k}{\omega} \wedge E$. Using Maxwell's equations, establish a corresponding relation between **E** and $\dfrac{\hat{k}}{\omega} \wedge B$. Check if your answer correlates properly with the properties of the wave being electromagnetic.

5.12   Show that the one-dimensional wave equation

$$\frac{\partial^2 \Psi(x,t)}{\partial x^2} = \frac{1}{v^2}\frac{\partial^2 \Psi(x,t)}{\partial t^2}, \text{ having solutions}$$

$$\Psi_1(x,t) = A\cos(kx - \omega t), \quad \text{and} \quad \Psi_2(x,t) = B\sin(kx - \omega t),$$

should also have $(\Psi_1 + \Psi_2)$ as a solution. That is,

$$\Psi = \Psi_1 + \Psi_2.$$

5.13 Show that for a plane wave $E(x,t) = E_o \hat{z} \cos(kx - \omega t + \varphi)$ traveling along x, the **B** field satisfies Maxwell's fourth equation, $\nabla \wedge \mathbf{B} = \mu_o \varepsilon_o \dfrac{\partial \mathbf{E}}{\partial t}$. Comment on the relevance of your answer.

5.14 Given the following EM wave $E(x, t) = (55 \ V/m)\hat{z} \cos\left[\dfrac{2\pi}{\lambda}x - \omega t + \pi/2\right]$ and letting $\lambda = 550$ nm,

   (a) Write the wave form of the magnetic field.
   (b) Find the irradiance of this wave.
   (c) Find the number of photons per meter squared per second crossing a surface over one second.

5.15 Use the entry data in the previous problem to calculate the irradiance of the wave if it propagates through

   (a) A cubic glass filled water of 1.0 m a side ($n_{water} = 1.33$)
   (b) A cubic glass filled alcohol of 1.0 m a side ($n_{alcohol} = 1.1$)

6

Light Waves, Coherence, Superposition, and Interference

We know only one source which directly reveals scientific facts – our senses.

Ernst Mach (1838–1916)

6.1 INTRODUCTION

The wave equations for both the electric and magnetic fields were derived in Chapter 5. Each is a second-order linear differential equation consisting of two differential terms, one spatial and the other temporal. The differentials, one on each side of the equation, are connected through a factor that relates to the speed of light in the medium of propagation; the speed is c in vacuum and is v in a medium of an index of refraction n (=c/v). For the electric field wave equation, the solution can be taken as a plane light wave polarized along the y axis, traveling in vacuum along both the positive and the negative x axis, having the forms

$$E_+(x,t) = E_1(x - ct), \tag{6.1}$$

$$E_-(x,t) = E_2(x + ct). \tag{6.2}$$

Mathematically speaking, each of the two separate solutions satisfies the wave equation, and a linear combination of the two solutions also satisfies the wave equation. That is, a wave with an electric field

$$E = E_+(x,t) + E_-(x,t) \tag{6.3}$$

is also a solution. This simple example facilitates the essence of an important principle in physics known as the superposition principle, which can be applied to determine the resultant of two waves that cross each other at a certain instant of time. At the instant of their overlap, their fields add algebraically, so that the resultant, in principle, is easily determined. Depending on the direction of propagation of the waves, direction of oscillation of their electric fields, and their phases, the overlap can be constructive, destructive, or an intermediate state of interference.

A clear understanding of interference may start with addressing the superposition of waves that have the same plane of vibration and direction of propagation. In such a choice, the complexity of involving the vector nature of light can be minimized, because the electric fields of these waves are treated as scalars.

There are several ways by which one can determine the resultant of two or more waves. Among these are two simple but informative methods: one analytic, employing harmonic waves and adding them algebraically, and the other is called the *phasor diagram method*. This method is most helpful when the superimposed waves have the same frequency.

6.2 SUPERPOSITION OF TWO WAVES

In this section, two simple but interesting cases of superposition are discussed: one is based on only the difference between the phase constants of the superimposed waves, and the other is based on the difference between their frequencies. Of course, a case based on both is another case of interest. In the following, a treatment of the first two cases is presented.

6.2.1 Superposition of Two Waves of the Same Frequency

As explained earlier, there are three function forms by which an electric wave is expressed: a cosine, a sine, or an exponential form. Using the cosine form, consider the following two waves:

$$E_1(x, t) = E_{o1} \cos(kx - \omega t + \varphi_1), \tag{6.4}$$

$$E_2(x, t) = E_{o2} \cos(kx - \omega t + \varphi_2), \tag{6.5}$$

where k is the propagation constant along the x axis, the same for both waves, and φ_1 and φ_2 are the phase constants of waves 1 and 2, respectively, measured at t = 0 from a reference point that can be considered the origin O, where x = 0 and t = 0.

Adding the individual waves as expressed in Equations 6.4 and 6.5 gives the electric field E(x, t) resulting from the superposed waves. This is

$$E_R(x, t) = E_1(x, t) + E_2(x, t)$$

$$E_R(x, t) = E_{o1} \cos(kx - \omega t + \varphi_1) + E_{o2} \cos(kx - \omega t + \varphi_1). \tag{6.6}$$

For convenience, let

$$k x + \varphi_1 = \delta_1, \quad k x + \varphi_2 = \delta_2.$$

Equation 6.6 becomes

$$E_R(x, t) = E_{o1} \cos(\delta_1 - \omega t) + E_{o2} \cos(\delta_2 - \omega t). \tag{6.7}$$

6.2.1.1 Two Waves with a Phase Difference of Integer Multiples of 2π Traveling in the Same Direction

That is,

$$\delta_2 - \delta_1 = 0, 2\pi, 4\pi, \dots 2n\pi; \quad n = 0, 1, 2, \dots.$$

In this case, $\cos(\delta_1 - \omega t) = \cos(\delta_2 - \omega t)$, and Equation 6.7 becomes

$$E_R(x, t) = E_{o1} \cos(\delta_1 - \omega t) + E_{o2} \cos(\delta_1 - \omega t)$$

or

$$E_R(x, t) = (E_{o1} + E_{o2}) \cos(\delta_1 - \omega t) \tag{6.8}$$

or

$$E_R(x, t) = (E_{oR})\cos(\delta_1 - \omega t) \tag{6.9}$$

where E_{oR} is the amplitude of the resultant wave, that is,

$$E_{oR} = E_{o1} + E_{02}. \tag{6.10}$$

The superposition of the two waves, E_1 and E_2, and the resultant E_R are shown in Figure 6.1.

As the intensity of a wave is determined by the square of its amplitude, the intensity of the resultant in Equation 6.10 is

$$I_R = (E_{oR})^2 = (E_{o1} + E_{o2})^2 = (E_{o1})^2 + (E_{o2})^2 + (2E_{o1}E_{o2}).$$

If the constituent amplitudes are equal (Figure 6.2),

$$E_{o1} = E_{o2} = E_o,$$

then

$$I_R = 4(E_o)^2$$

$$I_R = 4I_o. \tag{6.11}$$

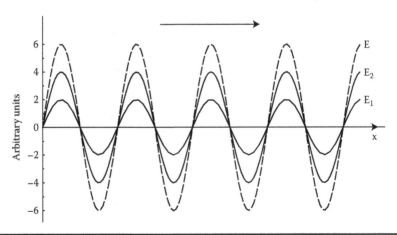

FIGURE 6.1   The figure demonstrates the superposition of two waves, E_1 and E_2; the waves are in phase but of different amplitudes; E_R is their resultant.

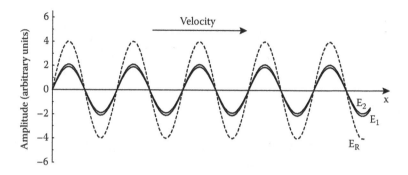

FIGURE 6.2   The figure demonstrates the resultant E_R of the superposition of two waves, E_1 and E_2, that are in phase but of slightly different amplitudes. The imposed slight difference in amplitude was intentional to clearly get them recognized in the plot.

6.2.1.2 Two Waves with a Phase Difference of Odd Integer Multiples of π

6.2.1.2.1 Waves Propagating in the Same Direction

That is,

$$\delta_2 - \delta_1 = \pi, 3\pi, \dots (2n+1)\pi; \quad n = 0, 1, 2, \dots,$$

giving

$$\delta_2 = (\delta_1 + \pi), (\delta_1 + 3\pi), \dots \left(\delta_1 + (2n+1)\pi\right).$$

Consider the addition of two waves:

$$E_R(x, t) = E_{o1} \cos(\delta_1 - \omega t) + E_{o2} \cos(\delta_2 - \omega t)$$

Since $\cos(\delta_1 + (2n+1)\pi - \omega t) = -\cos(\delta_1 - \omega t)$, Equation 6.7 becomes

$$E_R(x, t) = E_{o1} \cos(\delta_1 - \omega t) - E_{o2} \cos(\delta_1 - \omega t).$$

That is,

$$E_R(x, t) = (E_{o1} - E_{o2}) \cos(\delta_1 - \omega t)$$

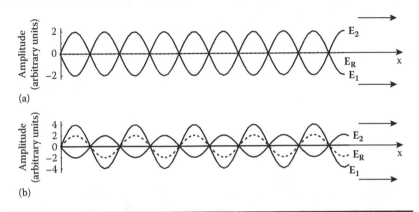

FIGURE 6.3 Superposition of two waves that are 180° out of phase, when (a) the waves are equal in amplitude, and (b) the amplitude of E_2 is half that of E_1. Both waves are propagating in the same direction.

or

$$E_R(x, t) = (E_{oR})\cos(\delta_1 - \omega t) \tag{6.12}$$

where

$$E_{oR} = E_{o1} - E_{o2} \tag{6.13}$$

is the amplitude of the resultant of the two waves, E_1 and E_2. The intensity of the resultant wave is

$$I_R = (E_{oR})^2 = (E_{o1} - E_{o2})^2 = (E_{o1})^2 + (E_{o2})^2 - (2 E_{o1}E_{o2}).$$

If the constituent amplitudes are equal, that is, $E_{o1} = E_{o2} = E_o$, then

$$I_R = 0. \tag{6.14}$$

This kind of superposition is called destructive interference and is best demonstrated in Figure 6.3, where two cases are displayed: (a) the amplitudes E_{o1} and E_{o2} are equal, and (b) $E_{o2} = 0.5\, E_{o1}$.

6.2.1.2.2 Waves Propagating in Opposite Directions: Standing Waves

Another simple but interesting case is the superposition of two waves that have the same frequency but are of a phase difference that varies with time, both traveling in opposite direction. A special case of relevance is when the two waves have the same amplitude and speed. The resultant can be determined from the component waves expressed in cosine, sine, or complex forms. Taking the exponential forms, consider the two waves

$$E_1 = E_o e^{-i(\omega t - kx)}. \tag{6.15}$$

$$E_2 = E_o e^{-i(\omega t + kx)} \tag{6.16}$$

$$E_R = E_1(x, t) + E_2(x, t),$$

or

$$E_R = E_o e^{-i(\omega t - kx)} + E_o e^{-i(\omega t + kx)}$$

that reduces to

$$E_R = e^{-i\omega t} \left[E_o \left(e^{ikx} + e^{-ikx} \right) \right]$$
$$= \left(2E_o \cos kx \right) e^{-i\omega t} \tag{6.17}$$

As can be noted, the resultant is a wave in the form of a product of two parts, one spatial, $2E_o \cos(kx)$, and the other solely time dependent, $e^{-i\omega t}$, that defines the phase of the wave as being the same for all positions at a defined instant of time. The real part of the above form is

$$E_R = \left(2E_o \cos kx \right) \cos \omega t. \tag{6.18}$$

This is called a standing wave. It is clear from Equation 6.18 that the spatial term defines the amplitude of the resultant wave at any position, x, for any given time. Calculating the resultant at all positions for a certain time interval would give the shape of this wave. Repeating that for several values of t would give the range of modes of oscillations that the wave has. Figure 6.4a and b demonstrates the situation for the standing wave of two component waves, E_1 and E_2, the amplitude of each of which was assumed a value of 3.0 units. The plots can be considered as snapshots taken at t = 0, T/8, 2T/8, 3T/8, 4T/8, where T is the period of each wave; the plots in Figure 6.4a correspond to phase differences of $\pi/2$, $\pi/4$, 0, $-\pi/4$, $-\pi/2$, respectively, as sketched in Figure 6.4b. It is worth noting that the form obtained in Equation 6.18 could have been obtained in different but equivalent expressions, depending on how one chooses to express the component waves as sines, cosines, or complex forms.

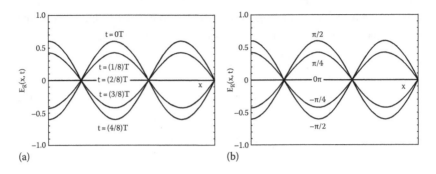

FIGURE 6.4   Standing waves at different positions of two waves E_1 and E_2 of equal amplitudes and equal frequency. (a) Snapshots taken at instances t = 0, T/8, 2T/8, 3T/8, 4T/8, where T is the period of each wave. (b) Snapshots of the two waves of phase differences π/2, π/4, 0, –π/4, –π/2, that correspond to the instances 0, T/8, 2T/8, 3T/8, 4T/8, respectively.

6.3 SUPERPOSITION OF MULTIPLE WAVES OF ARBITRARY PHASES

Through each of these methods, we have observed how one can determine the amplitude and phase of the resultant. The general nature of these methods makes them helpful in addressing all cases of interest including the two special cases we treated above.

6.3.1 ANALYTIC ALGEBRAICALLY BASED METHOD

Consider the following three waves:

$$E_1(x,t) = E_{o1} \cos(kx - \omega t + \varphi_1) \tag{6.19}$$

$$E_2(x,t) = E_{o2} \cos(kx - \omega t + \varphi_2) \tag{6.20}$$

$$E_3(x,t) = E_{o3} \cos(kx - \omega t + \varphi_3). \tag{6.21}$$

Introducing alternative phase factors $\delta_1 = (kx + \varphi_1)$, $\delta_2 = (kx + \varphi_2)$, and $\delta_3 = (kx + \varphi_3)$ in a manner similar to that defined earlier and adding Equations 6.19 and 6.21 gives the resultant as

$$E_R(x,t) = E_{o1} \cos(\delta_1 - \omega t) + E_{o2} \cos(\delta_2 - \omega t) + E_{o3} \cos(\delta_3 - \omega t) \tag{6.22}$$

or

$$E_R(x,t) = E_{o1}\left[\cos\delta_1\cos\omega t + \sin\delta_1\sin\omega t\right]$$
$$+ E_{o2}\left[\cos\delta_2\cos\omega t + \sin\delta_2\sin\omega t\right]$$
$$+ E_{o3}\left[\cos\delta_3\cos\omega t + \sin\delta_3\sin\omega t\right].$$

Factoring out (cos ωt) and (sin ωt) from the corresponding terms

$$E_R(x,t) = \left(E_{o1}\cos\delta_1 + E_{o2}\cos\delta_2 + E_{o3}\cos\delta_3\right)\cos\omega t$$
$$+ \left(E_{o1}\sin\delta_1 + E_{o2}\sin\delta_2 + E_{o3}\sin\delta_3\right)\sin\omega t. \qquad (6.23)$$

For Equation 6.23 to be a harmonic wave, the r.h.s. then must take the form

$$E_R(x,t) = E_{oR}\cos\delta_R\cos\omega t + E_{oR}\sin\delta_R\sin\omega t \qquad (6.24)$$

where

$$E_{oR}\cos\delta_R = E_{o1}\cos\delta_1 + E_{o2}\cos\delta_2 + E_{o3}\cos\delta_3 \qquad (6.25)$$

$$E_{oR}\sin\delta_R = \left(E_{o1}\sin\delta_1 + E_{o2}\sin\delta_2 + E_{o3}\sin\delta_3\right). \qquad (6.26)$$

Dividing Equation 6.26 by 6.25 gives

$$\tan\delta_R = \frac{E_{o1}\sin\delta_1 + E_{o2}\sin\delta_2 + E_{o3}\sin\delta_3}{E_{o1}\cos\delta_1 + E_{o2}\cos\delta_2 + E_{o3}\cos\delta_3} \qquad (6.27)$$

$$\delta_R = \tan^{-1}\left[\frac{E_{o1}\sin\delta_1 + E_{o2}\sin\delta_2 + E_{o3}\sin\delta_3}{E_{o1}\cos\delta_1 + E_{o2}\cos\delta_2 + E_{o3}\cos\delta_3}\right]. \qquad (6.28)$$

Upon squaring the two sides of each of Equations 6.25 and 6.26 and adding them, we get

$$E_{oR}^2 = \left(E_{o1} \cos \delta_1 + E_{o2} \cos \delta_2 + E_{o3} \cos \delta_3 \right)^2 + \left(E_{o1} \sin \delta_1 + E_{o2} \sin \delta_2 + E_{o3} \sin \delta_3 \right)^2.$$

The above reduces to

$$
\begin{aligned}
E_{oR}^2 = \left(E_{o1}^2 + E_{o2}^2 + E_{o3}^2 \right) &+ [(2 E_{o1} E_{o2} \cos \delta_1 \cos \delta_2 + 2 E_{o1} E_{o3} \cos \delta_1 \cos \delta_3 \\
&+ 2 E_{o2} E_{o3} \cos \delta_2 \cos \delta_3) + (2 E_{o1} E_{o2} \sin \delta_1 \sin \delta_2 \\
&+ 2 E_{o1} E_{o3} \sin \delta_1 \sin \delta_3 + 2 E_{o2} E_{o3} \sin \delta_2 \sin \delta_3)].
\end{aligned}
\tag{6.29}
$$

With some arrangement and trigonometric identities, Equations 6.29 and 6.30 can be rewritten as

$$
\begin{aligned}
E_{oR}^2 = \left(E_{o1}^2 + E_{o2}^2 + E_{o3}^2 \right) &+ [2 E_{o1} E_{o2} \cos(\delta_1 - \delta_2) \\
&+ 2 E_{o1} E_{o3} \cos(\delta_1 - \delta_3) + 2 E_{o2} E_{o3} \cos(\delta_2 - \delta_3)].
\end{aligned}
\tag{6.30}
$$

In a generalized form that expresses the intensity of the superposition of a large number of waves, N, the above in a reduced form becomes

$$E_{oR}^2 = \sum_i^N E_{oi}^2 + \left[2 \sum_{j>i}^N \sum_i^N E_{oi} E_{oj} \cos\left(\delta_i - \delta_j \right) \right].
\tag{6.31}$$

As it can be noted, the second term on the r.h.s. of the above is critically dependent on the phase difference between the superimposed waves; the condition $j > 1$ in the first double summation ensures no repetition of count of the same pair of waves. When $i = 1$, $j = 2, 3..., N$, and for $i = 2$, $j = 3, 4, 5...$ N, and so on.

Equations 6.28 and 6.31 are direct expressions for the phase and intensity, respectively, of the resultant for any finite number of superimposed waves.

EXERCISE

Apply the above expression to two waves for (a) $\delta_1 - \delta_2 = 2\pi$, (b) $\delta_1 - \delta_2 = \pi$, and contrast your results against those arrived at in the previous two cases (see Equations 6.11 and 6.14).

EXAMPLE 6.1

Consider the following two waves

$$E_1(x, t) = E_{o1} \cos(k x - \omega t + \varphi_1)$$

$$E_2(x, t) = E_{o2} \cos(k x - \omega t + \varphi_2)$$

that have equal amplitudes, $E_{o1} = E_{o2} = 4.0$ m, and are of a constant phase difference of $\pi/2$. (a) Determine the amplitude and phase of their resultant, and (b) describe the resultant by a proper wave form.

Solution

(a) From Equation 6.31, the amplitude

$$E_{oR}^2 = \sum_i E_{oi}^2 + \left[2 \sum_{j>i}^{2} \sum_{i}^{2} E_{oi} E_{oj} \cos(\delta_i - \delta_j) \right].$$

Thus,

$$E_{oR}^2 = E_{o1}^2 + E_{o2}^2 + \left[2 \sum_{j=2}^{2} \sum_{i=1}^{2} E_{o1} E_{o2} \cos(\delta_1 - \delta_2) \right].$$

Letting the phase of the first wave be zero, the second would be of $\pi/2$. Thus,

$$E_{oR}^2 = (4.0)^2 + (4.0)^2 = 32.$$

Therefore,

$$E_{oR} = 5.7 \, \text{m}.$$

The phase of the resultant is given by

$$\delta_R = \tan^{-1} \left[\frac{4.0 \sin 0 + 4.0 \sin(\pi/2)}{4.0 \cos 0 + 4.0 \cos(\pi/2)} \right] = \tan^{-1}(1.0).$$

Thus,

$$\delta_R = \pi/4 \, \text{rad}.$$

(b) The resultant can be described by the equation

$$E_R(x, t) = 5.7 \cos(kx - \omega t + \pi/4).$$

6.3.2 Phasor Diagram

In this method, the amplitudes of EM waves are treated as vectors that have magnitudes and directions. The direction of any of the superimposed waves is defined by its phase that is represented for a positive angle by a rotation measured in a counterclockwise rotation relative to the zero phase line. The amplitude of the wave represents the length of the phasor. To superimpose several waves, the amplitudes are treated as vectors; each is along a line that makes an angle with the zero line equal to the phase angle of the wave. This can be demonstrated through the following example.

EXAMPLE 6.2

Use the phasor method to determine the resultant of the following two waves:

$$E_1(x, t) = 3.0 \cos\left(\frac{\pi}{6} - \omega t\right)$$

$$E_2\left(x, st\right) = 4.0 \cos\left(\frac{2\pi}{3} - \omega t\right)$$

where the amplitudes are measured in meters, m, and phase in radians.

Solution

As noted, the two waves have the following phasors, which in the same order are

$$\text{Phasor 1: length } E_{o1} = 3.0\,\text{m} \quad \text{and} \quad \text{phase } \delta_1 = \pi/6 \text{ rads} \quad (6.32)$$

$$\text{Phasor 2: length } E_{o2} = 4.0\,\text{m} \quad \text{and} \quad \text{phase } \delta_2 = 2\pi/3 \text{ rads} \quad (6.33)$$

The two phasors E_{o1} of 3 m and E_{o2} of 4 m are drawn head to tail such that E_{o1} is drawn along OA, making an angle of 30° with the horizontal zero phase line Ox, and E_{o2} starting at the tip of E_{o1} is drawn along AB, making an angle of 120° with the horizontal direction. The line connecting the starting point O and the end point B represents the amplitude of the resultant. The angle δ_R is the phase of the resultant. To a good approximation, measuring OB gives a length of 5 m, and measuring δ_R gives a value of about 83° (=4π/9). These values suggest that the resultant can be expressed by the equation (see the following diagram)

$$E_R(x, t) = 5.0 \cos\left(\frac{4\pi}{9} - \omega t\right).$$

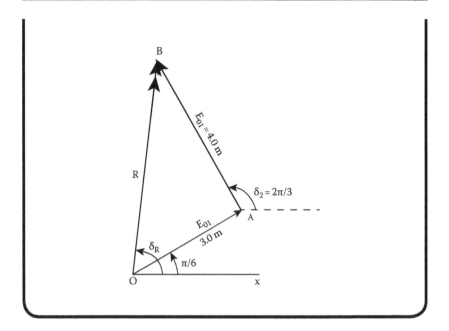

EXERCISE

Use geometry to support the above answers for the resultant. Also, one may use Equations 6.27, 6.28, and 6.31 to determine the amplitude and phase of the resultant. Check if your geometry-based answers confirm those obtained in the solution of the previous example.

COMMENT

The same procedure can be followed to determine the resultant of multiple waves of the same frequency.

6.4 SUPERPOSITION OF TWO WAVES OF A SLIGHTLY DIFFERENT FREQUENCY: GROUP VELOCITY

A variety of waves of an exact same frequency is too particular of a demand to fulfill. However, waves of a slight difference in wavelength are quite possible. As will be seen, this is not only more real of a case but is of a special importance, because its treatment unfolds a variety of useful concepts about the propagation of waves, their modulation, their velocity as a group, and the speed of a particular phase of the wave. In the general case that we will be addressing, the waves are assumed to be propagating in a nondispersive medium. Therefore, the speeds of the interfering waves are equal.

We will proceed in this argument in an approach similar to that followed in Section 6.2.1. Let us consider two waves that have the same amplitude E_o, but are slightly different in frequency, and accordingly different in wavelength; thus they are slightly different in the angular frequency ω and wave number k (Figure 6.5a). These differences imply that the wave's phases are different.

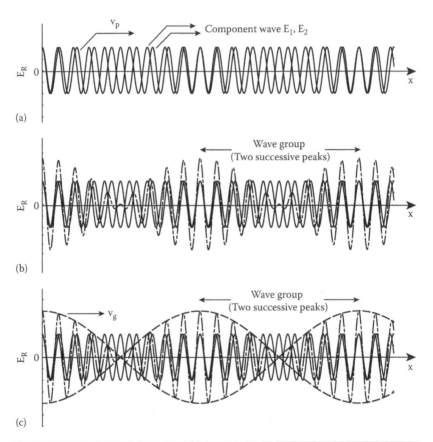

FIGURE 6.5   (a) Addition of two waves that are slightly different in frequency (Equations 6.34 through 6.36), (b) plot of $\cos\left(\dfrac{\bar{k}\,x}{2} - \dfrac{\bar{\omega}\,t}{2}\right)$, and (c) plot $\cos\left(\dfrac{(\Delta k)\,x}{2} - \dfrac{(\Delta\omega)\,t}{2}\right)$ showing modulati on of the two waves producing an envelope (dashed line) that travels with the group velocity.

Therefore, there is no need to include any additional difference in the phase constants φ_1 and φ_2, and we can let $\varphi_1 = \varphi_2 = 0$:

$$E_1(x, t) = E_o \cos(k_1 x - \omega_1 t) \tag{6.34}$$

$$E_2(x, t) = E_o \cos(k_2 x - \omega_2 t) \tag{6.35}$$

where k_1 and k_2 are the propagation constants along the x axis and ω_1 and ω_2 are the angular velocities of waves 1 and 2, respectively. The reference point can be considered an origin, O, where x = 0 and t = 0.

The slight differences in wave number and angular velocity imply that

$$k_1 \approx k_2 \approx \frac{k_1 + k_2}{2}, \quad \omega_1 \approx \omega_2 \approx \frac{\omega_1 + \omega_2}{2}; \tag{6.36}$$

Also, the difference between wave numbers and angular frequencies Δk, $\Delta \omega$, respectively, are very small:

$$\Delta k = k_1 - k_2 \quad \Delta \omega = \omega_1 - \omega_2$$

Adding the individual waves expressed in Equations 6.34 through 6.35 gives the resultant E(x, t), that is, the electric field resulting from these superposed waves is

$$E_R(x, t) = E_o \cos(k_1 x - \omega_1 t) + E_o \cos(k_2 x - \omega_2 t). \tag{6.37}$$

Letting

$$k_1 x - \omega_1 t = \delta_1, \quad k_2 x - \omega_2 t = \delta_2, \tag{6.38}$$

Equation 6.37 becomes

$$E_R(x, t) = E_o \left(\cos \delta_1 + \cos \delta_2 \right). \tag{6.39}$$

From trigonometry,

$$\cos(\delta_1 + \delta_2) = \cos \delta_1 \cos \delta_2 - \sin \delta_1 \sin \delta_2;$$

$$\cos(\delta_1 - \delta_2) = \cos \delta_1 \cos \delta_2 + \sin \delta_1 \sin \delta_2.$$

Adding these gives

$$\cos(\delta_1 + \delta_2) + \cos(\delta_1 - \delta_2) = 2\cos\delta_1 \cos\delta_2$$

$$= 2\left[\cos\frac{(\delta_1 + \delta_2) + (\delta_1 - \delta_2)}{2}\cos\frac{(\delta_1 + \delta_2) - (\delta_1 - \delta_2)}{2}\right].$$

Substituting for δ_1, δ_2 from expressions (6.38) reduces Equation 6.39 to

$$E_R(x, t) = 2\left[\cos\frac{\left(k_1 x - \omega_1 t\right) + \left(k_2 x - \omega_2 t\right)}{2}\cos\frac{\left(k_1 x - \omega_1 t\right) - \left(k_2 x - \omega_2 t\right)}{2}\right]$$

or

$$E_R(x, t) = 2\left[\cos\frac{\left(k_1 + k_2\right)x - \left(\omega_1 + \omega_2\right)t}{2}\cos\frac{\left(k_1 - k_2\right)x - \left(\omega_1 - \omega_2\right)t}{2}\right] \quad (6.40)$$

or

$$E_R(x, t) = 2\cos\left\{\bar{k}x - \bar{\omega}t\right\}\cos\left\{\frac{\Delta kx - \Delta\omega t}{2}\right\} \quad (6.41)$$

where

$$\frac{\left(k_1 + k_2\right)}{2} = \bar{k}. \quad (6.42)$$

and

$$\frac{\left(\omega_1 + \omega_2\right)}{2} = \bar{\omega}; \quad (6.43)$$

$\bar{k}$ and $\bar{\omega}$ are the average wave number and angular velocity, respectively

Let us write Equation 6.41 in an equivalent form:

$$E_R(x, t) = 2\cos\left(\frac{(\Delta k)x}{2} - \frac{(\Delta\omega)t}{2}\right)\cos\left(\bar{k}x - \bar{\omega}t\right). \quad (6.44)$$

One can notice in Equation 6.44 that the second cosine term represents a wave of an average propagation constant and average angular frequency that is similar in form to each of the individual waves in describing a propagating wave of velocity $v = \bar{\omega}/\bar{k}$. However, for the first cosine form, the propagation constant (Δk) is just the small difference between the two k's. Analysis of only the spatial part of the above wave can be done by letting $t = 0$. From the fact that $\bar{k} \gg \Delta k$, the cosine term $\cos\dfrac{\Delta kx}{2}$ will vanish for values of x much larger than those of x at which the term $\cos\dfrac{\bar{k}x}{2}$ vanishes. For reasons that will become clear later, let us denote x in the product (Δk) x by x_g and x in the product $\bar{k}x$ by x_p. The wave in Equation 6.41 is zero if

$$\text{(i)} \quad \cos\frac{\Delta kx_g}{2} = 0$$

or

$$\text{(ii)} \quad \cos\bar{k}x_p = 0$$

For case (i),

$$\frac{\Delta kx_g}{2} = \frac{\pi}{2}, \frac{3\pi}{2}, \frac{5\pi}{2}\ldots(2n+1)\frac{\pi}{2},$$

that is,

$$x_g = \frac{\pi}{\Delta k}.$$

For case (ii),

$$\frac{\Delta kx_p}{2} = \frac{\pi}{2}, \frac{3\pi}{2}, \frac{5\pi}{2}\ldots(2n+1)\frac{\pi}{2},$$

that is,

$$x_p = \frac{\pi}{2\bar{k}}.$$

The fact that $\bar{k} \gg \Delta k$ conditions the values of x_g to be much larger than x_p at which the wave $E_R(x, t) = 0$. This implies that the first cosine term is of a very slow variation compared to the second cosine term, and hence the period of the first cosine term embraces a large number of periods of the second cosine term. The slow variation of this first term in the product (6.44) ends up at any point chopping off all values of the second cosine term that are above or below the value of the envelope describing the first term. Figure 6.5c illustrates the above situation, where the amplitude of the resultant $E_R(x, t)$ in Equation 6.44 has the form

$$\text{Resultant amplitude} = 2\cos\left(\frac{(\Delta k)x}{2} - \frac{(\Delta\omega)t}{2}\right). \tag{6.45}$$

As demonstrated in the figure, the second term $\cos\left(\dfrac{\bar{k}x}{2} - \dfrac{\bar{\omega}t}{2}\right)$, resembling any of the individual propagating waves, is said to be grouped or modulated in amplitude by the first term, producing a periodic wave form of an envelope labeled on the diagram as the wave group that moves with a velocity v_g given by

$$v_g = \frac{\Delta\omega}{\Delta k} \tag{6.46}$$

However, the velocity of each of the individual waves, called wave velocity or phase velocity v_p, is given by

$$v_p = \frac{\bar{\omega}}{\bar{k}}. \tag{6.47}$$

If in the above discussion, the two interfering waves were equal in frequency, then from Equation 6.36, $\omega_1 = \omega_2 = \bar{\omega}$, or in general terms just ω. The same applies to $\bar{k} = k_1 = k_2 = k$. Thus, Equation 6.34 becomes

$$v_p = \frac{\omega}{k}. \tag{6.48}$$

For infinitesimally small differences in $\Delta\omega$, Δk, Equation 6.46 becomes

$$v_g = \frac{d\omega}{dk}, \tag{6.49}$$

which upon using Equation 6.48 gives

$$\begin{aligned} v_g &= \frac{d\omega}{dk} \\ &= \frac{d}{dk}(kv_p) \\ &= v_p + k\frac{d}{dk}(v_p). \end{aligned} \tag{6.50}$$

The phase velocity v_p represents the velocity of a point on any of the individual waves that composed the interference, and is a measure of the index of refraction of the medium in which the wave is propagating, where

$$v_p = c/n; \qquad (6.51)$$

c is the speed of light in vacuum

n is the index of refraction of the medium

Using Equation 6.51 in 6.50 gives

$$v_g = v_p + k\frac{d}{dk}\left(\frac{c}{n}\right)$$

$$v_g = v_p - \left(\frac{c}{n^2}\right)k\frac{dn}{dk} \qquad (6.52)$$

or

$$v_g = v_p\left(1 - \left(\frac{k}{n}\right)\frac{dn}{dk}\right). \qquad (6.53)$$

COMMENT

The group velocity and phase velocity are equal when $\dfrac{dn}{dk} = 0$. This is the case when n is independent of wavelength, implying that all the superimposed are waves having identical frequency.

EXAMPLE 6.3

The ω–k dispersion relation for a hypothetical pulse is shown below.

(a) Calculate the group velocity and phase velocity of this wave at the point P.

(b) Comment on the relevance of the value of the ratio v_g/v_p.

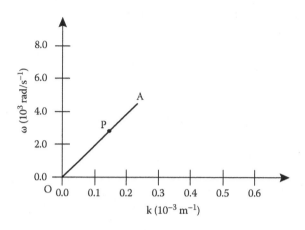

Solution

(a) Using Equation 6.48

$$v_p = \frac{\omega}{k}.$$

Substituting for $\omega = 2.5 \times 10^3$ rad/s and $k = 0.15 \times 10^{-3}$ m^{-1} (see the diagram), v_p becomes

$$v_p = \left(2.5 \times 10^3 \text{ rad/s}\right) / \left(0.15 \times 10^{-3} \text{ m}^{-1}\right)$$

$$= 1.7 \times 10^7 \text{ m/s}$$

Using Equation 6.49,

$$v_g = \frac{d\omega}{dk},$$

which is the slope of the line OA.

(b) Since the group velocity is equal to the phase velocity, $v_g/v_p = 1$, implying that all superimposed waves have the same frequency.

6.5 COHERENCE IS A MUST CONDITION FOR SUSTAINABLE INTERFERENCE

We notice that in the three cases discussed in the previous section, the resulting superposition depended on the difference in phase between the superimposed waves. It is important to mention a couple of points here: First, the effect of superposition applied to light waves is commonly known as interference. Second, as light waves are the result of atomic transitions from higher to lower atomic states, in a light wave of certain frequency emitted by a source, the electric field would have all directions of oscillation with no correlation in phase to waves of the same frequency emitted by another source nearby. For this reason, two such sources are called noncoherent, and interference of two waves emitted by the two sources is not possible due to the random nature of their phase differences. To sustain an interference pattern, the difference in phase between the interfering waves must be constant. This is the essence of *coherence* between the two waves. It is possible to provide a variety of cases, the simplest of which is having a light wave incident on a double-slit obstacle that acts to split the incident wave into two. As the two waves come from one source, any change in phase in one would automatically happen to the second. Another less restrictive condition for clear and distinct interference patterns is that the sources have the same wavelength. This will become clearer when interference is covered later in Chapter 7.

The significance of coherence or lack thereof can be demonstrated by rechecking Equation 6.31 for the interference of N light waves under the following conditions:

1. $\delta_i - \delta_j$ = constant, that is, independent of time. And assuming, for convenience, that this difference in phase is either zero or 2π and that all are of equal amplitude, E_o, Equation 6.31 becomes

$$E_{oR}^2 = \sum_i^N E_{oi}^2 + \left[2\sum_{j>i}^N \sum_i^N E_{oi}\, E_{oj} \right] \tag{6.54}$$

or

$$E_{oR}^2 = \sum_i^N (NE_o)^2. \tag{6.55}$$

Note There are several ways to show how the r.h.s. in Equation 6.54 reduces to the r.h.s. in Equation 6.55. The simplest is to demonstrate it for a limited number of waves: N = 2, N = 3, and N = 4.

$$\text{For } N = 2, \quad \sum_{i}^{2} E_{oi}^2 + \left[2 \sum_{j>i}^{2} \sum_{i}^{2} E_{oi}\, E_{oj} \right] = \left(2E_o\right)^2 = 4E_0^2.$$

$$\text{For } N = 3, \quad \sum_{i}^{N} E_{oi}^2 + \left[2 \sum_{j>i}^{N} \sum_{i}^{N} E_{oi}\, E_{oj} \right] = \left(3E_o\right)^2 = 9E_o^2.$$

$$\text{For } N = 4, \quad \sum_{i}^{N} E_{oi}^2 + \left[2 \sum_{j>i}^{N} \sum_{i}^{N} E_{oi}\, E_{oj} \right] = \left(4E_o\right)^2 = 16E_o^2.$$

Thus, (6.55) becomes

$$E_{oR}^2 = N^2 E_o^2. \tag{6.56}$$

That is, the intensity of the resultant

$$I_R = N^2\, I_o. \tag{6.57}$$

This shows how intense the resultant is compared to the individual wave intensities.

2. $\delta_i - \delta_j$ = random, oscillating—practically speaking—infinite number of values. The infinite number of waves will then have their cosine sum up to zero. Thus, the second term on Equation 6.54 vanishes, and Equation 6.55 becomes

$$E_{oR}^2 = \sum_{i}^{N} E_{oi}^2 \tag{6.58}$$

$$E_{oR}^2 = NE_o^2. \tag{6.59}$$

That is, the intensity of the resultant

$$I_R = N\, I_o. \tag{6.60}$$

PROBLEMS

6.1 Given a complex wave function, $E_+(x,t) = E_o e^{i(k_x x - \omega t)}$, find the change, if any, in the phase of the wave function upon multiplying it by (a) i, (b) −i, (c) −i².

6.2 Consider the two waves, $E_1(x,t) = E_{o1} \cos(kx - \omega t)$ and $E_2(x,t) = E_{o2} \cos\left(kx - \omega t + \dfrac{\pi}{4}\right)$, that have equal amplitudes, $E_{o1} = E_{o2} = 6.00$ m.

(a) Determine the amplitude and phase of their resultant at a certain value of x.

(b) Express the resultant wave in a proper representative wave form.

6.3 Redo the previous problem using the phasor method.

6.4 Consider the wave $E_{R+}(x,t) = 5.0 \cos\left(\dfrac{4\pi}{9} - \omega t\right)$ that was obtained in Example 6.2 as the resultant of two other waves.

(a) Write down an equation that describes a wave, call it E_{R-}, such that E_{R-} and E_{R+} have the same amplitude, but 180° out of phase.

(b) Determine the resultant of E_{R-} and E_{R+} using the analytic method.

6.5 Given the two waves,

$$E_+(x,t) = E_o e^{i(k_x x - \omega t)}, \quad E_-(x,t) = E_o e^{i(k_x x + \omega t)},$$

show that at any value of x, the resultant E_R of the waves is equivalently expressed by either the real part or the imaginary part of the complex form of the resultant.

6.6 Two coherent plane waves, $E_1 = 6 \sin(\omega t)$, $E_2 = 8 \sin(\omega t + \pi/2)$, polarized along the y direction are traveling together along the same line, the x direction; the amplitudes are in meters and t is in seconds.

(a) Sketch an approximate phasor diagram that displays the two waves and their resultant. Show on the diagram all quantities relevant to the waves such as amplitudes and phase constants.

(b) Write a mathematical expression for the resultant wave in the form $E_R = E_o \sin(\omega t + \delta)$.

6.7   Redo the previous problem analytically
  (a)   To determine the resultant of the above two waves. Relate the answers that you get analytically with those offered by the phasor diagram. Point out all aspects of consistency, if any, between the two methods.
  (b)   To write a mathematical expression for the resultant wave in the form $E_R = E_o \sin(\omega t + \delta)$.

6.8   Considering the following three waves, $E_1(x, t) = 3.0\cos\left(\dfrac{\pi}{6} - \omega t\right)$, $E_2(x, t) = 4.0\cos\left(\dfrac{2\pi}{3} - \omega t\right)$, $E_3(x, t) = 5.0\cos(\pi - \omega t)$, determine the amplitude and phase angle of their resultant E_R. (*Hint*: Check Example 6.2 that may be helpful.)

6.9   Use the phasor method to show that
  (a)   The addition of two waves, equal in amplitude, frequency, and phase, is simply twice either of them. Comment on the phase of the resultant in this case.
  (b)   The addition of two waves equal in amplitude and frequency but of 180° difference in phase is a wave of zero amplitude. Comment on the phase of the resultant obtained in this case.

6.10   Consider the triangle whose sides represent three phasors, E_{o1}, E_{o2}, and E_{o3}, drawn head to tail as depicted below. Knowing that the lengths of these sides are 1.0, 1.2, and 0.86, respectively, determine the resultant of E_{o1} and E_{o2}. (*Hint*: Use the trigonometric relation regarding the length of the side OB; OB = $[(OA)^2 + (AB)^2 - 2(OA)(AB) \cos \varphi)]^{1/2}$.

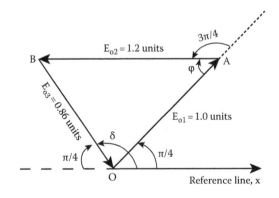

6.11   In the figure below, you have two ω–k dispersion plots for a hypothetical wave group for each.
  (a)   Calculate the group velocity and phase velocity at points P and
        Q in the following figure.
  (b)   Which plot (a) or (b) could be a realistic one?

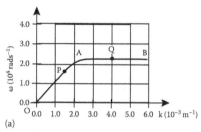

(a)

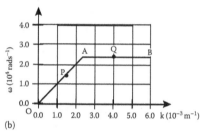

(b)

6.12   The dependence of the index of refraction n on wavelength in the
visible was expressed by Cauchy in the approximate form $n(\lambda) = A +
B/\lambda^2$, where A and B are constants. Find the phase and group veloci-
ties at 7.00×10^2 nm for one type of glass for which $A = 1.45$ and $B =
2.5 \times 10^{-14}$ m^2 and $n = 1.45$.

6.13   Find the group velocity of EM waves if the phase velocity is $v_p = a/\lambda$,
where a is constant.

6.14   There are several ways by which v_g and v_p are related. Benefit from
one or more of Equations 6.45 through 6.53 to show that

$$v_g = v_p - \lambda\left(dv_p/d\lambda\right).$$

6.15   Following the steps you may have used in the previous problem, show
that $v_g = v_p\left[1 + \dfrac{\lambda}{n}\dfrac{dn}{d\lambda}\right]$.

6.16   Given the dispersion relations, (i) $\omega = bk$ and (ii) $\omega^2 = bk^2$, determine
$(d\omega/dk)$ and use the derivation to establish a relation between the
group velocity and the phase velocity.

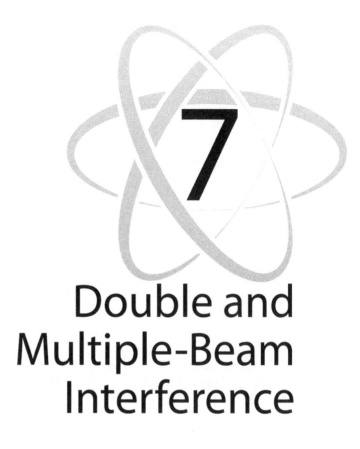

Double and Multiple-Beam Interference

The happiness of your life depends on the quality of your thoughts.

Marcus Aurelius (121–180)

7.1 INTRODUCTION

In this chapter, a special kind of superposition is explored. It is the interference of light waves subject to particular conditions of wavelength, amplitude, phase, and coherence. Most important among these is the coherence of the interfering waves and direction of oscillation of their electric fields; the former is critical for the sustainability of the interference, and the latter makes the analysis simpler, enabling better understanding of the properties of the resulting interference. Taking advantage of the algebra and analysis introduced in Chapter 6, we treat here a variety of methods followed in understanding interference. For interference of two waves to happen a difference in phase that remains constant between them is fundamental. Interference may happen between two beams or multiple beams. Young's double-slit setup that provides a two-beam interference was the earliest experiment used to produce an interference pattern. It generated more interest and helped to empirically support the wave nature of light, which was established theoretically several decades later by Maxwell in his electromagnetic theory of light.

7.2 YOUNG'S DOUBLE-SLIT EXPERIMENT

Young's double-slit experiment consists of a point source S of fixed frequency/ wavelength, that is, monochromatic, a lens, and an obstacle AB with two narrow slits separated by a small distance d with a screen placed a long distance $D \gg d$ away from the slits. The lens facing the source is placed in front of the obstacle at a distance from the source that equals its focal length (Figure 7.1).

With this arrangement, the spherical waves incident on the lens emerge as plane waves illuminating the two slits. Thus, the waves arriving at the slits S_1 and S_2 are all in phase. Taking into consideration one wave from each slit directed at any given point on the screen, we have two waves that would travel forward along different paths, S_1P and S_2P, labeled in the sketch as r_1 and r_2, respectively. For a small slit separation d of the order of 10^{-6} m, the separation D between the obstacle AB and the screen, being in the range of 0.6–1.0 m, is practically infinite compared to d (Figure 7.1 is not to scale). Therefore, the two paths, r_1 and r_2, are considered parallel lines, meeting at an infinitely distant point, P. A line S_1N dropped perpendicular to path r_2 is then perpendicular to both paths r_1 and r_2, making the two segments S_1P and

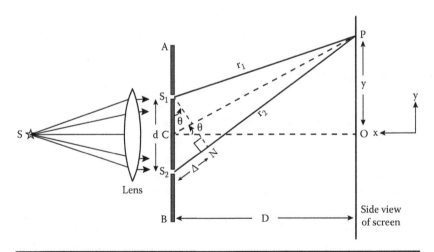

FIGURE 7.1 An illustration of Young's double-slit interference setup. The spherical waves emitted by the source emerge from the lens as plane waves.

NP equal. Thus, S_2N is the path difference between r_1 and r_2. Denoting this difference by Δ, we have

$$r_2 - r_1 = \Delta \tag{7.1}$$

where, from the geometry,

$$\Delta = d \sin \theta. \tag{7.2}$$

Following the argument laid out in Section 6.3.1, the two waves, having the same frequency and of phase constants δ_1 and δ_2, may be described by the equations

$$E_1(r, t) = E_{o1} \cos(kr - \omega t + \delta_1),$$

$$E_2(r, t) = E_{o2} \cos(kr - \omega t + \delta_2),$$

where k is the propagation constant along r, the same for both waves. Their resultant, established in Equation 6.31, is

$$E_{oR}^2 = \sum_i^2 E_{oi}^2 + \left[2 \sum_{j>i}^2 \sum_i^2 E_{oi} E_{oj} \cos(\delta_i - \delta_j) \right]. \tag{7.3}$$

That is, for the two waves, the resultant at a point P is given by

$$E_{oR}^2(r, t) = E_{o1}^2 + E_{o2}^2 + 2E_{o1}E_{o2}\cos\delta,$$

where δ is the angular phase difference between the two waves at P. As the amplitudes of the two waves originating at S_1 and S_2 are equal, $E_{o1} = E_{o2} = E_o$, the amplitude of the resultant becomes

$$E_{oR}^2(r, t) = 2E_o^2 + 2E_o^2 \cos\delta$$

or

$$E_{oR}^2(r, t) = 2E_o^2(1 + \cos\delta),$$

or

$$E_{oR}^2(r, t) = 4E_o^2 \cos^2 \frac{\delta}{2}. \tag{7.4}$$

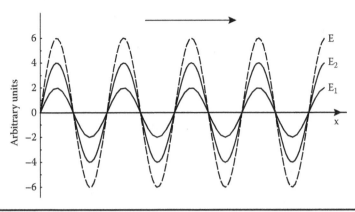

FIGURE 7.2   Demonstration of the resultant E_R of the superposition of two waves, E_1 and E_2, that are of zero phase difference.

As the point P is an arbitrary point on the screen along y, there are two sets of positions of point P where the phase difference δ is of particular interest.

I.   *First Case: Positions of Constructive Interference at P*
    Constructive interference at point P occurs for

$$\delta = 0, \ \pm 2\pi, \ \pm 4\pi, \dots \pm 2m\pi; \quad m = 0, 1, 2, 3, \dots \tag{7.5}$$

The waves are perfectly in phase; thus, they interfere in a *total constructive interference* (Figure 7.2). From Equation 7.4, the resultant amplitude would be twice the amplitude of either wave, and the resultant intensity is four times the intensity of any of the interfering waves.

For m = 0, the phase difference between the two waves is zero. That is, from Equation 7.2, the path difference Δ between the two waves is zero. This corresponds to point O on the screen where r_1 and r_2 are equal. From an intensity perspective, the corresponding locations of intensity are called bright bands. The integer value of m refers to the order of the band; m = 0 is the zeroth or central bright band, m = 1 is the first bright band, and so on for the higher values of m.

II.   *Second Case: Positions of Destructive Interference at P*:

$$\delta = \pm \pi, \ \pm 3\pi, \dots \pm 2\left(m + \frac{1}{2}\right)\pi; \quad m = 0, 1, 2, 3, \dots \tag{7.6}$$

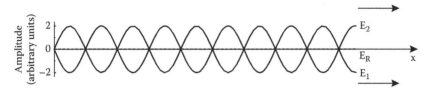

FIGURE 7.3   Superposition of two waves that are 180° degrees out of phase for two waves of equal amplitude.

The two waves would be out of phase with each other; one wave would have a minimum amplitude at the time the other wave had a maximum amplitude. Again, as the interfering two waves coming from the slits S_1 and S_2 are of equal amplitudes, the resultant amplitude would be zero, and the resultant intensity would also be zero (Figure 7.3). This descriptive layout is what Equation 7.4 provides for the positions of zero light intensity at their corresponding values of δ. This is called *total destructive interference*; $m' = 0, 1, 2, 3, ...$ defines the order of the dark band, the first, second, third dark bands, respectively.

Figure 7.4 is a sketch of the intensity versus the factor $\cos^2 \dfrac{\delta}{2}$ showing the maxima and minima of the intensity at the corresponding values of $\delta = 0, \pm 2\pi, \pm 4\pi, ... \pm(2m\pi)$ for the maxima and $\delta = \pm\pi, \pm 3\pi, ... \pm 2\left(m' - \dfrac{1}{2}\right)\pi$ for the minima.

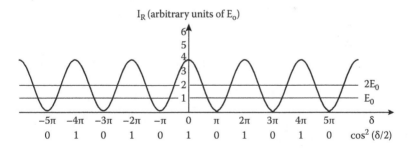

FIGURE 7.4   An illustration of the intensity as a function of δ. The maxima and minima of the function $\cos^2 \dfrac{\delta}{2}$ correspond to the maxima and minima of diffraction.

7.2.1 ANALYSIS OF YOUNG'S DOUBLE-SLIT INTERFERENCE PATTERN

From Equation 7.1, the path difference between the two waves as they arrive at P is

$$r_2 - r_1 = \Delta.$$

The angular phase difference δ that corresponds to this path difference is

$$\delta = k\Delta. \tag{7.7}$$

Substituting for $k = \dfrac{2\pi}{\lambda}$ and the condition for constructive interference, $\delta = 2m\pi$, reduces the above equation to

$$2m\pi = \frac{2\pi}{\lambda}\Delta,$$

while for destructive interference, $\delta = (2m + 1)\pi$, Equation 7.7 reduces to

$$(2m+1)\pi = \frac{2\pi}{\lambda}\Delta.$$

After a little rearrangement of terms in the above two equations, they become

1. *Condition for constructive interference*

$$\Delta = m\lambda; \quad m = 0, \pm1, \pm2, \pm3... \tag{7.8}$$

2. *Condition for destructive interference*

$$\Delta = \left(m' - \frac{1}{2}\right)\lambda; \quad m' = \pm1, \pm2, \pm3... \tag{7.9}$$

Substituting in Equation 7.8 for $\Delta = d\sin\theta$ for constructive interference gives

$$d\,\sin\theta_m = m\lambda, \quad m = 0, \pm1, \pm2,... \tag{7.10}$$

or

$$\sin\theta_m = m\frac{\lambda}{d} \quad m = 0, \pm1, \pm2,... \tag{7.11}$$

For small angles, $\tan\theta \cong \sin\theta = \theta$; thus, we have

$$\theta_m = m\frac{\lambda}{d} \quad m = 0, \pm1, \pm2,... \tag{7.12}$$

Equations 7.11 and 7.12 give the angular positions of the mth bright fringe.

Also, for a coordinate system with an origin at the center of the screen, the x axis perpendicular to the plane of the screen and the y axis in the plane along the vertical direction make

$$\tan\theta \cong \sin\theta = \frac{y_m}{D}.$$

Equation 7.10 becomes

$$d\frac{y_m}{D} = m\lambda,$$

giving *the height of the mth bright fringe* on the screen to be

$$y_m = m\frac{D\lambda}{d}; \quad m = 0, \pm 1, \pm 2,... \tag{7.13}$$

Using Equation 7.13, the height of the (m + 1)th bright fringe is then

$$y_{m+1} = (m+1)\frac{D\lambda}{d}.$$

Thus,

$$y_{m+1} - y_m = (m+1)\frac{D\lambda}{d} - m\frac{D\lambda}{d}.$$

Thus, denoting the lateral distance between the centers of two successive bright fringes $y_{m+1} - y_m$ by Δy makes

$$\Delta y = \frac{D\lambda}{d}. \tag{7.14}$$

Thus, all bright fringes have the same width and so do the dark fringes. Thus, Equation 7.14 applies for both bright and dark fringes. Students should verify this latter statement for themselves. Also note that in the way m and m′ we just defined, m = 0, 1, 2, 3,... refer to the zeroth, first, second, third bright fringes, and so on, while m′ = 1, 2, 3,... refer to the first, second, third dark fringes, and so on.

EXAMPLE 7.1

In a Young's double-slit experiment, the two slits are 0.20 mm apart and the screen is 1.452 m away from the slits. If the fourth minimum is 14.52 mm from the center of the screen

(a) Determine the wavelength of the incident light.
(b) In the above pattern, what is the ratio between the intensities of the first and second fringes?

Solution

(a) Referring to the order of interference by m', the condition for a minimum is

$$d \sin\theta = (m' - 1/2)\lambda.$$

The fourth minimum implies that m' = 4. Therefore,

$$d\frac{y}{D} = (7/2)\lambda; \quad (0.20 \times 10^6 \text{ nm})(14.52 \times 10^6 \text{ nm})/(1.452 \times 10^9 \text{ nm})$$

$$= (7/2)\lambda.$$

This gives λ = 571.4 nm.
(b) The interference fringes are equal in intensity. Thus, the ratio between the intensities of the first and second fringes is one (see Equation 7.4 and consider δ).

7.2.2 FURTHER ANALYSIS OF YOUNG'S DOUBLE-SLIT INTERFERENCE PATTERN

7.2.2.1 Height of Bright Fringes

Figure 7.5 illustrates Young's setup and the interference pattern as formed on a screen according to the intensity distribution, Equation 7.4. Equation 7.13 expresses the dependence of the lateral position of a bright fringe y_m on the wavelength of light employed in the experiment.

In a fixed setup, the position of a fringe of certain order would appear farther away from the center O if light of longer wavelength was used. An important side of this feature is that using a light of two component wavelengths, similar fringe orders in the two patterns would be slightly shifted from each other. *However, the central zeroth fringe for both will coincide.*

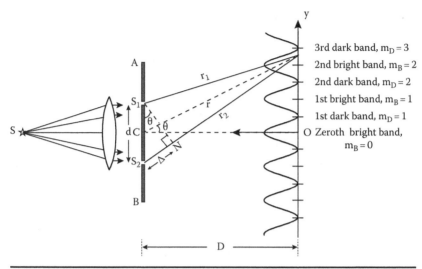

FIGURE 7.5 A demonstration of the double-slit diffraction pattern. The maxima and minima are displayed on the screen.

7.2.2.2 Width of the Bright Fringes

As described earlier in Equation 7.14, besides showing that the width Δy of all fringes is the same, it inversely depends on the slit separation d. A setup with a smaller separation, d, gives a wider width of the fringes and bigger overall screen-covered area. Figure 7.6 illustrates two such patterns obtained

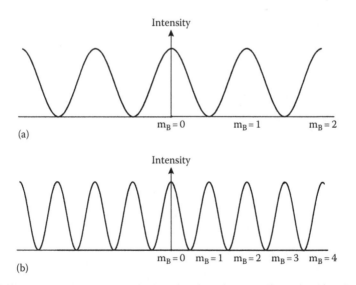

FIGURE 7.6 An illustration of the dependence of the width of Fraunhofer diffraction fringes on the slit separation: (a) slit separation d and (b) slit separation 2d.

in a Young's double-slit experiment, one obtained for a slit separation, d, and another having double the slit separation, 2d.

EXAMPLE 7.2

In a Young's double-slit experiment, yellow light of wavelength 589.2 nm illuminates two slits that are 0.200 mm apart. If the screen is 1.75 m away from the slits, determine

(a) The spacing between the interference fringes
(b) The lateral position of the fourth maximum fringe from the center of the screen

Solution

(a) The spacing is given by Equation 7.14:

$$\Delta y = \frac{D\lambda}{d}.$$

Therefore,

$$\Delta y = \frac{(1.75\,\text{m})\left(589.2\times10^{-9}\,\text{m}\right)}{\left(0.200\times10^{-3}\,\text{m}\right)},$$

giving

$$\Delta y = 516\times10^{-6}\,\text{m} = 5.16\,\text{mm}.$$

(b) The condition for the fourth maximum is

$$d\sin\theta = (m)\lambda; \quad m = 4.$$

Therefore,

$$d\frac{y}{D} = \left(\frac{7}{2}\right)\lambda$$

This gives y = 20.6 nm.

7.3 LLOYD'S MIRROR

This setup consists of a source S, a mirror, and a screen (Figure 7.7). In the arrangement known as Lloyd's mirror, a flat mirror is laid out horizontally along a line perpendicular to a screen that stands facing the incident light. The mirror at one end does not have to touch the screen, though it could be demonstrating a special case of interest. Meanwhile, consider two beams, ray 1 incident on the screen at P and ray 2 incident on the mirror with an angle that makes its reflected ray 2' at N. The two rays 1 and 2' meeting at P interfere with each other, producing a bright or dark spot or a spot whose intensity is somewhat in between bright and dark. By intuition, checking the resultant at other positions as one sights down the screen, conditions of maxima and minima are found until arriving at C. Also consider two additional rays, ray 3 and 3', slightly displaced from one another headed toward C. Ray 3 just misses the edge of the mirror and so is not reflected. Ray 3' hits the farthest edge point and is reflected. The reflected ray intersects ray 3 at C. The condition for constructive interference would produce a bright spot at C. However, experiments showed that the spot at C was dark. The only explanation was that ray 3' must have changed phase by π upon reflection. Thus, it is now a foregone conclusion that a light wave originating in a medium of index of refraction n incident on a boundary of a medium whose index of refraction is higher than n changes phase by π. However, the converse does not happen. A light wave originating in a medium of refraction n incident on a boundary of a medium of lower index of refraction does not change phase upon reflection. This observation was crucial for the understanding of reflection from

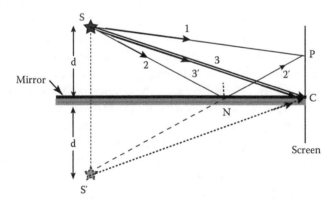

FIGURE 7.7 Lloyd's mirror. The source S and its image S' act as two coherent sources. Any ray reflected from the mirror can be considered as being emitted by the source image S'.

surfaces of single thin films and multilayer coatings. In contrast to reflection, a light wave endures no phase change during transmission across different media.

As the diagram illustrates, the reflected ray 2′ extended backward along NS′ looks as if it were originating from the image source S′. Thus, the situation is equivalent to having two coherent sources, S and S′, from which there are two rays reaching point P producing an interference pattern. Except for two features, fringes on the screen in Lloyd's mirror look similar to those appearing in Young's double-slit setup. One difference is that the center of the screen in Lloyd's mirror setup is dark, and the second is that the interference in Lloyd's on the screen is not symmetric around the center. Besides, an interesting physical feature of Lloyd's experiment is that the spacing between the two sources here is twice the height of S above the screen.

7.4 NEWTON'S RINGS

This setup consists of a plano (top surface)-convex (bottom surface) lens whose spherical surface is placed on top of a flat glass plate such that the two surfaces share a very narrow area of contact. The schematic in Figure 7.8 shows a light ray (1) incident almost normally on the lens from above gets partially reflected from the spherical surface along (1′) and partially transmitted toward the plate. This part of light gets reflected from the plate's top surface; ray 2′ passes through the lens along 2″ back toward the observation position.

The figure shows a situation at P_1. The ray reflected from the plate changes phase by π, while the first ray, (1′), does not. Therefore, as the two rays meet at O, there is a phase difference of π along with another phase difference due to a path difference between them, and ray (2) is traveling through double the thickness of the thin air film. Depending on whether the total difference in phase is 2π, 4π, 6π, ... or π, 3π, 5π, ..., the interference viewed at O would produce constructive (bright) or destructive (dark) circles, respectively; these are called fringes. For a monochromatic light wave, the circles are more like narrow concentric rings with a center at P. The fringe at P is a dark spot formed from light incident normal to the lens surface at P; this is because the two rays 1 and 2″ have equal path lengths, thus having a phase difference of π only. The path difference between two successive dark fringes would equal one wavelength, λ. With the central fringe labeled as the zeroth fringe, a number of m dark fringes are observed between M_2 and M_1. This means that there are m + 1 fringes between M_2 and M_1. TM_2 in the sketch would represent the radius of the mth fringe.

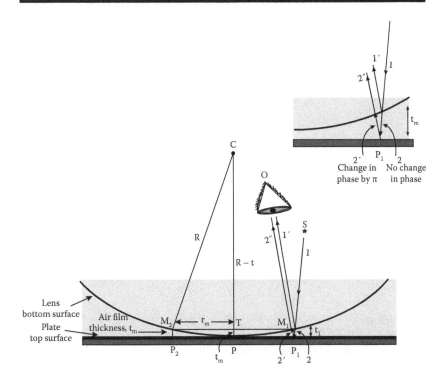

FIGURE 7.8   An illustration of Newton's rings apparatus. In the figure, a lens of radius R and a glass plate are 1′ and a transmitted part 2. 2 is reflected off the plate as 2′ that transmits through the lens as 2″. The interference occurs between 1′ and 2″.

Labeling the center of the lens spherical surface as C and its radius as R, the triangle CTM_2 is a right-angled triangle. Thus,

$$R^2 = r_m^2 + (R - t_m)^2,$$

which reduces to

$$2Rt_m = r_m^2 + t_m^2,$$

giving

$$R = (r_m^2 + t_m^2)/2t_m. \tag{7.15}$$

Since r_m is measurable by a traveling microscope, having a lens of predetermined radius of curvature R allows the thickness t_m to be determined.

From the condition of interference for the mth *dark fringe,*

$$2t_m = m\lambda. \tag{7.16}$$

This equation would give the wavelength of the light used in the experiment. Obviously, for the mth *bright fringe,* the above condition becomes

$$2t_m = \left(m + \frac{1}{2} \right)\lambda. \tag{7.17}$$

The discussion has been focused on fringes as seen from an observation point, O, that lies above the setup, a setting called the reflection mode. If the arrangement is made to see the interference fringes from below, the setting is in the transmission mode. In this mode, the dark and bright fringes switch appearances; the central becomes bright, followed by a series of dark–bright fringes. In Figure 7.9 are the patterns as they would look in the two modes.

In the transmission mode, Equations 7.16 and 7.17 would still be used, but with a switch in their roles, Equation 7.16 for the bright fringe and (7.17) for the dark fringe.

It is worth noting that once the thickness is calculated from (7.15), Equations 7.16 and 7.17 may be used to determine the index of refraction of a transparent solution that may be released into the space between the lens and the plate. Care should be observed to whether the solution index of refraction is lower or higher than the lens and the plate indices of refraction. Thus, the appropriate interference conditions should be used. If n is lower than the indices of refraction of the lens and the plate, Equations 7.16 and 7.17 after including the index of refraction of the film become

$$2nt_m = m\lambda \tag{7.18}$$

$$2nt_m = \left(m + \frac{1}{2} \right)\lambda. \tag{7.19}$$

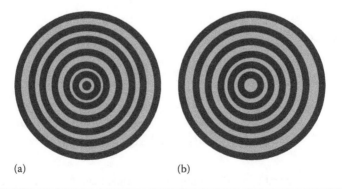

(a)   (b)

FIGURE 7.9   A sketch (not a photograph) of Newton's rings as seen in (a) reflection mode and (b) transmission mode.

EXAMPLE 7.3

Newton's rings are formed between the spherical surface of a lens and an optical flat. A clear liquid (acetone) of an index of refraction n = 1.20 fills the gap between the lens and the flat plate. For a lens of 20.0 cm radius of curvature, Newton's rings are observed with a red light of 700.0 nm wavelength. Determine the number of fringes that are contained within a fringe of 1.00 mm radius.

Solution

   (a) From $r_m^2 = 2Rt_m$, $t_m = r_m^2/2R = 1.00 \times 10^{-6}\,m^2/(2 \times 0.200\,m) = 2.5 \times 10^{-6}\,m$.

   (b) Now $t_m = m\lambda/2n_f$,

      or

$$m = 2n_f t_m/\lambda = 2(1.2)(2.5 \times 10^{-6}\,m)/(700 \times 10^{-9}\,m) = 8.57.$$

Thus, eight dark fringes would be visible within the 1.00 mm radius.

7.5 INTERFERENCE OF LIGHT IN THIN FILMS

The schematic in Figure 7.10 illustrates a dielectric thin film of an index of refraction n and thickness t. The figure shows a light ray incident along path 1 on the top surface of thin film, its reflection along path 1', and its refraction travels through the film along path 2. Ray 2 reflects off the lower boundary at B along path 2', which refracts at boundary C as shown, emerging into the original medium. Although for clarity, the diagram shows a rather large angle of incidence, θ, the analysis that follows is based on light incident normal to the film surface.

At the lower boundary, part of ray 2 refracts at B along path 4, emerging into the last medium, air in this case. As discussed earlier, ray 1' changes phase by π upon reflection at A, while rays 2, 2', and 3 do not encounter such change. However, ray 3 has covered a longer path than ray 1'. This excess in path is equal to AB + BC. Therefore, there is a path difference between rays 1' and 3, resulting in the two rays interfering constructively or destructively according to the net phase difference between these rays as they arrive at the observation point O.

A total phase difference of 2π, 4π, 6π, etc., would be the condition for constructive interference. Due to the phase difference of π that ray 1' has carried upon reflection, the condition for constructive interference of rays 1' and 3 would be

$$AB + BC = p\frac{\lambda'}{2},$$

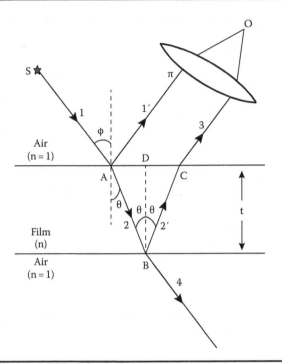

FIGURE 7.10 Interference through a thin film.

where p = 1, 3, 5, ... and λ' is the wavelength within the film which is λ/n. This may be written as

$$n(AB+BC)=\left(m+\frac{1}{2}\right)\lambda,$$

where m = 0, 1, 2, 3, ... and [n(AB + BC)] is the optical path difference between the two rays. From the geometry, for an almost normal incidence of light, θ and θ' are very small, and cos θ' may be approximated to unity. Therefore, (AB + BC) = 2t. This makes the condition for a *constructive interference* of light incident normally on the film

$$2nt =\left(m+\frac{1}{2}\right)\lambda; \quad m=0, 1, 2, 3,... \qquad (7.20)$$

where m is called the order of interference. It can be seen that the *destructive interference* would be satisfied if

$$2nt =(m)\lambda; \quad m=1, 2, 3,.... \qquad (7.21)$$

It has to be noted that if the film was deposited on a substrate of index of refraction n_s (Figure 7.11), the above conditions would change if $n_s > n$,

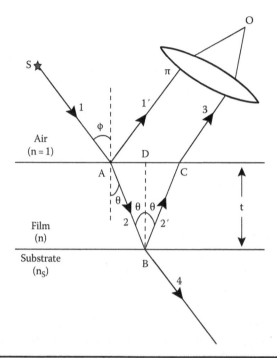

FIGURE 7.11   A film deposited on a substrate of index of refraction n_s. The condition of interference depends on whether $n_s > n$ or $n_s < n$.

because there is an additional phase shift of pi at the second boundary. However, there will be no change in the above two expressions if $n_s < n$. Thus, for normal incidence and for $n_s > n$, the conditions (7.20 and 7.21) become

(i)   *Constructive interference*

$$2nt = (m+1)\lambda; \quad m = 0, 1, 2, 3,... \qquad (7.22)$$

(ii)   *Destructive interference*

$$2nt = \left(m + \frac{1}{2}\right)\lambda; \quad m = 0, 1, 2, 3,... \qquad (7.23)$$

Note that if the beam was white and destructive interference was observed for a particular color, then for that wavelength,

$$2nt = \lambda/2,$$

giving

$$nt = \lambda/4. \qquad (7.24)$$

This is a case of antireflection, and the thin deposited layer is called antireflection coating, used to minimize undesired reflections of that particular wavelength. Since nt is the optical thickness of the layer, the film is called a quarter-wave plate. More generally, in case light is incident with an arbitrary angle, θ', on a film surrounded by air on both sides, Equations 7.22 and 7.23 would be

(i) *Constructive interference*

$$2nt\cos\theta' = \left(m + \frac{1}{2}\right)\lambda; \quad m = 0, 1, 2, 3,\dots \tag{7.25}$$

(ii) *Destructive interference*

$$2nt\cos\theta' = (m)\lambda; \quad m = 0, 1, 2, 3,\dots \tag{7.26}$$

Depending on the value of θ', constructive and destructive interference would become harder or impossible to observe if light is coming from a point source. This is due to unknown variations in path differences associating different incident rays and their multiple reflections prior to getting to the observation position. That is why an extended source becomes a necessity for those fringes to be seen. Figure 7.12 illustrates this situation. The figure shows three rays emanated from the source S. However, for clarity, only one ray was pursued for phase analysis.

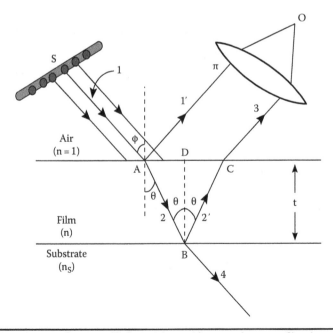

FIGURE 7.12 Interference of light incident with an angle φ on a film deposited on a substrate of index of refraction n_s. The light source is an extended source.

EXAMPLE 7.4

Consider an oil film (n = 1.40) on a glass (n = 1.45) slide on which a beam of white sunlight is shining. It was observed that a strong first-order reflection of yellow light, λ = 600 nm wavelength, occurs. Assuming normal incidence, determine the thickness of the oil film.

Solution

Due to having a change in phase to each of the rays, the condition is a maximum reflection. Thus,

$$2n_f t = m\lambda.$$

That is,

$$2(1.40)t = (1)(600\,nm),$$

giving t = 214 nm.

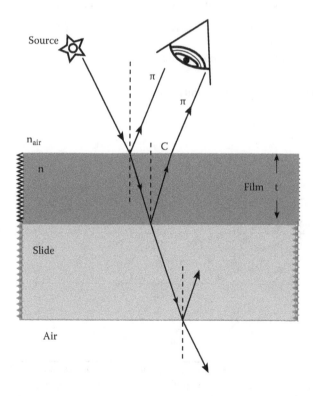

7.6 MULTIPLE-BEAM INTERFERENCE

Treatment of interference in the thin film (Figure 7.12) where the first refracted ray (ray 2) encounters multiple reflections within the film gets more complicated. In this case, multiple rays exist on each side of the film as demonstrated in Figure 7.13. The set of rays above the first boundary constitutes the reflected part of light, and the set below constitutes the transmitted part. Calling the intensities of these parts **R** and **T**, respectively, and having the incident amplitude normalized to unity, a dielectric film satisfies the following relation:

$$R + T = 1. \tag{7.27}$$

Except for ray 1′ each of the reflected rays 2′, 3′, 4′, ... travels through a path of multiples of 2nt longer than that traveled by 1′; n is the film index of refraction. As none of 2′, 3′, 4′, etc., experience any phase change during reflections, aside from the path difference, they are all in phase. However, 1′ carries a phase change of π due to reflection. Therefore, if 2nt = mλ, then 1′ and the rest combined would give a destructive interference.

The contrasting case of a constructive interference would occur if $2nt = \left(m + \dfrac{1}{2}\right)\lambda$; m in both cases is an integer that marks the order of the interference. In brief, these conditions are

(i)   *Destructive*, that is, condition for minima

$$2nt = m\lambda; \quad m = 1, 2, 3, 4,... \tag{7.28}$$

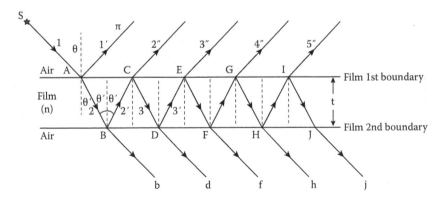

FIGURE 7.13   Multiple-beam interference. Ray 1 is the incident beam. The side above the film first boundary is the reflection side, and that below the film second boundary is the transmission side.

(ii)   *Constructive*, that is, condition for maxima

$$2nt = \left(m + \frac{1}{2}\right)\lambda, \quad m = 0, 1, 2, 3, 4,\dots \tag{7.29}$$

7.7 FRINGES OF EQUAL INCLINATION: FIZEAU FRINGES

Multiple-beam interferometry is used in measuring the thickness of thin films and air film confined between two wedge-forming glass slides (Figure 7.14). Using an extended source and monochromatic light is fundamental in generating visible fringes in the area of the confined air film. For normal or near-normal incidence, a set of fringes form, each of which corresponds to a particular optical path difference attributed in each case to a fixed angle of incidence; that is why they are called fringes of equal inclination. The path difference due to one ray traveling in the air film of index n_f is twice the film optical thickness, that is, $2n_f t$. Accounting for a π change in the phase of one of the rays and realizing that each value of m corresponds to a different thickness of the wedge-shaped film, the conditions for the resulting interference fringes are as follows:

(i)   *Constructive interference*

$$2n_f t = \left(m + \frac{1}{2}\right)\lambda; \quad m = 0, 1, 2, 3,\dots \tag{7.30}$$

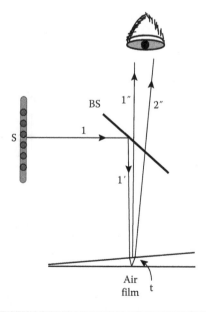

FIGURE 7.14   Interference through a wedge of air formed between two glass plates.

(ii)   *Destructive interference*

$$2n_f t = (m)\lambda; \quad m = 0, 1, 2, 3,... \tag{7.31}$$

Another method that employs Fizeau fringes to measure the thicknesses of thin layers is that developed by Tolansky. This is shown in Figure 7.15, where the thin film, deposited on a substrate AB, extends from C to D.

The setup consists of an extended coherent source, usually a sodium lamp, a beam splitter, and a microscope. The film and the exposed areas of the substrate are covered by depositing in vacuum a heavy metallic coating, silver in most cases. The step D′D″ of the coating material reproduces the step on the underlying material. A partially silvered platelet EF is placed on the step forming a wedge that extends slightly beyond the step to facilitate coverage of the step.

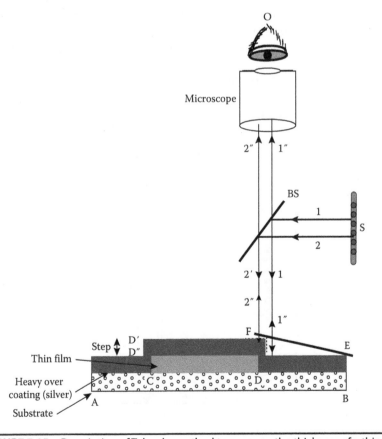

FIGURE 7.15   Description of Tolansky method to measure the thickness of a thin film.

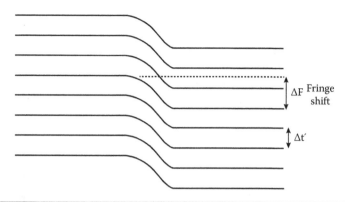

FIGURE 7.16 Shift of interference fringes due to a step that corresponds to the film thickness shown in Figure 7.15.

Consider two light rays, 1 and 2, from the source incident on the beam splitter that reflects these down along 1′ and 2′. These in turn travel down toward the platelet, such that 1′ is incident slightly before the step and 2′ is incident slightly beyond the step. The two rays are reflected back along 1″ and 2″; each is experiencing the same optics described earlier for the wedge. Thus, two straight line sets of fringes are generated with a shift in the fringes of one set off those of the other set (Figure 7.16).

This shift is due to the step, which can be related to the thickness of the film by the relation

$$2n_f t = m\lambda. \tag{7.32}$$

Consider the physical distance between two adjacent fringes, which is equivalent to a thickness change of the film wedge, $\Delta t = \lambda/2n_f$ (i.e., $\Delta m = 1$) on both the top and bottom of the step. That can be used as a scaling factor for the vertical distances in Figure 7.16. The physical shift of the fringe pattern using adjacent fringes can be determined by measuring the physical distance between adjacent fringes in the figure, call it $\Delta t'$, and the physical distance of the shift of either of the fringes passing over the step. Call this ΔF. Therefore, the actual step height $\Delta H = (\Delta F/\Delta t')\Delta t$ and for a film of air $n_f = 1.0$, so $\Delta H = (\Delta F/\Delta t')\lambda/2$.

7.8 MICHELSON INTERFEROMETER

The Michelson interferometer is an important double-beam interference device designed by Albert A. Michelson (1887); the interferometer was crucial in an interference experiment carefully set to establish that the speed of light in vacuum was constant, which helped dismiss the prevalent notion of ether as a medium necessary for the propagation of light.

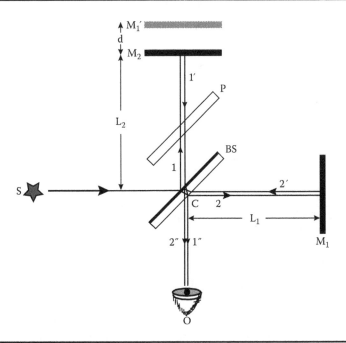

FIGURE 7.17   Michelson interferometer.

A schematic of Michelson's interferometer is shown in Figure 7.17. It consists of a source, S; a 50% transmission-/reflection-coated glass plate called a beam splitter, BS; and two mirrors, M_1 and M_2, positioned along two directions that are perpendicular to each other. The mirrors are at equal distances from the position of the beam splitter at C.

Each of the mirrors is attached to a back support controlled by a microm-eter capable of fine linear adjustments. Controlled tilting of the mirrors is also possible to improve visibility at O. A plate P (a compensator) is placed parallel to the beam splitter in the path between the splitter and M_2. The com-pensator P is a glass plate identical in thickness to the beam splitter that is placed in the path of beam 1, so that by the time beam (1-1'-1") reaches point O, it would have crossed through the plate thickness three times (two times through P and once through BS), the same number of times as the beam 2-2'-2" crosses the plate thickness. A beam from the source incident on the beam splitter splits into two beams at the coated surface, a reflected beam (1) and a transmitted beam (2). Beam (1) gets reflected off mirror M_2 along path (1') and passes through the beam splitter traveling along 1" toward O. Beam (2) reflects off M_1 along (2') that reflects off the beam splitter along 2" toward O. Thus, the two beams 1" and 2" meet at O, interfere, and form interference fringes. These fringes are very sensitive to the setting of the mirrors. At the

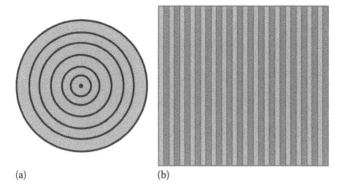

(a)   (b)

FIGURE 7.18 Fringes observed in a Michelson interferometer (a) Circular fringes for M_1 perpendicular to M_2 and (b) almost parallel fringes for M_1 tilted slightly from the perpendicular position.

very strict setting of M_2 being perpendicular to M_1 when the beam splitter is set at 45° with M_1, the fringes are circular with their center at the middle of the field. Let M_1' represent the image of M_1, where M_2 should be located for equal optical path lengths of the transmitted and the reflected beams. For any displacement d of M_2 closer to BS or away from it, the circular fringes will expand or contract, collapsing at the center. With a slight tilting of M_1, the lines would be curved in one of two directions, depending on the sense (positive or negative tilt) of the angle, and may appear parallel at large curvature. Figure 7.18 shows in Figure 7.18a the circular fringes and in Figure 7.18b the parallel fringes.

For any displacement d of one of the mirrors, a path difference of 2d would then be generated between the virtual sources of the two beams, 1″ and 2″, at the observation position O; this is due to the reflection of the rays at the mirror on its path. If the viewed spot happens to be at a bright fringe, then the condition is a *constructive interference*. As both beams change phase by π upon reflection (1) at the BS and the respective mirrors, the difference in the optical path Δ (= path difference in air) would fulfill the following condition:

$$2d = m\lambda; \quad m = 0, 1, 2, 3,\ldots, \tag{7.33}$$

m being the order of the interference fringe being viewed.

However, the condition for dark fringes, that is, *destructive interference*, would be

$$2d = \left(m + \frac{1}{2}\right)\lambda; \quad m = 0, 1, 2, 3,\ldots. \tag{7.34}$$

In case M_1 is slightly tilted through an angle, θ, the difference in the optical path would then be $2d \cos \theta$, and the above conditions for destructive and constructive interference would be as follows:

(i)   *Constructive interference*

$$2d\cos\theta = m\lambda; \quad m = 0,\ 1,\ 2,\ 3,\dots \tag{7.35}$$

(ii)   *Destructive interference*

$$2d\cos\theta = \left(m + \frac{1}{2}\right)\lambda \quad m = 0,\ 1,\ 2,\ 3,\dots \tag{7.36}$$

REMARK

From Equation 7.35,

$$\cos\theta = m\lambda/2d.$$

For a low-order dark interference fringe, $m = 0$ makes $\cos\theta = 0$, implying that $\theta = 90°$, while for the smallest angle, $\theta = 0$, $\cos\theta = 1$. Since $\cos\theta$ being 1 is the largest possible value for the cosine, the order m on the right-hand side of the equation implies that m assumes the largest value for the fringe observed at $\theta = 0$. To overcome this inconvenience, introduce a labeling that starts with the central fringe as the zeroth order and the farthest as the maximum order through a new integer, s, such that

$$s = m_{max} - m; \quad m = 0,\ 1,\ 2,\dots m_{max} - 1,\ m_{max}$$

This then makes $s = 0$ when $m = m_{ma}$.

EXAMPLE 7.5

Consider a Michelson interferometer with arms 22.0 cm long accommodating two gas-filled identical cylindrical glass cells, one in each arm. The gas index of refraction $n = 1.00003$ and the cells are 11.0 cm each. If the setup was illuminated with light of 450.0 nm wavelength, and without disturbing one cell, the gas in the other was pumped out, determine how many fringes will shift during the evacuation process.

Solution

As the index of refraction of vacuum is $n = 1.00000$, the optical path difference between the two cells is $\Delta = 2 \times 11 (1.00003 - 1.00000)$ cm.
  Thus, $2 \times (0.110 \text{ m}) (1.00003 - 1.00000) = m(450.0 \times 10^{-9})$.
  Thus, $m = 14.7$, which implies that 14 fringes have passed.

PROBLEMS

7.1   Light falls on two parallel slits separated by 0.0200 mm. If the interference fringes on a screen 75.0 cm away have a spacing of 2.20 mm, what is the wavelength of the light used?

7.2   In an interference experiment of the Young's type, the distance between the slits is 0.500 mm. The wavelength of the light is 600.0 nm. If it is desired to have a fringe spacing of 1.00 mm at the screen, what is the corresponding screen distance from the slits?

7.3   Interference fringes from two slits in a double-slit setup are observed, using light of wavelength λ_1 = 400.0 nm and λ_2 = 600.0 nm.

(a)   At what values for the orders, if any, do the bright bands of the two wavelengths coincide? Refer to interference orders by m_1 for the first wavelength and m_2 for the second wavelength.

(b)   Do you expect that some orders of the dark fringes would coincide?

7.4   Using light of wavelength λ = 513.3 nm, interference fringes from two slits in a double-slit setup are observed on a screen at a distance of 1.75 m from the slits. For a spacing of the fringes equal to 3.86 mm, determine

(a)   The spacing between the slits

(b)   The separation between the first minima on the two sides of the central fringe

7.5   In a Young's double-slit experiment, one finds that by introducing a thin sheet of polymer of an index of refraction n = 1.430 in the path of the beam, the central fringe shifts to the position that was originally occupied by the tenth dark fringe. If the wavelength of light incident on the slits is 589.3 nm, determine the thickness of the sheet.

7.6   In a replication of Lloyd's experiment, 31.0 cm away from a screen is a source 0.400 mm above a mirror placed horizontally. The source is emitting a yellow light of wavelength λ = 589.2 nm toward the mirror whose far end is touching the screen at C. Determine

(a)   The height of the fourth maximum fringe from the center of the screen

(b)   The spacing between the interference fringes

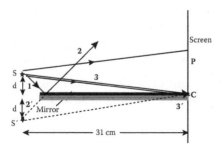

7.7   Newton's rings are formed between the spherical surface of a lens and an optical glass slide. If the radius of the tenth bright ring of light of wavelength 598.6 nm is 1.20 mm in diameter, what is the radius of curvature of the lens surface?

7.8   Newton's rings are formed between the spherical surface of a lens of radius R = 0.760 m and an optical slide.
   (a)   What would be the thickness of the air film at which one would view the ninth dark fringe of light having wavelength $\lambda = 600.0$ nm?
   (b)   What is the radius of the ninth fringe?

7.9   Redo the previous problem, assuming that the space between the flat plate and the lens is filled with water (n = 1.33).

7.10   Consider an oil film (n = 1.42) spread on the surface of a water-filled pool (n =1.33). The film of thickness d is illuminated by white light from a source overhead. For an observer looking down on the film, a strong first-order reflection of red light, $\lambda = 660.0$ nm wavelength, is observed. Assuming a normal incidence case
   (a)   Determine the thickness of the film rounded off to one decimal place.
   (b)   State how a white sheet of paper placed a few millimeters beneath the film would look.

7.11   When mirror M_1 of a Michelson interferometer is moved toward M_2 a distance d, a shift of 450 fringes past the center was viewed. If the device was illuminated with a light beam of 500.0 nm wavelength, calculate the distance d.

7.12   In the previous problem, Michelson interferometer is of two identical arms each of which is a cylindrical glass tube, 8.00 cm long. One arm, provided with a valve hooked to a pump, is filled with Helium (n = 1.00000349). Find the number of fringes that shift from the center of observation during this process.

7.13   A silicon monoxide thin coating of 108.0 nm deposited on the surface of a lens (n = 1.45) resulted in complete antireflection when normally illuminated with a wavelength of 589.2 nm. What is the index of refraction of the coating?

7.14   In the previous problem, use the value of the index of refraction of the coating to find what wavelength would result in a minimum reflection if the angle of incidence was 30°. Comment on how realistic that is as far as the calculated wavelength is visible.

7.15   Using sodium light of 589.2 nm wavelength, a pattern of fringes formed at a step of a thin film deposited on a flat glass showed a displacement of 5.0 mm. For a spacing of 1 mm between the fringes, determine the thickness of the film.

Diffraction I
Fraunhofer Diffraction

The book of nature is written in the language of mathematics.

Galileo Galilei (1564–1642)

8.1 INTRODUCTION

Diffraction is the result of any wave (electromagnetic and/or mechanical) deviating from its normal straight path due to the presence of an obstacle or aperture of a size comparable to the wavelength of the incident wave. The sharp edges of the obstacle or aperture cause the diffraction effect. The projection of the diffracted wave, after passing around or through the obstacle or aperture, onto a screen forms a real image that is called a diffraction pattern. The features of a diffraction pattern depend on the physical geometry and kind of the obstacle. It is a daily life experience for sound waves. For a student sitting in a classroom with a door that opens to a corridor, he or she could hear the chat of two people standing in the corridor close to the door even though they are out of sight of the student. The sound, propagating in waves, diffracts through the door because its dimensions are comparable to the wavelength of sound waves. Light waves have this property, and as they diffract, they create a pattern of successive maxima and minima of light distribution due to the interference phenomenon. As will be discussed in detail in this chapter, these maxima and minima arise from the interference of a very large number of wave fronts experiencing diffraction just as they encounter an obstacle or pass through an opening. Thus, interference of diffracted waves is inherently involved in the making of a diffraction pattern. Among the simplest setups to observe such a phenomenon are a source of light, an obstacle with a slit, a screen, and a lens or lenses. The arrangement of these elements is based on one of two considerations: one that allows for the diffraction of plane waves focused on a screen via a lens; thus, the source and the screen are effectively treated as if they were at infinite distances from each other; this is known as Fraunhofer diffraction. The other allows for diffraction of spherical waves received on a screen at a finite distance from the source; this is called Fresnel diffraction. This chapter presents a detailed discussion of Fraunhofer diffraction.

8.2 SETUP OF SINGLE-SLIT DIFFRACTION

A typical arrangement for a single-slit diffraction consists of a coherent source, S; a slit, AB; two lenses, L_1 in front of the slit and L_2 behind it; and a screen placed at the secondary focal plane of L_2, Figure 8.1.

The slit, rectangular in shape and of a narrow width, a, is cut in a solid opaque plane. The arrangement secures illuminating the slit with plane waves by having the source at the primary focal point of L_1. The second lens focuses the resulting diffraction pattern on the screen.

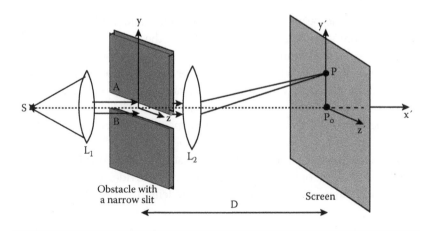

FIGURE 8.1   A typical arrangement of single-slit diffraction setup.

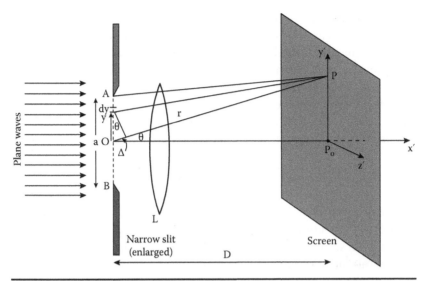

FIGURE 8.2 An enlarged side view of a rectangular aperture in a single-slit diffraction setup.

In Figure 8.2, the incident waves were plane, and accordingly L_1 is not shown. The obstacle is chosen to be in the yz plane with an origin, O, at the middle point of the slit. Parallel to slit plane is the screen plane y′z′. P_o is at the center of the screen lying along the horizontal line of symmetry that extends from O, and P is an arbitrary point chosen along the y′ axis on the screen. The line OP defines the direction of propagation of a wavelet emanated at O with angle of inclination θ. This is detailed in Figure 8.2, where the slit AB is magnified for clarity.

8.2.1 THEORY OF SINGLE-SLIT DIFFRACTION

The foundation for understanding diffraction is based on the Huygens–Fresnel principle, which assumes that as plane waves reach the slit, its opening acts as a line of a large number of point sources, each of which emits spherical wavelets at every point of the wave front reaching the slit. These, leaving the slit parallel to the axis OP_o, are focused on the screen. Except for the point P_o, each observation point along the y′ axis on the screen receives a large number of the wavelets with a direction of propagation that makes an angle, $0 < \theta \leq 90°$. Thus theoretically speaking, the pattern on the screen spreads along the y axis from 0 to infinity in the form of fringes symmetrically distributed with P_o at the center of the pattern. The angle θ for wavelets travelling along OP_o is zero. The wavelets entering lens L parallel to the axis of symmetry are focused on the screen at P_o. Summing over all the amplitudes of the wavelets arriving at any point of the screen gives the overall resultant amplitude at those points. Out of computational convenience, we divide the slit width into a large number of infinitesimally small elements, each of width dy. Letting the constant amplitude per unit length of the wavelet at any point on the slit be dE_a, the amplitude of the wavelet coming out of each element would be $dE_a dy$. For an element situated at the origin, a spherical wave from it arriving at point P will have an electric field of amplitude dE_p inversely proportional to the distance r. That is

$$dE_p = \left(\frac{dE_a dy}{r} \right) \sin(kr - \omega t). \tag{8.1}$$

For an element, dy, at a point, y, away from O, a spherical wave arriving at point P will travel through a distance different from that coming from point O by a small segment, Δ. This results in a phase difference between the two wavelets arriving at P. From the geometry

$$\Delta = y \sin \theta.$$

The amplitude contributed by this wave at P would be

$$dE_p = \left(\frac{dE_a dy}{r} \right) \sin\left[k(r - \Delta) - \omega t \right]$$
$$dE_p = \left(\frac{dE_a dy}{r} \right) \sin\left[k(r - y \sin \theta) - \omega t \right]. \tag{8.2}$$

The total amplitude from the slit is the integration of dE_p in Equation 8.2 over the width a of the slit. That is

$$Ep = \int_{-a/2}^{+a/2} dEp$$
$$Ep = \int_{-a/2}^{+a/2} \left(\frac{dE_a dy}{r} \right) \sin\left[k(r - y \sin \theta) - \omega t \right], \tag{8.3}$$

or

$$Ep = \int_{0}^{a/2} \frac{dE_a dy}{r} \left\{ \sin\left[k(r - y \sin \theta) - \omega t \right] + \sin\left[k(r + y \sin \theta) - \omega t \right] \right\}. \tag{8.4}$$

Applying the trigonometric identity

$$\sin(A+B) + \sin(A-B) = 2 \sin\frac{(A+B)+(A-B)}{2} \cos\frac{(A+B)-(A-B)}{2}$$
$$= 2 \sin A \cos B \tag{8.5}$$

or equivalently

$$\sin \phi_1 + \sin \phi_2 = 2 \sin \frac{(\phi_1 + \phi_2)}{2} \cos \frac{(\phi_1 - \phi_2)}{2} \tag{8.6}$$

to the bracketed quantity in Equation 8.4 reduces the equation to

$$E_p = \int_0^{a/2} \left(\frac{dE_a dy}{r} \right) 2 \sin[kr - \omega t] \cos ky \sin \theta$$

or

$$E_p = 2 \left(\frac{dE_a}{r} \right) \sin(kr - \omega t) \int_0^{a/2} \cos \left(ky \sin \theta \right) dy$$

$$E_p = 2 \left(\frac{dE_a}{r} \right) \sin(kr - \omega t) \frac{\sin \left(ky \sin \theta \right)}{k \sin \theta} \Bigg|_0^{a/2} \tag{8.7}$$

$$E_p = a \left(\frac{dE_a}{r} \right) \sin(kr - \omega t) \frac{\sin \left(\frac{1}{2} ka \sin \theta \right)}{\frac{1}{2} ka \sin \theta}.$$

The above may be reduced to the form

$$E_p = \left(E_o \frac{\sin \alpha}{\alpha} \right) \sin(kr - \omega t). \tag{8.8}$$

where

$$E_o = \frac{a \, dE_a}{r}, \tag{8.9}$$

and

$$\alpha = \frac{1}{2} ka \sin \theta. \tag{8.10}$$

The factor $\left(E_o \frac{\sin \alpha}{\alpha} \right)$ in Equation 8.8 represents the resultant amplitude at point P and is a function of θ. Thus, Equation 8.8 can be re-expressed as

$$E_p = E(\theta) \sin(kr - \omega t), \tag{8.11}$$

where

$$E(\theta) = E_o \frac{\sin \alpha}{\alpha}. \tag{8.12}$$

The intensity of the resultant at point P is

$$E^2(\theta) = E_o^2 \left(\frac{\sin \alpha}{\alpha} \right)^2,$$

that is,

$$I(\theta) = I_o \frac{\sin^2 \alpha}{\alpha^2}, \tag{8.13}$$

where I_o is the maximum value that the intensity $I(\theta)$ can have, which occurs when $\left(\dfrac{\sin \alpha}{\alpha} \right) = 1$, that is, at $\alpha = 0$, or equivalently, $\theta = 0$. Thus, $I(\theta)|_{\theta=0} = \max$. This gives the intensity of the diffraction pattern at the center of the screen, that is, at point P_o.

8.2.2 ANALYSIS OF SINGLE-SLIT DIFFRACTION

For all zero values of $\sin \alpha$ other than $\alpha = 0$, the intensity $I(\theta)$ is equal to zero, and that gives the angular positions of the minimum intensity values of the diffraction pattern. Thus, a set of minima in the intensity of the diffraction pattern is generated. The presence of successive minima in the intensity distribution implies the presence of maxima in between them. The positions of the minima and maxima characterize positions of what is usually called destructive and constructive diffraction, respectively. These can be analyzed as follows.

8.2.2.1 Destructive Diffraction

The positions of the successive minima in the diffraction pattern occur for

$$\alpha = m_\alpha \pi; \quad m_\alpha = 1,2,3,\ldots, \tag{8.14}$$

where

m_α is called the order of the minimum band
$m_\alpha = 1$ denotes the first minimum
$m_\alpha = 2$ is the second minimum, and so on

From Equation 8.10,

$$\frac{1}{2}ka\sin\theta = m_\alpha\pi; \quad m_\alpha = 1,2,3,\ldots$$

or

$$\frac{1}{2}\left(\frac{2\pi}{\lambda}\right)a\sin\theta = m_\alpha\pi,$$

which gives

$$a\sin\theta = m_\alpha\lambda; \quad m_\alpha = 1,2,3,\ldots \tag{8.15}$$

As $\sin\theta \cong \theta = \dfrac{y'}{D}$, then from the above and the geometry of the setup

$$a\frac{y'}{D} = m_\alpha\lambda,$$

that is,

$$y' = m_\alpha\left(\frac{D}{a}\right)\lambda; \quad m = 1,2,3,\ldots \tag{8.16}$$

The separation $\Delta y'$ between any two successive minima becomes

$$\Delta y' = \frac{D\lambda}{a}, \tag{8.17}$$

independent of the orders of these minima. Note that the treatment of diffraction is for waves incident normal to the plane of the slit. In case light waves from the source were inclined along a direction that makes an angle of φ with OP_o, the above Equation 8.15 would be

$$a\left(\sin\theta + \sin\varphi\right) = m_\alpha\lambda; \quad m_\alpha = 1,2,3,\ldots \tag{8.18}$$

QUESTION

What happens if an aperture's opening a is very large, practically infinity?!

ANSWER

In this case, the separation between the diffraction bands vanish. Thus, the whole area in front of the slit is illuminated as geometrically predicted. The geometric optics in this case is dominating, meaning that light is traveling in straight lines.

8.2.2.2 Constructive Diffraction

The positions of the successive maxima in the diffraction pattern can be found from Equation 8.12 by differentiating it with respect to α and equating that to zero, that is, $\dfrac{\partial}{\partial \alpha} E(\theta) = 0$. Thus

$$\frac{\partial}{\partial \alpha}\left(E_o \frac{\sin \alpha}{\alpha}\right) = \frac{\alpha \cos \alpha - \sin \alpha(1)}{\alpha^2} = 0,$$

which yields

$$\alpha = \tan \alpha \qquad\qquad (8.19)$$

The above equation represents the condition on α for which the intensity of diffraction at point P would be maximum. Introducing a new variable, y, Equation 8.19 can be written as a coupling of two parametric equations:

$$y = \alpha \qquad\qquad (8.20)$$

$$y = \tan \alpha \qquad\qquad (8.21)$$

The first represents a straight line of a slope equal to 1, while the second generates tangent plots for all α. The intersections of the line $y = \alpha$ and plots of $\tan \alpha$ give the required values of α at which diffraction is maximum. The plots are shown, Figure 8.3.

Careful calculations show that the positions of these maxima occur at $\alpha = 1.430\pi$, 2.459π, 3.471π, 4.477π, and so on. The intensity of these maxima can be found from the ratio

$$\frac{I(\theta)}{I_o} = \frac{\sin^2 \alpha}{\alpha^2},$$

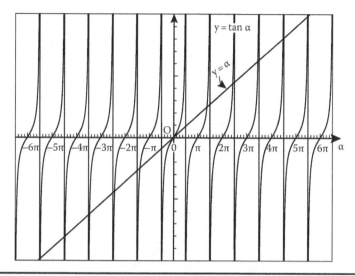

FIGURE 8.3 A plot of Equations 8.20, 8.21, and 8.22: $y = \tan \alpha$ and $y = \alpha$. The intersections of these plots give the angular positions of the maxima in Fraunhofer diffraction.

that is

$$I(\theta) = I_o \frac{\sin^2 \alpha}{\alpha^2}. \tag{8.22}$$

For example, the intensities of the first two maxima ($\alpha = 1.430\pi$, $\alpha = 2.459\pi$) compared to the central maximum are

$$\frac{I(\theta)_{1st\,max}}{I_o} = \frac{\sin^2(1.43\pi)}{(1.43\,\pi)^2} = \frac{0.952}{20.2} = 0.0472,$$

$$\frac{I(\theta)_{2nd\,max}}{I_o} = \frac{\sin^2(2.46\,\pi)}{(2.46\,\pi)^2} = \frac{0.984}{59.7} = 0.0164.$$

Thus, the intensity of the first-order maximum is 4.7% of the central maximum, and the second is only 1.6%. Obviously, they are very weak compared to the central maximum. Almost all the energy in the single-slit diffraction pattern goes into the central maximum. That is why the first-, second-, third-, etc., order diffraction maxima are called *secondary diffraction maxima*, and from a practical perspective, these are of little or no relevance. Figure 8.4

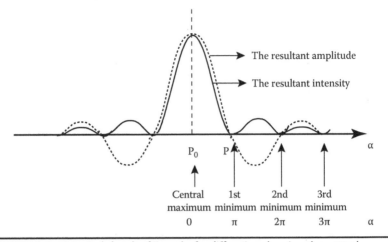

The resultant amplitude

The resultant intensity

P_0 P

Central   1st   2nd   3rd
maximum minimum minimum minimum

0   π 2π 3π α

FIGURE 8.4   A typical sketch of Fraunhofer diffraction showing the central maximum and two secondary maxima on each side.

illustrates the diffraction pattern of a single slit. We note that they are positioned on the right and left of the central maximum.

In the discussion of the diffraction grating in later chapters, we find what are *called subsidiary maxima*, and they should not be mistaken for the single-slit diffraction secondary maxima.

EXAMPLE 8.1

A narrow slit of width $a = 4.00 \times 10^{-6}$ m is illuminated by a yellow monochromatic light of $\lambda = 600$ nm. Knowing that $D = 0.800$ m, find the height along the screen of (a) the first- and (b) second-order minima in the single-slit diffraction pattern.

Solution

(a) Using Equation 8.15,

$$m_\alpha \, \lambda = a \sin \theta,$$

or

$$\sin \theta = \frac{m_\alpha \lambda}{a}.$$

Substitution for the given values gives

$$\sin \theta = \frac{(1)(6.00 \times 10^{-7} \text{m})}{(4.00 \times 10^{-6} \text{m})}$$

or

$$\sin\theta = 0.150;$$

$$\theta = \sin^{-1}(0.150) = 8.63^{\circ}.$$

From the geometry

$$\tan\theta = \left(\frac{y'}{D}\right)$$

Thus,

$$y'_{1st} = D\tan\theta$$
$$= (0.800\,\text{m})\tan(8.63^{\circ}) = (0.800\,\text{m})(0.153)$$
$$= 0.121\,\text{m}$$

(b) Following the same steps, for the second-order minimum

$$\sin\theta = 0.300.$$

Thus

$$\theta = \sin^{-1}(0.300) = 17.5^{\circ}.$$

And

$$y'_{2nd} = (0.800\,\text{m})\tan(17.50^{\circ}) = (0.800\,\text{m})(0.300)$$
$$= 0.240\,\text{m}$$

Analysis

Notice that the linear separation from P_0 of the second-order minimum diffraction is very close to being twice the height of the first order. It is not quite twice, because the sine of an angle does not equal two times the sine half of the angle.

8.2.3 FURTHER ANALYSIS OF THE SINGLE-SLIT DIFFRACTION: THE VIBRATION CURVE

In this analysis, we will apply the phasor diagram method on a large number of wavelets arriving at different points along the y′ axis of the screen. Consider a slit of width a, where two sets of rays, one (regular) leaving the slit parallel and incident normal to the slit plane, all reaching P_o and another (italic) leaving the slit parallel but inclined with an angle θ, all reaching point P.

Let us divide the slit into a large number of identical segments, N. All wavelets originating at the slit have the same amplitude and phase, but upon reaching point P, they will be different in phase. This is because the wavelets travel different distances between the slit and point P. Except for the center of the screen, P_o, there is a finite difference in phase at P between each pair of two successive waves emanated at the slit; let that be δ, and let the phase difference between a wavelet coming from the top source at A and the lower source at B be 2α, Figure 8.5a. Out of convenience, let the slit be comprised of a limited number of identical segments, 11, for example, where the center of each represent a secondary source so that the whole make 11 identical coherent sources that emit wavelets toward the screen. Based on the construction of the phasor diagram presented in Chapter 6, the phasor diagram of wavelets at P would look like those shown in Figure 8.5a.

8.2.3.1 Central Maximum

Considering the solid set of wavelets in Figure 8.6, the amplitudes of the wavelets arriving at P_o are equal, and all are in phase. For determining their

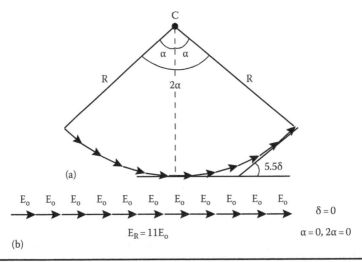

FIGURE 8.5   (a) A typical phasor diagram of 11 phasors, each of which is different in phase from the adjacent by an angle δ. In this sketch, the difference between the top phasor and the lowest one is 2α, and (b) phasors are all in phase, δ = 0.

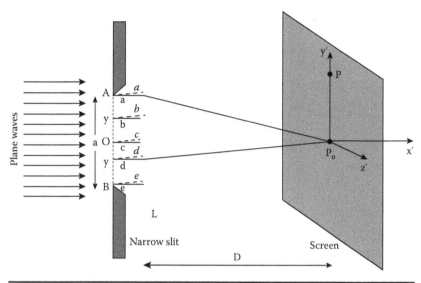

FIGURE 8.6   A sketch of a single-slit diffraction setup.

resultant, such phasors may be sketched head to tail along one line giving a resultant amplitude E_R, as shown in Figure 8.5b.

The intensity of the diffracted light at P_o, denoted by I_o, is the square of its amplitude. Thus

$$I_o = \left(E_R\right)^2 = \left(\sum_i E_i\right)^2 = \left(11E_o\right)^2. \tag{8.23}$$

One observes that the resultant amplitude E_R ($=11E_o$) in this case is the maximum value that such a sum could have. But our pick of 11 phasors was an arbitrarily chosen number, and I_R ($=(11E_o)^2$) followed this choice. Thus, we may consider the resultant intensity at P_o as unity to which the resultant $I(\theta)$ at any other point P on the screen can be defined through the ratio $I(\theta)/I_o$.

8.2.3.2 First Minimum

The vibration curve should close into itself such that the phase difference between the first and the last phasor $2\alpha = 2\pi$. This means that the tip of the last phasor in the vibration curve closes on the tail of the first phasor, forming a polygon of 11 equal sides. Of course had we had the slit divided into thousand segments, the polygon that would look more like a circle would be obtained, and an ideal circle for an infinite number of segments (Figure 8.7b). The resultant amplitude at point P is then zero, that is

$$E_R = 0, \quad I_{1st\,min} = 0.$$

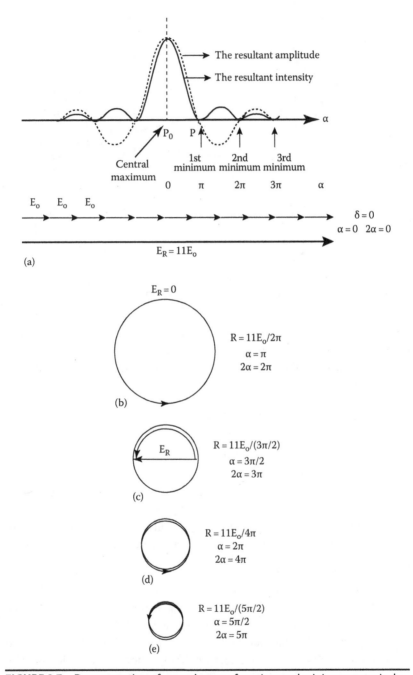

FIGURE 8.7 Demonstration of several cases of maxima and minima successively displayed on the screen for a single-slit diffraction. To make the repetitive rounds distinct, the vibration curves after completing the first full circle in cases c, d, and e were intentionally slightly shifted from the first circle.

8.2.3.3 First Secondary Maximum

For the first maximum to occur, the total length of phasors combined should, after completing one full circle, still have enough to curve along an additional half a circle. The length of the total vibration curve becomes equal to the (1½) circumferences of the circle shown in Figure 8.7c, that is, the phasors sweeping through an angle of $2\alpha = 3\pi$. Thus

$$(3\pi)(E_R/2) = \text{Total length of the curve} = \sum_i E_i = E_{Ro}$$

This gives

$$E_R = \frac{2}{3\pi} E_{Ro}.$$

Thus, the intensity for the *first secondary maximum*, exhibit (c), would be

$$I_{1st\,max} = \left(\frac{2}{3\pi} E_{Ro}\right)^2$$

or

$$I_{1st\,max} = \left(\frac{2}{3\pi}\right)^2 I_o.$$

For the *second minimum diffraction*, the phasors would be completing two circles, sweeping an angle of 4π. Thus, $2\alpha = 4\pi$, and the radius would be $E_R/2 = (11\,E_o/4\pi)$; this is demonstrated in case (d).

In an argument similar to that presented for the first maximum in case (b), the *second secondary maximum*, case (e), would occur when the vibration curve has swept through 2 ½ circles, that results in angle of $2\alpha = 5\pi$. The intensity would then be

$$I_{2nd\,max} = \left[(2/5\pi)^2\right] I_o$$

In conclusion, the condition for *diffraction minima* is

$$2\alpha = 2\pi,\, 4\pi,\, 6\pi,\, 8\pi,\ldots$$

or

$$\alpha = \pi,\, 2\pi,\, 3\pi,\, 4\pi,\ldots$$

or

$$\alpha = m_\alpha \pi; \quad m_\alpha = 1, 2, 3, 4, \ldots$$

While the positions of maxima as determined by the vibration curve method $2\alpha = (3/2)\pi$, $(5/2)\pi$, each may seemingly occur at half way between two minima, they in fact are slightly off; this feature was analytically determined earlier. The intensities of these secondary maxima, however, are in full agreement with the analytical values (see Equation 8.13).

EXAMPLE 8.2

Using the same data in the previous example, find the linear spacing on the screen between the first two successive minima, (m_α = 1, 2) in the diffraction pattern.

Solution

Using Equation 8.17

$$\Delta y'_{min} = \frac{D\lambda}{a} = \frac{(0.800 \, \text{m})(6.00 \times 10^{-7} \, \text{m})}{(4.00 \times 10^{-6} \, \text{m})}$$

$$= 0.120 \, \text{m},$$

i.e.,

$$\Delta y' = 12 \, \text{cm}$$

8.3 DIFFRACTION FROM RECTANGULAR APERTURES

Let us consider Figure 8.8 that illustrates a rectangular aperture with a height a and width b at a distance D from a screen; D is large compared to the setup dimensions. Also, take an elemental rectangular area, dS (=dydz), of the aperture at a distance, r, from a point of observation, P, on the screen; P is at distance r_o from the center of the aperture.

In parallel with definitions introduced for the single-slit diffraction, we assume that dE_a is the contribution per unit area of a source point of the aperture. Each of the slit dimensions a and b produces a diffraction, one in the x'y' from the height y and another in the x'z' from the width b.

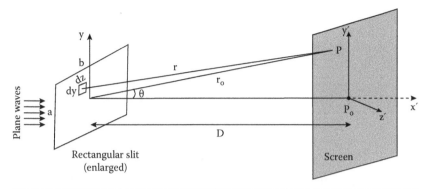

FIGURE 8.8   A setup showing a rectangular slit and a screen at a distance D from the slit. A distant source (not shown) is emitting waves that are received by the slit as plane waves.

As established for the single-slit diffraction in Equation 8.7

$$E_p = \left(\frac{dE_a}{r}\right)\sin(kr - \omega t)\int_{-a/2}^{a/2}\cos\left(ky\sin\theta_a\right)dy,\qquad(8.24)$$

in which the integral can be more conveniently written in a complex exponential form. Thus absorbing all in a constant, C, we then have

$$E_p = C\int_{-a/2}^{a/2}e^{i\left(ky\sin\theta\right)}dy.\qquad(8.25)$$

In the above,

$$\sin\theta_a = \tan\theta_a = \frac{y'}{D};$$

θ_a is the angle for the diffraction in the x'y' plane. Thus, the above equation becomes

$$E_p = C\int_{-a/2}^{a/2}e^{i\left(ky\frac{y'}{D}\right)}dy.\qquad(8.26)$$

Introducing a new wave vector, $k_y = k\dfrac{y'}{D}$, reduces the above to the following:

$$E_p\left(k_y\right) = C\int_{-a/2}^{a/2}e^{i\left(k_y y\right)}dy\qquad(8.27)$$

Equation 8.27 can be exploited further by defining a function that represents the aperture, single slit in this case. Denoting that by $A(y)$ such that $A(y) = 1$ for all points on the aperture where $|y| < a/2$ and zero otherwise, reforms (8.27) to

$$E_p\left(k_y\right) = C \int_{-a/2}^{a/2} A\left(y\right) e^{i\left(k_y y\right)} dy. \tag{8.28}$$

Equation 8.28 is the interesting Fourier transform of $A(y)$ discussed in more detail in Chapter 12. Obviously this is a one-dimensional transform that describes the pattern obtained by a single slit. For a two-dimensional aperture, the argument can be pursued for the width of the aperture that creates a diffraction in the $x'z'$ plane with an angle θ_b given by

$$\sin\theta_b = \tan\theta_b = \frac{z'}{D}$$

The path r would be given by the approximation

$$r = r_0 - \frac{y'}{D} = \frac{z'}{D}. \tag{8.29}$$

Introducing the definition for a wave vector $k_z = k\dfrac{z'}{D}$, the diffraction integral in (8.27) for the two-dimensional aperture would now become

$$E_p\left(y', z'\right) = C' \int_{-b/2}^{b/2} \int_{-a/2}^{a/2} e^{ikr} dA = C' \int_{-b/2}^{b/2} \int_{-a/2}^{a/2} e^{ik\left(r_0 + \frac{yy'}{D} + \frac{zz'}{D}\right)} dA \tag{8.30}$$

$$E_p\left(y', z'\right) = C' e^{ikr_0} \int_{-b/2}^{b/2} \int_{-a/2}^{a/2} e^{i\left(k_y y + k_z z\right)} dy dz. \tag{8.31}$$

In a manner similar to defining an aperture function for the one-dimensional single-slit diffraction in Equation 8.28, we can define the aperture function for the rectangular function as $A(y, z)$, so that the two-dimensional diffraction integral in Equation 8.31 may be re-expressed as

$$E_p\left(y', z'\right) = C' e^{ikr_0} \int_{-\infty}^{\infty} \int_{-\infty}^{\infty} A\left(y, z\right) e^{i\left(k_y y + k_z z\right)} dy dz. \tag{8.32}$$

In Equation 8.32, $A(y, z)$ is the Fourier transform of $E_p(y', z')$, and $E_p(y', z')$ is the Fourier transform of $A(y, z)$.

From the relations connecting k_y and k_z, to y' and z', more appropriately, we write $E_p(y', z')$ as $E_p(k_y, k_z)$. This reduces Equation 8.32 to

$$E_p\left(k_y, k_z\right) = C'e^{ikr_0} \int\limits_{-\infty}^{\infty} \int\limits_{-\infty}^{\infty} A\left(y, z\right) e^{i\left(k_y y + k_z z\right)} dy dz. \qquad (8.33)$$

As noted, $A(y, z)$ is the Fourier transform of $E_p(k_y, k_z)$, and $E_p(k_y, k_z)$ is the Fourier transform of $A(y, z)$, that is

$$A\left(y, z\right) = C'e^{-ikxr_0} \int\limits_{-\infty}^{\infty} \int\limits_{-\infty}^{\infty} E_p\left(k_y, k_z\right) e^{i\left(k_y y + k_z z\right)} dy dz. \qquad (8.34)$$

8.4 DOUBLE-SLIT DIFFRACTION

Figure 8.9 shows two narrow identical slits that can be used to produce what is called double-slit diffraction. The process may be explained on the basis of diffraction of light through each slit, combined with interference of wavelets that are also being diffracted. Thus, the mechanism consists of diffraction and interference happening simultaneously with no means of separating the two from each other as long as the dimensions of the slits qualify for diffraction and the separation between them allows for the interference (see Chapter 7).

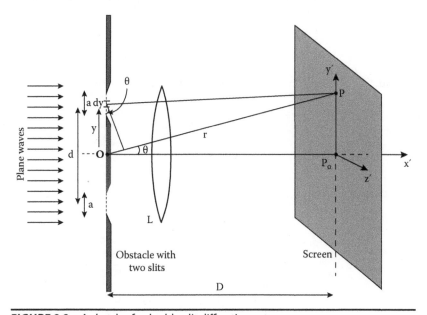

FIGURE 8.9 A sketch of a double-slit diffraction setup.

Thus, summing, or more specifically integrating, over all the amplitudes of the wavelets emanated by the sources at the two slits and arriving at any point of interest on the screen gives the overall resultant amplitude at that very point, producing a diffraction pattern.

Let us consider two identical elements, dy+ on the top slit and dy− on the lower slit, both at equal distances y from O. Letting the contributions to the resultant amplitude per unit length of the wavelet at any point on the two slits be dE_{a+} and dE_{a-}, respectively, the spherical waves arriving from the slits at a point P, r away from O, will have an electric field, dE_p, of amplitude that is inversely proportional to the distance r, that is, for any element ds, the contribution at P (see Equation 8.1) is

$$dE_p = \left(\frac{dE_a dy}{r}\right) \sin(kr - \omega t).$$

For the contributions from the two elements on dE_{p+}, dE_{p-} the two slits are $dEp = dE_{p+} + dE_{p-}$. Thus

$$dE_p = \left(\frac{dE_{a+} dy_+}{r}\right)\sin(kr - \omega t) + \left(\frac{dE_{a-} dy_-}{r}\right)\sin(kr - \omega t) \qquad (8.35)$$

or

$$dE_p = \left(\frac{dE_{a+} dy_+}{r}\right)\sin\left[k(r-\Delta) - \omega t\right] + \left(\frac{dE_{a-} dy_-}{r}\right)\sin\left[k(r+\Delta) - \omega t\right], \qquad (8.36)$$

or

$$dE_p = \left(\frac{dE\,dy}{r}\right)\left[\sin[k(r - y\sin\theta) - \omega t] + \sin\left[k(r + y\sin\theta) - \omega t\right]\right],$$

where

$$\Delta = y \sin\theta$$

and magnitude wise,

$$dy_+ = dy_-, \quad dE_+ = dE_- = dE.$$

Thus,

$$E_p = \left(\frac{dE\,dy}{r}\right)[2\sin[kr - \omega t]\cos\left[ky\sin\theta\right].$$

Integrating the above with the limits of integration being from $(d - a)/2$ to $(d + a)/2$, we get

$$E_p = \int_{(d-a)/2}^{(d+a)/2} \left(\frac{dEdy}{r} \right) [2 \sin[kr - \omega t] \cos\left[k\, y \sin\theta \right]$$

$$E_p = \left(\frac{2dE}{r} \right) \frac{ky \sin\theta}{k \sin\theta} \Big|_{(d-a)/2}^{(d+a)/2} \sin[kr - \omega t],$$

which gives

$$E_p = \left(\frac{2dE}{r\, k\, \sin\theta} \right) \left[k \frac{(d+a)}{2} \sin\theta - k \frac{(d-a)}{2} \sin\theta \right] \sin(kr - \omega t). \quad (8.37)$$

Using the trigonometric identity

$$\sin(A+B) - \sin(A-B) = 2\cos \frac{(A+B)+(A-B)}{2} \sin \frac{(A+B)-(A-B)}{2}$$

$$= 2\cos A \sin B \qquad (8.38)$$

or, equivalently,

$$\sin\phi_1 - \sin\phi_2 = 2\cos \frac{(\phi_1 + \phi_2)}{2} \sin \frac{(\phi_1 - \phi_2)}{2} \qquad (8.39)$$

Equation 8.37 reduces to

$$E_p = \left(\frac{2dE}{r\, k\, \sin\theta} \right) \left[2\sin\left(\frac{ka \sin\theta}{2} \right) \cos\left(\frac{kd \sin\theta}{2} \right) \right] \sin[kr - \omega t].$$

With some rearrangement, the above equation becomes

$$E_p = \frac{2\, a\, dE}{r} \left(\frac{\sin\left(\frac{ka \sin\theta}{2} \right)}{\frac{1}{2} k a \sin\theta} \right) \cos\left(\frac{kd \sin\theta}{2} \right) \sin[kr - \omega t] \qquad (8.40)$$

Define

$$E_o = \frac{2\,a\,dE}{r}, \tag{8.41}$$

$$\alpha = \frac{ka\sin\theta}{2}, \tag{8.42}$$

$$\beta = \frac{kd\sin\theta}{2}. \tag{8.43}$$

Now comparing the double-slit diffraction in Equation 8.40 with the diffraction of a single slit and using Equations 8.12 and 8.41 makes it possible to write Equation 8.30 as

$$E_p = 2\,E_o\left(\frac{\sin\alpha}{\alpha}\right)\cos\beta\,\sin\left[kr - \omega t\right]. \tag{8.44}$$

The above equation shows that the maximum value of E_p for the double-slit diffraction pattern is

$$E_p = 2E_o\left(\frac{\sin\alpha}{\alpha}\right)\cos\beta. \tag{8.45}$$

And upon squaring Equation 8.45, the intensity in the double-slit pattern diffraction becomes

$$I_p = 4I_o\left(\frac{\sin\alpha}{\alpha}\right)^2\cos^2\beta \tag{8.46}$$

or

$$I_p = I_{max}\left(\frac{\sin\alpha}{\alpha}\right)^2\cos^2\beta, \tag{8.47}$$

where $I_{max} = 4I_o$; I_o is the maximum intensity obtained in the single-slit diffraction. As noted, the maximum intensity obtained for the double-slit diffraction pattern is four times the maximum intensity of the diffraction pattern for a single slit. In the product of Equation 8.47, the factor $\left(\frac{\sin\alpha}{\alpha}\right)^2$ is what we had in Equation 8.12 for the diffraction of a single slit, Equation 8.12, and the factor $\cos^2 \beta$ is what we had in the interference obtained in Young's double slit. Thus, we may write Equation 8.47 as

$$I_p = 4\,I_o\,\boldsymbol{D\,I}, \tag{8.48}$$

where

$$\boldsymbol{I} = \cos^2 \beta \tag{8.49}$$

$$\boldsymbol{D} = \left(\frac{\sin \alpha}{\alpha}\right)^2. \tag{8.50}$$

From Equation 8.47, the double-slit diffraction is governed by its dependence on the values of both the interference and diffraction factors, $\cos^2 \beta$ and $\left(\dfrac{\sin \alpha}{\alpha}\right)^2$, respectively, and on the product of these factors. This can be observed in the plots of Equations 8.49 and 8.50 for a range of values of β and α in Figure 8.10a and b. Figure 8.11 is a plot of the product $I_p = 4I_o\left(\dfrac{\sin \alpha}{\alpha}\right)^2 \cos^2 \beta$. While I_{max} is normalized to unity, a choice for $\beta = 3\alpha$ is displayed.

8.4.1 ANALYSIS OF DOUBLE-SLIT DIFFRACTION

From Equations 8.42 and 8.43,

$$\alpha = \frac{ka\sin\theta}{2}, \quad \beta = \frac{kd\sin\theta}{2}$$

Dividing the two equations gives

$$\frac{\beta}{\alpha} = \frac{d}{a}.$$

As multiples of π for the values of α, that is, $m_\alpha\pi$, (m_α = 1, 2...) for α define the minima positons for diffractions, while multiples of π for values of β, that is, $m_\beta\pi$, (m_β = 1, 2,...) for β define positions of interference maxima, we may introduce an integer p.

That is

$$\frac{\beta}{\alpha} = p, \quad \text{where } p(\text{an integer}) = 1,2,3,\dots,$$

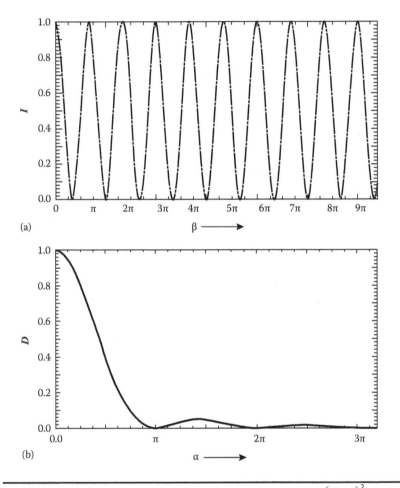

(a)

(b)

FIGURE 8.10 The figure displays plots of (a) $I = \cos^2\beta$ and (b) $D = \left(\dfrac{\sin\alpha}{\alpha}\right)^2$.

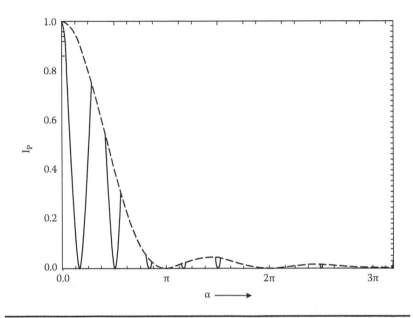

FIGURE 8.11   The figure illustrates a plot of $I_p = 4I_o \left(\dfrac{\sin \alpha}{\alpha} \right)^2 \cos^2 \beta$; I_o is in arbitrary units.

or

$$\beta = p\,\alpha. \tag{8.51}$$

Thus for p = 3, that is, $\beta = 3\alpha$ the first minimum on the diffraction pattern coincides with the third maximum interference fringe obliterating it almost completely. We say almost because the variation of the diffraction plot $\left(\dfrac{\sin \alpha}{\alpha} \right)^2$ near the $\alpha = \pi$ curve is not as steep as the interference $\cos^2\beta$ at the corresponding values of β near multiples of π. Hence, an insignificant remnant of the third interference fringe on its two sides survives. The same can be said about the second diffraction minimum obliterating the sixth interference fringe and so on (Figure 8.12). The obliterated interference fringes are called missing interference orders. If p = 2, $\beta = 2\alpha$, the second, fourth, sixth... orders of the interference fringes would be missing. The interference maxima that survive are called principal maxima.

QUESTION

What would happen if p = 1, that is, $\beta = \alpha$? Use the above argument to justify your answer in a manner that is relevant to the change in the physical setup when $\beta = \alpha$.

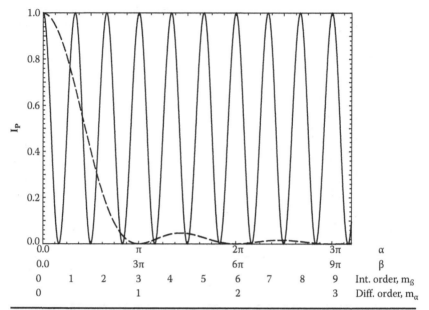

FIGURE 8.12 The figure illustrates the missing interference maxima $\cos^2 \beta$ caused by the diffraction plot $\left(\dfrac{\sin \alpha}{\alpha}\right)^2$ being zero at those maxima.

EXAMPLE 8.3

Consider a double-slit setup where their separation is three times the aperture opening, $d = 3a$, $a = 4.00 \times 10^{-7}$ m. For a light wave of 600.0 nm incident on the slits, calculate the intensity of the first and second principal maxima that follow the central maximum.

Solution

Since $d = 3a$, the third, sixth, and ninth principal interference fringes (peaks) are obliterated due to the diffraction envelope being minimum at $m_\alpha = 1, 2, 3$ for which $\alpha = \pi, 2\pi, 3\pi$, etc.

The first and second subsidiary maxima occur at $\alpha = (1/3)\pi$ $\alpha = (2/3)\pi$, while $\beta = \pi, 2\pi$.

$$I = \cos^2 \beta, \quad D = \left(\frac{\sin \alpha}{\alpha}\right)^2$$

Substituting in Equation 8.47, $I_p = I_{max} \left(\dfrac{\sin \alpha}{\alpha}\right)^2 \cos^2 \beta$, we find the following:

For the first principal maxima,

$$I_p = I_{max} \left(\frac{\sin(1/3)\pi}{(1/3)\pi} \right)^2 \cos^2(\pi) = I_{max} \left(\frac{0.866}{(1/3)\pi} \right)^2 \cos^2(\pi) = 0.683\, I_{max}$$

For the second principal maxima,

$$I_p = I_{max} \left(\frac{\sin(2/3)\pi}{(2/3)\pi} \right)^2 \cos^2(2\pi) = I_{max} \left(\frac{0.866}{(2/3)\pi} \right)^2 \cos^2(\pi) = 0.171 I_{max}$$

8.5 DIFFRACTION GRATINGS

A diffraction grating is basically a slab of solid transparent plate or sheet on which a large number of slits are generated via parallel grooves much like parallel straight lines etched on the plate, creating a multi slit instrument. Figure 8.13 is a typical sketch of a setup employing a lens and a screen positioned at a relatively long distance from the grating. A parallel beam incident normally to the grating face gets diffracted by each of the slits. The diffracted wavelets from a large number of slits, N, interfere, producing a diffraction pattern on a screen. The mechanism of deriving the contributions of such a large number of slits to the pattern on the screen is very similar to what has been presented for the double-slit diffraction, Equation 8.37.

The intensity of light at any point on the screen is obtained by summing over all the slits in the grating. There are several ways to carry out this procedure, where all bring out the same final answer. One is carrying an integration over all the open space of the grating. An alternative, much simpler method is to combine, that is, sum, a large number of waves, each of which is different in phase from the preceding one by a constant angle, β, that corresponds to a path difference, $d \sin \theta$, previously defined by the relation $\beta = \dfrac{kd\sin\theta}{2}$.

Following the symbolism adopted in presenting the single-slit diffraction, we assume that the difference in phase between waves from two

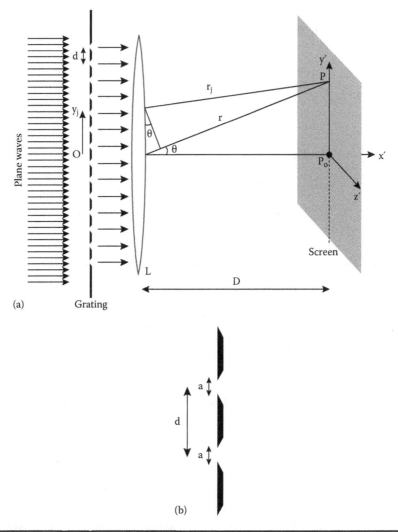

FIGURE 8.13 A sketch of a diffraction grating setup showing plane waves shone on the grating and received on a screen. The wavelets emitted at the slits are also plane. The lens L focuses the diffraction pattern on the screen.

adjacent sources as they arrive at a point P is 2β. Letting the origin be located at the bottom of the grating, and as the amplitude $E(\theta) = E_{o}\dfrac{\sin\alpha}{\alpha}$ of all waves becomes identical, the summation of these waves takes the form

$$E(\theta) = E_{o}\frac{\sin\alpha}{\alpha}(1 + \sin 2\beta + \sin 4\beta + \cdots \sin(N-1)2\beta. \qquad (8.52)$$

The sum in the bracketed terms $(1 + \sin 2\beta + \sin 4\beta + \cdots \sin(N-1)2\beta$ can be determined by considering the complex series

$$S = \sum_{j=1}^{N} e^{i\varphi j} = A\left(e^{i\varphi} + e^{2i\varphi} + e^{3i\varphi} + \cdots e^{i(N-1)\varphi}\right), \tag{8.53}$$

where

$$\varphi = 2\beta.$$

Equation 8.53 is a geometric series that has the sum

$$S = A\frac{1 - e^{i(N)\varphi}}{1 - e^{i\varphi}}.$$

Thus, $E(\theta)$ becomes

$$E(\theta) = E_0 \frac{\sin\alpha}{\alpha} \frac{\left(1 - e^{iN\varphi}\right)}{\left(1 - e^{i\varphi}\right)}.$$

The intensity of the resultant at the point P is $I(\theta) = |E(\theta)|^2$, resulting in

$$I(\theta) = I_0\left(\frac{\sin^2\alpha}{\alpha^2}\right)\frac{(1 - \cos N\varphi)}{(1 - \cos\varphi)}$$

$$I(\theta) = I_0\left(\frac{\sin^2\alpha}{\alpha^2}\right)\frac{\left(\sin^2 N\,\varphi/2\right)}{\left(\sin^2 \varphi/2\right)}$$

$$I(\theta) = I_0\left(\frac{\sin^2\alpha}{\alpha^2}\right)\frac{\left(\sin^2 N\beta\right)}{\left(\sin^2\beta\right)} \tag{8.54}$$

$$I(\theta) = I_0\,D\,I, \tag{8.55}$$

where

$$I = \frac{\left(\sin^2 N\beta\right)}{\left(\sin^2\beta\right)}, \tag{8.56}$$

$$D = \frac{\sin^2 \alpha}{\alpha^2} \qquad (8.57)$$

are the terms responsible for interference and diffraction, respectively. Equations 8.56 and 8.57 are separately plotted in Figure 8.14a and b. Figure 8.15 shows a plot of I(θ) as described by Equation 8.54.

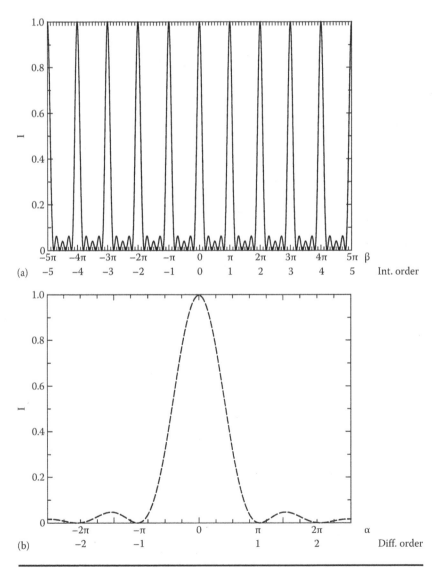

FIGURE 8.14   The figure illustrates (a) a plot of $I = (\sin^2 N\,\beta)/(\sin^2\beta)$ showing the interference orders and (b) a plot of $D = \sin^2\alpha/\alpha^2$ showing the diffraction orders.

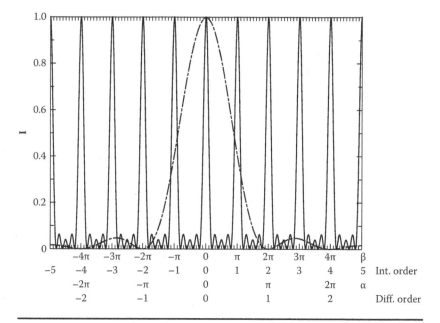

FIGURE 8.15   A plot of $I(\theta) = I_o \dfrac{\sin^2\alpha}{\alpha^2}\dfrac{\left(\sin^2 N\beta\right)}{\left(\sin^2\beta\right)}$, on which all relevant interference and diffraction orders are displayed.

QUESTION

Using the above relation for a diffraction grating for which N = 1, check if Equation 8.41 reduces to a single slit expression!

ANALYSIS

The argument is almost trivial. Try it!

8.5.1 ANALYSIS OF THE GRATING DIFFRACTION PATTERN

Let us have another look at the interference term $I = \dfrac{\left(\sin^2 N\beta\right)}{\left(\sin^2\beta\right)}$ in Equation 8.54. This term is maximum when the numerator and denominator are both zero. This happens when $N\beta = s\pi$ for $s = 0, \pm1, \pm2, \pm3,....$ As for the denominator, it becomes zero when $\beta = m_\beta\pi$ for $m_\beta = N, 2N,....$ Thus, between each two interference principal maxima, there are (N − 1) values for β, where the interference term is zero. These (N − 1) minima accommodate N − 2 maxima that are called subsidiary maxima. This is clearly noticed in Figure 8.13. As N = 5, there are four minima separated by three secondary maxima.

Thus, the number of subsidiary maxima between successive principal maxima increases as the number of the slits increases. However, their intensity decreases to an almost vanishing order of magnitude. Following this argument, one can see that the subsidiary maxima, being equally spaced between two successive integer multiples of N, their angular separation is $\frac{\pi}{N}$. Hence, these minima occur at

$$\beta_{subsidiary\ min} = \pm\frac{\pi}{N}, \pm\frac{2\pi}{N}, \pm\frac{3\pi}{N}, \ldots, \pm\frac{(N-1)\pi}{N}, \pm\frac{(N+1)\pi}{N}, \ldots, \pm\frac{(2N-1)\pi}{N} \quad (8.58)$$

Assuming that each subsidiary maxima is midway between two successive minima, then on either side of the central maximum, the positions of any subsidiary maximum would be midway between two successive values of the β values stated in Equation 8.58, that is, starting from the central maximum, the position of the first is at

$$\beta_{1st\ subsidiary\ max} = \frac{\pi}{N} + \frac{1}{2}\left(\frac{2\pi}{N} - \frac{\pi}{N}\right) = \frac{3\pi}{2N}$$

and for the second,

$$\beta_{2nd\ subsidiary\ max} = \left[\frac{3\pi}{2N} + \frac{1}{2}\left(\frac{3\pi}{N} - \frac{\pi}{N}\right)\right] = \frac{5\pi}{2N}$$

and so on for the positions of the rest generating the sequence

$$\beta_{subsidiary\ max} = \frac{3\pi}{2N}, \frac{5\pi}{2N}, \frac{7\pi}{2N}, \frac{9\pi}{2N} \quad (8.59)$$

In Figures 8.14, a plot of $\frac{\sin^2\alpha}{\alpha^2}$ and the product $\frac{\sin^2\alpha}{\alpha^2}\frac{(\sin^2 N\beta)}{(\sin^2\beta)}$ are shown. It can be noted that the $\frac{\sin^2\alpha}{\alpha^2}$ function chops significant portions of the function $\frac{(\sin^2 N\beta)}{(\sin^2\beta)}$ at particular values of α and β that share the same value of θ. This commonality determines the ratio $\frac{\beta}{\alpha}$ as follows:

$$\frac{\beta}{\alpha} = \frac{(d\sin\theta)/2}{(a\sin\theta)/2} = \frac{d}{a} \quad (8.60)$$

Introducing integer parameters m_β and m_α that label interference maxima and diffraction minima, respectively, then a ratio, $\dfrac{d}{a} = 2$, means that $\dfrac{\beta}{\alpha} = \dfrac{m_\beta \pi}{m_\alpha \pi} = 2$, that is, $\beta = 2\alpha$ and $m_\beta = 2\, m_\alpha$. Thus, the second-order maximum interference fringe (i.e., second principal maxima) will lie at the position of the first minimum diffraction and get obliterated. Similarly, the fourth-order maximum interference fringe will lie at the position of the second minimum diffraction, and so on for $m_\alpha = 3$, $m_\beta = 6$. Thus the second, fourth, and sixth interference maxima are obliterated by the first, second, and third minima, respectively, in the diffraction. The second, fourth, sixth,… obliterated principal interference maxima are called missing orders; these are illustrated in Figure 8.15.

EXERCISE

What change would happen for the diffraction pattern of a grating if $\beta = \alpha$? Use the above argument to justify your answer in a manner that relates to the change in the physical setup when $\beta = \alpha$.

8.5.2 FURTHER ANALYSIS OF THE DIFFRACTION PATTERN FOR A GRATING

Keeping in mind that for all $\alpha \neq 0$, the diffraction $\dfrac{\sin^2 \alpha}{\alpha^2}$ continues to decrease, all surviving interference fringes are decreasing in intensity for all values of β other than $\beta = 0$. The integers $m_\beta = 1,\ 2,\ 3,\dots$ that lie within the central diffraction maximum define possible interference spectral lines (principal maxima) that happen to have their wavelengths satisfying the condition for a maximum interference; from the discussion in the previous section, this condition is

$$N\beta = s\pi; \quad s = 0, 1, 2, 3,\dots, N, (N+1),\dots(2N-1), (2N),\dots(3N),$$

that is, when

$$\beta = \frac{s\pi}{N}; \quad s = 0, 1, 2, 3,\dots, N, (N+1),\dots(2N-1), (2N),\dots(3N)$$

The principal interference maxima occur when

$$s = 0, N, 2N,\dots,$$

that is, the condition is

$$\frac{S}{N} = 0, 1, 2, 3, \ldots, m.$$ (8.61)

Substituting for $\beta = (d \sin \theta)/2$, the above reduces to

$$(kd \sin \theta)/2 = m \lambda; \quad k = 2\pi/\lambda$$

$$d \sin \theta = m \lambda$$ (8.62)

Equation 8.62 is known as the grating equation, and d is the grating spacing or the grating constant.

EXAMPLE 8.4

Consider a diffraction grating that has 6000 lines/cm. For a beam of white light incident normal to the grating surface, find the angular position of the first-order maxima of yellow light of 600 nm wavelength.

Solution

From the given dimensions of the grating, its spacing d is

$$d = \frac{1}{6.00 \times 10^3 \text{lines/cm}} = 1.67 \times 10^{-4} \text{cm/line.}$$

From the grating equation,

$$m \lambda = d \sin \theta,$$

that is,

$$\theta_{yellow} = \sin^{-1}\left[\frac{m\lambda}{d}\right],$$

which upon substitution for the given quantities gives

$$\theta_{yellow} = \sin^{-1}\left[\frac{(1)(6.00 \times 10^{-7} \text{m})}{1.67 \times 10^{-6} \text{m}}\right] = \sin^{-1}(0.359) = 21.1°.$$

8.6 ANGULAR DISPERSION AND POWER OF A GRATING

Among the interesting features of the diffraction pattern is the width of each principal maximum. From Equation 8.58, the adjacent minimum on both sides of a principal maximum has an angular extension

$$\Delta\beta = \left(\frac{\pi}{N}\right) - \left(-\frac{\pi}{N}\right) = \frac{2\pi}{N} \tag{8.63}$$

As for any interference maximum of order m

$$\beta = \left(kd \sin\theta_m\right)/2 = \pi d \sin\theta_m/\lambda,$$

then

$$\Delta\beta = \left(\pi d \cos\theta_m\right)(\Delta\theta)/\lambda. \tag{8.64}$$

Combining Equations 8.63 and 8.64 results in

$$\left(d\cos\theta_m\right)(\Delta\theta)/\lambda = \frac{2}{N},$$

giving

$$\Delta\theta = \frac{2\lambda}{Nd\cos\theta_m}. \tag{8.65}$$

As noted, the two factors increase in the number of the slits in the grating, and an increase in the width of the grating spacing d offers a narrower angular spread that implies sharper spectral lines.

The change of θ with λ can be determined from the grating equation, which for any order m is

$$d \sin\theta_m = m\lambda.$$

Thus

$$\frac{\Delta\theta}{\Delta\lambda} = \frac{m}{d\cos\theta_m}. \tag{8.66}$$

Equation 8.66 describes the angular spread for wavelengths that are slightly different from each other. Thus, for an order m, a minimum in the angular spread

$$\Delta\theta_{min} = \frac{m\,\lambda_{min}}{d\cos\theta_m}. \tag{8.67}$$

However, from Equation 8.65, for a two-component light wave

$$\Delta\theta = \frac{2\bar\lambda}{Nd\cos\theta_m},$$

where $\bar\lambda$ is the average wavelength of the wavelengths of the two waves.

The resolving power R of a grating becomes relevant when two light components that are slightly different in wavelengths are incident on a grating and being separated from each other. The resolving power is defined as $R = \dfrac{\bar\lambda}{(\Delta\lambda)_{min}}$, where $(\Delta\lambda)_{min}$ is the least resolvable difference in the wavelengths being diffracted. From Rayleigh criterion, this least resolution occurs when the principal maximum (interference) coincides with the first minimum of the other, Figure 8.16. That is, the two maxima, one from each wavelength are separated by the angular spread of either principal maximum, that is

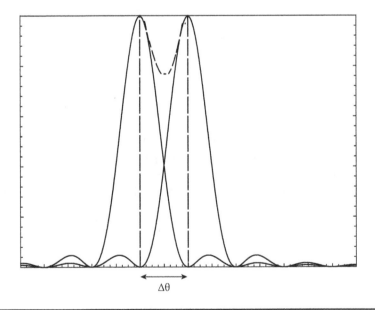

FIGURE 8.16 Rayleigh criterion for two wavelengths to be barely resolved: principal maximum of one falls on the first minimum of the other.

$$\Delta\theta_{min} = \frac{\Delta\theta}{2} = \frac{\bar{\lambda}}{Nd\cos\theta_m}. \tag{8.68}$$

Equating the two r.h.s. of Equations 8.67 and 8.68 and rearranging terms yields

$$\Delta\theta_{min} = \frac{\bar{\lambda}}{(\Delta\lambda)_{min}} = mN. \tag{8.69}$$

Thus, the resolving power R of a grating is mN.

EXAMPLE 8.5

Consider a diffraction grating of 6.0 cm wide having 6300 grooves per cm. Calculate its resolving power in the second-order diffraction of light of wavelength $\lambda = 600$ nm incident normally to the grating surface.

Solution

$$R = Nm = \left(6300\ cm^{-1} \times 6.0\ cm\right) \times 2 = 7600$$

PROBLEMS

8.1   In a single-slit diffraction pattern, find the ratio $I(\theta)/I_o$ for the second diffraction, also called secondary maximum. Comment on the relevance of the value of your answer.

8.2   For diffraction of a single slit of width a, illuminated with light of 600.0 nm, the distance between the second-order maxima on both sides of the $\alpha = 0$ line of symmetry is 2.50 mm. For a distance of 0.600 m between the screen and the slit, (a) determine the slit width a; (b) find the separation between the two third-order minima on the two sides of the central maximum.

8.3   A narrow slit of width a $= 4.00 \times 10^{-6}$ m is illuminated by a yellow monochromatic light of $\lambda = 600.0$ nm. Knowing that the screen is placed at the focal length of a lens of 0.800 m, find the linear spacing between (a) the first- and (b) third-order maxima in the single-slit diffraction pattern.

8.4   Using the same data in the previous problem, find the angular positions θ_1 and θ_3 of (a) the first- and (b) third-order maxima in the single-slit diffraction pattern.

8.5   Equation 8.54 describes diffraction for a double slit. Look into the equation to find what it becomes if d = a.

8.6   Given two identical narrow slits separated by 1.2×10^{-4} m and placed at 0.80 m from a screen is illuminated by a monochromatic light of 540.0 nm wavelength. The diffraction pattern on the screen showed four distinct interference principal maxima beyond which the pattern was very weak. Find the width of each slit.

8.7   Consider two narrow slits separated by 1.60×10^{-6} m, and each of width 4.00×10^{-7} m. The distance between the screen and the double slit is 0.80 m. For a green spectral line of wavelength $\lambda = 540.0$ nm, incident normally on the double slit, find the angular positions on the screen of the first and second principal maxima.

8.8   For the double slit described in the previous problem
   (a) Define the angular positions in terms of β and α of the first, second, and third missing orders of the green spectral lines in the double-slit diffraction.
   (b) Calculate the angular and linear width of the above missing orders, and comment on your answer.

8.9   Consider an obstacle consisting of six slits where d = 3a. Calculate the intensity of the first and second principal maxima that follow the central principal maximum.

8.10   Using the relation derived for a diffraction grating for which N = 2, check if Equation 8.54 reduces to a double-slit arrangement.

8.11   For two visible light sources separately used to shine normally on a diffraction grating, the second order of the diffraction pattern for one wavelength λ_1 overlaps with the third order of the diffraction pattern for λ_2. Find the ratio between those wavelengths. Could there be more than a pair of overlapping wavelengths?

8.12   For a grating, whose grating constant is 1.19×10^{-6} m, determine the difference between the angular positions of the first-order hydrogen lines of the wavelengths blue, $\lambda = 411.0$ nm, and red, 656.0 nm, observed on a screen 0.80 m away from the grating.

8.13   Use your answers for the previous problem to find the linear distance between the first- and second-order wavelengths stated in the problem.

8.14   Consider a grating of eight slits for which d = 5a and a screen placed in front of the slits 0.80 m away. Determine the intensity of the first, second, and third subsidiary maxima in the diffraction pattern obtained for this grating.

8.15   For a diffraction grating of 6.0 cm wide having 6300 grooves per cm, determine the minimum wavelength difference around an average wavelength of 600.0 nm that could be resolved in the first order by this grating.

9

Diffraction II
Fresnel Diffraction

Every body attracts every other with a force directly proportional to the product of their masses and inversely proportional to the square of the distance between them.

Isaac Newton (1642–1727)

9.1 INTRODUCTION

In Chapter 8, the Fraunhofer diffraction, also known as far-field diffraction, was discussed in detail. As noted then, the locations of the source and screen in the setup of Fraunhofer diffraction were arranged to be effectively at infinity from the diffraction device, a slit-like aperture, or a diffraction grating. Such a condition simplified the treatment significantly; it made it possible to treat the emitted waves reaching the aperture as plane waves, all having the same phase and amplitude. The aperture opening acts as a line of a large number of point sources each of which emits spherical wavelets at every point of the wave front reaching the slit. They differ in phase as they reach an observation point, P, on the screen, but all reach the point with the same amplitude. However, Fresnel diffraction, also known as near-field diffraction, is free from any particular limitation.

The setup requires a source, a diffracting device, and a screen. Thus, Fresnel diffraction involves a variety of variables that influence its formation.

Among the early observations of Fresnel diffraction were the effects of a sharp-edge obstacle on light illuminating the area behind the obstacle, where light spreads over an area beyond geometric limits and a series of bright and dark bands were observed. Fresnel offered convincing mathematical analyses for diffraction from straight edges, rectangular openings, and circular apertures.

9.2 LAYOUT AND ASSUMPTIONS: OBLIQUITY FACTOR

A fundamental consideration in Fresnel diffraction is the fact that waves emitted by a point source at a finite distance from an aperture or a straight-edge obstacle are spherical. This also applies to the wave fronts reaching points of the aperture opening or the edge of the obstacle. On the basis of Huygens' theory, all points on the wave front at the opening become secondary sources of spherical wavelets, and these, supposedly propagating in all directions, are oriented with angles ranging between 0° and ±180° (see Figure 9.1).

This means that light from the secondary sources at the aperture is traveling forward beyond the obstacle and backward in areas in front of the obstacle. As this is not what is observed in real situations, Fresnel supplemented Huygens' theory with what is called an obliquity factor $F(\theta)$, shown by Kirchhoff to be of the form

$$F(\theta) = \frac{1}{2}(1 + \cos \theta) \qquad (9.1)$$

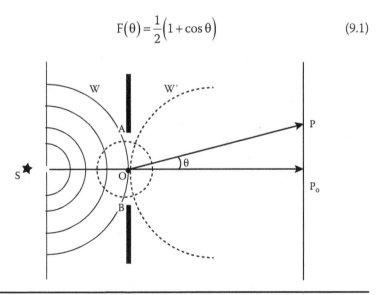

FIGURE 9.1 Huygens' secondary wavelets emitted at a narrow opening are impacted by the obliquity factor. In the figure, only one Huygens' wavelet is shown.

that affects the direction of propagation of the wavelets emitted by the secondary sources. In Equation 9.1, θ is the angle between a horizontal line through the secondary source and the line connecting the latter with the observation point. The factor $F(\theta)$, multiplied by the amplitude of a spherical secondary wavelet at the aperture, reduces it by a factor ranging between 1 and 0; the angle $\theta = 0°$ is for the forward direction, and $\theta = 180°$ is for the backward direction. As noticed, $F(\theta) = 1/2$ for $\theta = \pi/2$. This shows that all wavelets directed between $\theta = -\pi/2$ and $\theta = \pi/2$ contribute to the disturbance at point P. The obliquity factor $F(\theta)$, though lacking in physical reasoning, helped supplement Huygens' theory in explaining Fresnel diffraction.

Another fundamental aspect is that the finite distances from the aperture elements to a point, P, vary considerably and impact the wavelets' amplitudes and phases. In Fraunhofer diffraction, the geometry of infinite distances of the aperture from the source and the screen implied that equidistant points on the ray path from an aperture to a point, P, on the screen, measured from P, lie on a plane (Figure 9.2a, screen distance not to scale). However, in Fresnel diffraction, such equidistant points from P lie on a spherical surface (Figure 9.2b). Hence, the difference between the phase on a spherical surface of an original wave front and that of a wavelet generated by the wave at the aperture is critical for the overall contribution of all points of the aperture to the resultant diffraction.

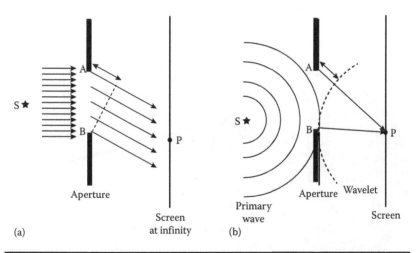

FIGURE 9.2 The figure in (a) is an illustration of wavelets at the opening of a Fraunhofer setup; it illustrates a linear change in phase across the opening, while in (b) the phase in Fresnel diffraction changes in a quadratic manner. (From Smith, F.G., King, T.A., and Wilkins, D., *Optics and Photonics*, John Wiley & Sons, Hoboken, NJ, 2007, p. 240.)

9.3 HUYGENS–FRESNEL DIFFRACTION

To develop a mathematical treatment of Huygens–Fresnel diffraction, we consider Figure 9.3, in which the arc W–W represents a primary wave with a spherical wave front of radius r' emitted by a source S, and P is an observation point on the screen. The arc W'–W' (shown in Figure 9.4) represents a wavelet of radius r emitted by a secondary source on the primary wave front at O. The line connecting S and P passing through the polar point O consists of the segments SO and OP, which are labeled r_o' and r_o, respectively. Assuming that E_o is the amplitude at unit distance from the source, the disturbance per unit area of the aperture at a point of the wave front lying on the aperture can be described by

$$E_A = \frac{E_o}{r'} e^{i(k r' - \omega t)}.$$ (9.2)

For convenience, designating the instant, when the wave front is at the selected arbitrary point as t = 0, the factor $e^{-i\omega t}$ would be unity and can be dropped from the expression above. The above impacted by the obliquity factor $F(\theta)$ becomes

$$E_A F(\theta) = \frac{F(\theta) E_o}{r'} e^{ikr'}.$$ (9.3)

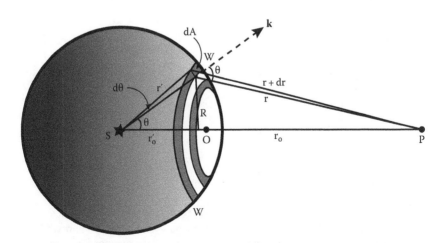

FIGURE 9.3 Illustration of a spherical wave front emitted by a source, S. A limited number of Fresnel zones are distinctively marked.

The wave in Equation 9.3 propagates toward the receiving point P as a secondary wavelet of an element of disturbance dE_P given by

$$dE_P = \frac{F(\theta)E_A}{r}e^{ik(r-\omega t)}.$$ (9.4)

Combining Equations 9.3 and 9.4 gives the disturbance at a point P due to the wavelets reaching P, and that is

$$dE_P = \frac{F(\theta)E_o}{rr'}e^{i[k(r+r')-\omega t]}.$$ (9.5)

To include contributions of all wavelets generated by the aperture points, the above reduces to

$$E_P = \int_A F(\theta)\frac{E_o}{rr'}e^{i[k(r+r')-\omega t]}dA.$$ (9.6)

The above is an integration over the aperture whose geometry differs from one application to another. As will be noted in the next few sections, it was Fresnel who offered an imaginative remedy to the problem through incorporating a successful structure of a primary spherical wave advancing toward the aperture where then different points on the primary wave would have different phases.

9.3.1 FRESNEL HALF-PERIOD ZONE STRUCTURE ASSUMPTIONS AND FEATURES

In an attempt to carry out the integration in Equation 9.6, Fresnel devised a special construction of the primary incident wave front that encounters the aperture as it propagates. With rationally founded assumptions and approximations, he then was able to account for the contribution from aperture diffraction at any receiving point of the screen. The Fresnel scheme is called Fresnel half-period zones. In this scheme, the incident wave front is divided into zones of circles that have a finite width centered on the pole O (see Figure 9.3). The circles are constructed such that the distance from the observation point P to the first (smallest) circle is $r + \lambda/2$; the rest follow, each distance from P being larger than the preceding one by $\lambda/2$. Let us call these zones $Z_1, Z_2, Z_3, ..., Z_m$.

The task then reduces to determining how many zones cross the aperture, counting the path differences and the corresponding phase

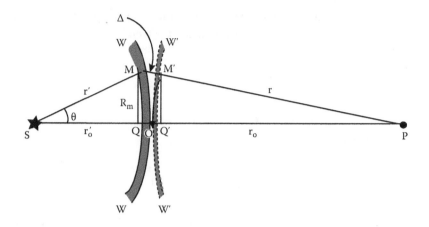

FIGURE 9.4 The path difference between secondary waves along SP and SMM′P is shown.

differences for all wavelets reaching P, and finally adding all the contributions that produces diffraction on the screen.

For a simple start of such a process, refer to Figure 9.4 in which a side view of an aperture, a segment of a Fresnel zone on a primary wave, and a wavelet generated at a point of the aperture are shown. Consider the two points O and M through which the paths SOP and SMP pass. The first is the shortest path of light between the source and receiving point P traversed by the primary wave and the secondary wavelet generated at O. The second is the path OM plus that of a wavelet generated at M, another point on the primary wave. The two paths, being not equal, are of a path difference Δ that is equal to

$$\Delta = (r + r') - (r_o + r_o'), \tag{9.7}$$

which by Fresnel approximation may be expressed as

$$\Delta = QO + Q'O. \tag{9.8}$$

The segments QO and Q′O are known as the sagittas of the arcs W–W and W′–W′. For a small height of points M and M′ above the line SOP, these heights can be shown to be equal to $QO = \dfrac{R^2}{2r_o'}$, $Q'O = \dfrac{R^2}{2r_o}$.

Equation 9.7 becomes

$$\Delta = \frac{R^2}{2r_o'} + \frac{R^2}{2r_o} \tag{9.9}$$

$$\Delta = \frac{R^2}{2}\left(\frac{r_o + r_o'}{r_o r_o'}\right).$$

(9.10)

According to Fresnel assumption, the path difference Δ of a zone m implies that

$$\Delta = m\frac{\lambda}{2}.$$

(9.11)

From Equation 9.10, for the mth zone, we have

$$m\frac{\lambda}{2} = \frac{R_m^2}{2}\left(\frac{r_o + r_o'}{r_o r_o'}\right),$$

(9.12)

or

$$R_m^2 = m\lambda\left(\frac{r_o r_o'}{r_o + r_o'}\right).$$

(9.13)

Also, referring to Figure 9.4, and since $rr_o + \Delta$, we have

$$R_m^2 = \left(r_o + \frac{m\lambda}{2}\right)^2 - r_o^2,$$

or

$$R_m^2 = r_o^2\left[\left(\frac{m\lambda}{r_o}\right) + \left(\frac{m\lambda}{2r_o}\right)^2\right].$$

For the condition $m\lambda/r_o \ll 1$, the second term on the right-hand side of the above can be dropped, resulting in

$$R_m^2 = mr_o\lambda.$$

That is

$$R_m = \sqrt{mr_o\lambda}.$$

(9.14)

Thus, for an aperture of radius a, whose center is on the line connecting the source and the observation point P, the number of zones N that pass through the aperture is calculated from

$$\pi a^2 = N A_m,$$

where A_m refers to the mth zone. Letting the central zone (the first zone) be labeled by m = 1, with $A_1 = \pi R_1^2$, the number of zones denoted by N that would fit into the aperture is

$$N = \frac{a^2}{R_1^2} = \frac{a^2}{r_o \lambda}. \tag{9.15}$$

The number N is used as a criterion that draws a distinction between Fraunhofer and Fresnel diffraction. For $N \ll 1$, the case is a Fraunhofer one, while for $N > 1$, it is a Fresnel diffraction case.

The area A_m of any zone, mth, is

$$A_m = \pi \left(R_m^2 - R_{m-1}^2 \right).$$

Using Equation 9.13, the above becomes

$$A_m = \pi \left[m\lambda \left(\frac{r_o \, r_o'}{r_o + r_o'} \right) \right] - \left((m-1)\lambda \left(\frac{r_o \, r_o'}{r_o + r_o'} \right) \right).$$

The above after a slight algebra gives

$$A_m = \pi\lambda \left(\frac{r_o \, r_o'}{r_o + r_o'} \right). \tag{9.16}$$

Equation 9.16 shows that the area of any zone is independent of the zone ranked order. Hence, within the approximated expressions for the sagittas, all zones have equal areas.

9.3.2 ANALYTIC TREATMENT OF FRESNEL DIFFRACTION—CIRCULAR APERTURE

As noted earlier, we are interested in summing over all points of the wave front as it enters the aperture, then letting that instant be the zero in time (t = 0), the factor $e^{-i\omega t}$ would become unity and can be dropped from Equation 9.6.

As r' is the radius of the primary wave front hitting the aperture, it is constant as far as the integral in the expression is concerned. It is r and θ that determine an element of area dA on the primary wave front. This reduces Equation 9.6 to

$$E_P = \frac{E_o e^{ikr'}}{r'} \int_A F(\theta) \frac{e^{ikr}}{r} dA, \tag{9.17}$$

where the element dA is

$$dA = r'^2 \sin\theta \, d\theta \, d\varphi. \tag{9.18}$$

From Figure 9.3 we have

$$r^2 = r'^2 + \left(r'_o + r_o\right)^2 - 2r'\left(r'_o + r_o\right)\cos\theta. \tag{9.19}$$

Thus

$$r \, dr = r'\left(r' + r_o\right)\sin\theta \, d\theta,$$

giving

$$\sin\theta \, d\theta = \frac{r \, dr}{r'\left(r' + r_o\right)}. \tag{9.20}$$

Substituting for $\sin\theta \, d\theta$ in Equation 9.18 makes dA equal to

$$dA = \frac{r' r \, dr}{\left(r' + r_o\right)} d\varphi, \tag{9.21}$$

and substituting for the above expression for dA in Equation 9.17 yields

$$E_P = \frac{2\pi E_o e^{ikr'}}{r' + r_o} F(\theta) \int_A e^{ikr} dr. \tag{9.22}$$

As noticed, we have only one variable of integration, and the contribution to the diffraction pattern at a point P from any zone, the mth, is going to be bounded between the borders of two successive zones. Let these be Z_{m-1} and Z_m, which correspond to $r_{m-1} = r_o + (m-1)\lambda/2$ to $r_m = r_o + m\lambda/2$. Thus

$$E_{P,m} = \frac{2\pi E_o e^{ikr'}}{r' + r_o} F_m(\theta) \int_{r_o+(m-1)\lambda/2}^{r_o+m\lambda/2} e^{ikr} dr.$$

The above, after some algebra, turns out to be

$$E_{P,m} = -C \frac{2\pi i \, E_o e^{ik(r'+r_o)}}{(r'+r_o)} F_m(\theta) \frac{e^{ik_m\lambda/2}\left(1 - e^{-ik\lambda/2}\right)}{k}. \tag{9.23}$$

Noting that $k = 2\pi/\lambda$, the part $\dfrac{e^{ik_m\lambda/2}\left(1 - e^{-ik\lambda/2}\right)}{k}$ in the above expression is

$$\frac{e^{ik_m\lambda/2}\left(1 - e^{-ik\lambda\,2}\right)}{k} = \frac{\lambda}{2\pi} e^{i\pi m}\left(1 - e^{-i\pi}\right) = \frac{\lambda}{2\pi}\left(\cos\pi_m\right)(2)$$

$$= \frac{\lambda}{\pi} e^{i\pi m}\left(1 - e^{-i\pi}\right) = \frac{\lambda}{\pi}(-1)^m,$$

Equation 9.23 becomes

$$E_{P,m} = -C \frac{2\pi i \, E_o e^{ik(r'+r_o)}}{(r'+r_o)} F_m(\theta) \frac{e^{ik_m\lambda/2}\left(1 - e^{-ik\lambda/2}\right)}{k}. \tag{9.24}$$

Equation 9.24 expresses the amplitude of the diffraction at point P contributed by zone m. The overall contribution of the diffracted wavelets consists of those of the zones entering the aperture, which is obtained by summing over all of them, that is, for a total number of zones N, the diffraction amplitude at P is

$$E_P = \sum_{m=1}^{N} E_{P,m}$$

giving

$$E_P = \frac{2\lambda i E_o e^{ik(r'+r_o)}}{(r'+r_o)} \sum_{m=1}^{N}(-)^{m+1} K_m(\theta) \tag{9.25}$$

where

$$K_m(\theta) = 2F_m(\theta) = \left(1 + \cos\theta_m\right); \quad 0 \le \theta_m \le \pi/2 \tag{9.26}$$

$$E_P = \frac{\lambda i E_o e^{ik(r'+r_o)}}{(r'+r_o)} \sum_{m=1}^{N}(-)^{m+1} K_m(\theta). \tag{9.27}$$

Obviously, the summation over all the zones, N, entering the aperture is what determines the resultant amplitude contributed by the whole incident primary wave. The summation consists of N terms that have alternating signs, +, −, ending with either a positive or negative last term depending on whether the number of the zones is odd or even. The difference in the magnitude of the amplitude due to changes in the angle of inclination θ for various points on a particular zone is negligible. However, the phasors that represent the amplitudes contributed by the successive zones continue to decrease very slowly as m increases. Thus, calling the resultant amplitude E_P, we then have N phasors summed as follows:

$$E_P = E_1 - E_2 + E_3 - E_4 + E_5 - E_6 + \cdots + (-1)^{N+1} E_N. \qquad (9.28)$$

The properties, magnitude, and directions of the phasors in Equation 9.28 may be grouped as follows:

$$E_P = \frac{E_1}{2} + \left(\frac{E_1}{2} - E_2 + \frac{E_3}{2}\right) + \left(\frac{E_3}{2} - E_4 + \frac{E_5}{2}\right) + \cdots + \left(\frac{E_{N-2}}{2} - E_{N-1} + \frac{E_N}{2}\right) + \frac{E_N}{2}.$$

For odd N, one can see that each of the bracketed terms is positive and is almost zero, leaving the sum to be

$$E_P = \frac{E_1}{2} + \frac{E_N}{2} \quad \left(N \text{ is odd}\right). \qquad (9.29)$$

However, for even N, one can see that each of the bracketed terms is positive and is almost zero, leaving the sum to be

$$E_P = \frac{E_1}{2} - \frac{E_N}{2} \quad \left(N \text{ is even}\right). \qquad (9.30)$$

For a large number of zones on a primary incident spherical wave, values of the angle θ would range between θ = 0° for the forward direction (first zone) and θ = 180° for the backward direction (the last zone). Since the obliquity factor F(θ) would be zero at 180°, the last zone would not contribute to the diffraction amplitude at P. Therefore, the resultant amplitude at P would be just half the amplitude attributed by the first zone, that is

$$E_P = \frac{E_1}{2}. \qquad (9.31)$$

Figure 9.5 demonstrates a phasor diagram for Fresnel half-period zones in which the first 15 phasors are distinctly marked. The diagram shows the resultant $E = E_1/2$ is the overall contribution (resultant) by an unobstructed wave front.

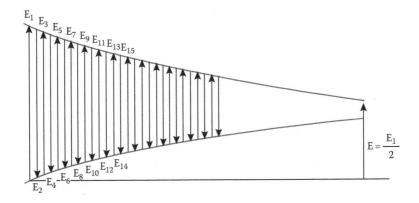

FIGURE 9.5 Phasor diagram for a finite number of Fresnel half-period zones; the first 15 of these are labeled.

Incorporating the cases expressed by Equations 9.29 and 9.30 in Equation 9.27 reduces the latter to the following:

$$E_P = i\lambda \left(K_1 \pm K_N \right) \frac{E_o e^{ik(r'+r_o)}}{\left(r' + r_o \right)}, \qquad (9.32)$$

where the upper sign is for N being odd and the lower sign is when N is even. The above expression describes the total diffraction at an arbitrary point, P, on the screen. In parallel with Equation 9.31, 9.32 becomes

$$E_P = i\lambda K_1 \frac{E_o e^{ik(r'+r_o)}}{\left(r' + r_o \right)}. \qquad (9.33)$$

9.3.3 Special Remarks on the Amplitude of Fresnel Diffraction

There are several important observations that one can make on the features of Fresnel diffraction as inferred from the above derivation:

1. As stated in Equation 9.31, the resultant amplitude of Fresnel diffraction generated on a screen by a circular aperture is half of the amplitude contributed by the first zone.
2. According to Equation 9.15, the radii of Fresnel zones increase with the order of the zone as follows: $R_m = \sqrt{m}$.

3. For the expressions in (9.32 and 9.33) to represent a spherical wave, the factor

$$i\lambda K_1 = 1.$$

That is,

$$K_1 = -\frac{i}{\lambda} = \frac{e^{-i\pi/2}}{\lambda}.$$

This implies that there is a change in phase of $\pi/2$ for the secondary wave from that of the primary one. This corresponds to one quarter of one wavelength. As each of Fresnel zones corresponds to half a period difference from neighboring zones, the secondary wavelet is one fourth of a period.

4. For the value of $K_1 = -\frac{i}{\lambda}$, the presence of λ in the denominator means that the amplitude of the secondary wavelet is $\left(\frac{1}{\lambda}\right)$ of the primary wave, implying that the intensity of the wavelet is extremely insignificant compared to the intensity of the primary wave.

5. Finally, substituting for $K_1 = -\frac{i}{\lambda}$ in Equations 9.32 and 9.33, we get the total amplitude of the diffraction pattern at P, becoming

$$E_P = \frac{E_o e^{ik(r'+r_0)}}{(r'+r_0)}. \tag{9.34}$$

EXERCISE

Verify the sum in Equation 9.30.

EXAMPLE 9.1

Consider a monochromatic light of wavelength $\lambda = 550.0$ nm incident on an aperture of radius a = 5.00 mm. The aperture was situated in the middle between the source and a point, P, on the screen along the line SOP (Figure 9.4). For S and P to be 10.0 cm apart, find

(a) The radius of the central zone
(b) The number of Fresnel zones that would be accommodated through the aperture

Solution

(a) From Equation 9.13, $R_m^2 = m\lambda\left(\dfrac{r_o r_o'}{r_o + r_o'}\right)$. Thus, to find the radius of

the first zone, $m = 1.00$, $R_1^2 = (1.0)\lambda\left(\dfrac{r_o r_o'}{r_o + r_o'}\right)$, which upon substi-

tuting for $r_o = 0.05$ m, $r_o' = 0.05$ m, and $\lambda = 550.0 \times 10^{-9}$ m, we get

$$R_1^2 = (1.00)\lambda\left(\frac{r_o r_o'}{r_o + r_o'}\right)$$

$$= (1.00)(550.0 * 10^{-9}\,\text{m})\left[\frac{(0.0500\,\text{m})(0.0500\,\text{m})}{(0.0500\,\text{m})+(0.0500\,\text{m})}\right].$$

The above reduces to $R_1^2 = 1.375 \times 10^{-8}$ m^2, giving $R_1 = 1.173 \times 10^{-4}$ m.

(b) The number of Fresnel zones that would pass through the aperture is N (to be denoted here by N_F), which from Equation 9.15 is

$$N_F = \frac{a^2}{R_1^2}.$$

Upon substitution for $a = 0.005$ m and $R = 1.173 \times 10^{-4}$ m, we get the number of Fresnel zones N_F from

$$N_F = \frac{(0.00500\,\text{m})^2}{(1.173\times 10^{-4}\text{m})^2} = 1820.$$

EXERCISE

How would your answers in Example 9.1 change if r_o' was infinity? Comment on the relevant change in the physics in this case.

9.4 ZONE PLATE

This is a two-dimensional plate placed to cover the aperture whose diffraction was treated in the previous section. The plate has a hole in its center that aligns with the aperture center. The hole has a radius that could be varied such that it has one-to-one correspondence with Fresnel zones structured on the primary wave, such that the hole radius coincides with an integer multiple of Fresnel zone m. The idea behind this special design comes from the amplitudes contributed by the successive zones described in Equation 9.28 that we rewrite in the following equivalent form:

$$E_P = \left[E_1 + E_3 + E_5 + \cdots (-1)^{N+1} E_N \right] - \left[\left(E_2 + E_4 + E_6 + \cdots \right) \right]; \quad N \text{ even} \quad (9.35)$$

$$E_P = \left[E_1 + E_3 + E_5 + \cdots \right] - \left[\left(E_2 + E_4 + E_6 + \cdots E_N \right) \right]; \quad N \text{ odd} \quad (9.36)$$

In brief, each of the above forms is just the sum of two square-bracketed sets, one set is the sum of the even zones and the other is the sum of the odd zones. Thus, if we manage one of these sets, the odd or the even terms set, the amplitude contributed by the unblocked zones would be much higher than what we derived it to be in Equations 9.29 and 9.30 or Equation 9.31. This is the function of a zone plate. According to Equation 9.14, the radius of any zone, the mth, is

$$R_m = \sqrt{m\, r_o\, \lambda}.$$

Thus, for an incident light of defined wavelength, the radius of any zone is proportional to $\sqrt{m}$. A white screen on which concentric black circles of radii $R_1 = \sqrt{r_o\, \lambda}$, $R_3 = \sqrt{3\, r_o\, \lambda}$, $R_5 = \sqrt{5 r_o\, \lambda} \ldots$, centered on the aperture center would do the elimination of the odd zones, and another with concentric circles of radii $R_2 = \sqrt{2\, r_o \lambda}$, $R_4 = \sqrt{4\, r_o \lambda}$, $R_6 = \sqrt{6\, r_o \lambda}$ would eliminate the even zones. In either case, as wavelets generated by points on the primary wave are all in phase, and each is one wavelength path difference larger than the neighboring shorter one, then all phasors representing one set of waves are in the same direction. That is why the resulting amplitude in either choice yields a much higher amplitude than having the contribution from all the zones. The diffraction passing through the unblocked segments of the plate is rather intense. Figure 9.6 illustrates the two cases we just described.

In another interesting connection, combining Equations 9.9 and 9.11 gives the value of Δ for the mth zone as

$$\Delta = m\frac{\lambda}{2} = \frac{R_m^2}{2}\left(\frac{1}{r} + \frac{1}{r_o'} \right),$$

that is,

$$\frac{1}{r} + \frac{1}{r_o'} = \frac{m\lambda}{R_m^2}, \quad (9.37)$$

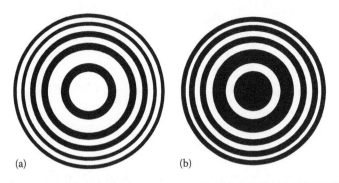

FIGURE 9.6 Zone plates. (a) The odd zones, 1st (central), 3rd, 5th,... are exposed, while in (b) the even zones 2nd, 4th, 6th,... are exposed.

which is an equation having special resemblance to a thin lens equation, $\frac{1}{r} + \frac{1}{r_o'} = \frac{1}{f}$, where we have a lens of focal length f, an object at distance of r_o', and an image at a distance r from the lens. Setting $\frac{1}{f} = \frac{m\lambda}{R_m^2}$, the resemblance becomes perfect. Thus, letting

$$f = \frac{R_m^2}{m\lambda},\qquad(9.38)$$

we can determine the position of diffraction at a point, p, along the symmetry axis of the plate for a given wavelength emitted by a source at a defined distance from the plate, r_o'.

We can say that the zone plate behaves like a lens that focuses the diffracted light at a point, P (Figure 9.4) on the screen. From Equation 9.16, the area of any zone is independent of m, that is, the areas of all zones are approximately equal, and $R_m = \sqrt{m}\, R_1$. Also, for an object at infinity, that is, a plane wave, $r_o' = \infty$, the focal length $f = r_o$. In Equation 9.38, the radius of the first zone is R_1, and the equation for m = 1 becomes

$$f = \frac{R_1^2}{\lambda}.\qquad(9.39)$$

As noted in Equation 9.39, due to the dependence of f on the square of R_1, an increase in R_1 results in a significant increase in f. From the above, we can determine the radius of the central zone when the focused position on the screen is given. This is given by

$$R_1 = \sqrt{f\lambda}.\qquad(9.40)$$

This f then represents the distance on the screen of the first bright circle from the center. Thus, the Fresnel zone plate behaves as a lens, and because it depends on the wavelength, a relevant chromatic aberration is expected.

9.5 FRESNEL DIFFRACTION FOR A RECTANGULAR APERTURE—FRESNEL ZONE STRUCTURE

In the diffraction for a circular aperture, on which a spherical wave is incident, spherical polar coordinates were applied to simplify the integral in Equation 9.6. In a rectangular aperture, however, we will use Cartesian coordinates because they work perfectly well in integrations over the aperture's straight edges (Figure 9.7). Also, the kind of a light source used in such an

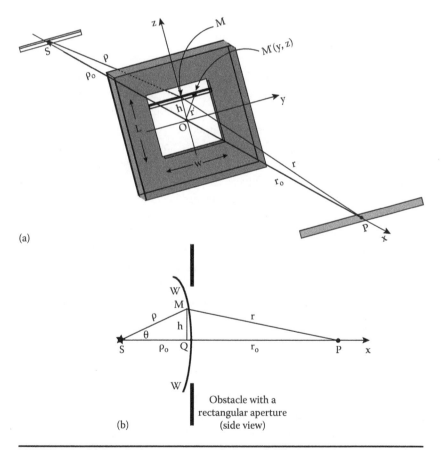

FIGURE 9.7 (a) A typical arrangement of a Fresnel diffraction using a rectangular slit. (b) A side view of the aperture and an incoming wave front.

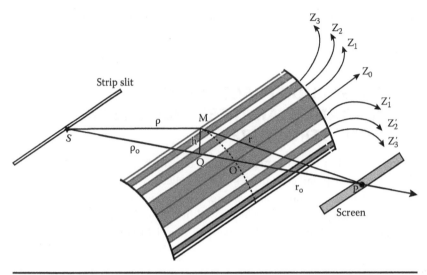

FIGURE 9.8 An arrangement for an extended source emitting cylindrical waves. Several half-period zones are marked on the cylindrical wave front.

application is important. As an elemental strip $d\ell$ chosen along the width W of the aperture would act as a single slit, a spherical wave incident on a very large number of these strips would face a problem in requiring the phase at all points of the strip to be constant, let alone requiring the same along the successive elemental strips. Thus, a cylindrical source parallel to the aperture plane would be our best fit so that all points of the cylindrical wave front would have the same phase. Hence, a slit is placed in front of the source at a distance, ρ_o, from the aperture. The aperture is centered on the origin O of the yz plane, and x is the wave direction of propagation. The Fresnel zone structure in this case would be marked as strips on a primary cylindrical wave front. The wave front is divided into successive strips whose edges are at distances that increase by one half wavelength away from the observation point P. Thus, the distance from the observation point to the edge of the first strip is $r_o + \lambda/2$; the rest follow in sequence. Figure 9.8 illustrates the boundaries Z_1, Z_2, Z_3,... and $Z_1', Z_2', Z_3',...$ between the successive strips $Z_1 Z_2$, $Z_2 Z_3$,... $Z_{m-1} Z_m$, and $Z_1' Z_2'$, $Z_2' Z_3',...Z_{m-1}' Z_m'$, respectively.

9.5.1 Analytic Solution

For the analysis of Fresnel diffraction for the current aperture, some of the basic ideas, assumptions, and derivations ending with Equation 9.6 can be used. Aside from the geometry of the aperture, which requires the source to be an extended slit so that cylindrical waves are considered, no constraints were applied. After accommodating the minor changes in the parameters as marked on Figures 9.7 and 9.8, Equation 9.6 becomes

$$E_P = \int_A F(\theta) \frac{E_o}{r\rho} e^{i[k(r+\rho)-\omega t]} dA \qquad (9.41)$$

where quantities in the equation have the same definitions and relevance introduced earlier.

As h is very small compared to r, r', these can be replaced by r_o, ρ_o. Thus, dropping the obliquity factor for its negligible effect, Equation 9.41 becomes

$$E_P = \frac{E_o}{r_o \rho_o} \int_A e^{i[k(r_o+\rho_o)-\omega t]} dA. \qquad (9.42)$$

The path SMP passing through a point M on the primary wave front related to path SOP that represents the shortest path between S and P can be written as

$$r+r' = \sqrt{\left(r_o^2 + h^2\right)} + \sqrt{\left(\rho_o^2 + h^2\right)} = r_o\sqrt{\left(1 + \frac{h^2}{r_o^2}\right)} + \rho_o\sqrt{\left(1 + \frac{h^2}{\rho_o^2}\right)},$$

where for a small h the square roots can be expanded, giving

$$r+r' = r_o\left(1 + \frac{h^2}{2r_o^2} + \cdots\right) + \rho_o\left(1 + \frac{h^2}{2\rho_o^2} + \cdots\right)$$

or

$$r+r' = r_o + \rho_o + \frac{h^2}{2}\left(\frac{1}{r_o} + \frac{1}{\rho_o}\right). \qquad (9.43)$$

To simplify the algebra, let us introduce the following quantity L such that

$$\frac{1}{L} = \left(\frac{1}{r_o} + \frac{1}{\rho_o}\right). \qquad (9.44)$$

With Equation 9.44, now Equation 9.42 becomes

$$E_P = \frac{E_o}{r_o \rho_o} \int_A e^{i\left[k\left(r_o+\rho_o+h^2/2L\right)-\omega t\right]} dA,$$

which can be rewritten as

$$E_p = \frac{E_o e^{i\left[k(r_o + \rho_o) - \omega t\right]}}{r_o \, \rho_o} \iint_A e^{ik \, h^2/2L} dA. \tag{9.45}$$

As the primary cylindrical wave front reaches the slit, only a segment along its surface of width W will be entering the slit, and it is that portion of the wave that will be diffracted (Figure 9.7a). Keep in mind that the inclination factor has been dropped. This portion at a height h, that is, height z above the y axis, secures an elemental area dA (=Wdz) that is exposed for diffraction by the aperture. The elemental area dA being equal to Wdz assumes that all points of this element along the width contribute equally. This is only approximately true. The more accurate choice is dA = dy dz taken at point M'. The error being accommodated within the obliquity factor makes an insignificant difference. Thus, Equation 9.45 becomes

$$E_p = \frac{E_o e^{i\left[k(r_o + \rho_o) - \omega t\right]}}{r_o \, \rho_o} W \int_{z1}^{z2} e^{ik \, z^2/2L} dz. \tag{9.46}$$

The above integral can be simplified by introducing a dimensionless quantity, w, such that

$$w = z\sqrt{\frac{2}{\lambda L}}, \tag{9.47}$$

that is,

$$z = w\sqrt{\frac{\lambda L}{2}}. \tag{9.48}$$

This reduces Equation 9.46 to the form

$$E_p = \frac{E_o e^{i\left[k(r_o + \rho_o) - \omega t\right]}}{r_o \, \rho_o} W \sqrt{\frac{L\lambda}{2}} \int_{w1}^{w2} e^{i\pi \, w^2/2} dw. \tag{9.49}$$

Letting

$$F(w) = e^{i\pi w^2/2}; \quad e^{i\pi w^2/2} = \cos\left(\pi w^2/2\right) + i\sin\left(\pi w^2/2\right), \tag{9.50}$$

the integral in (9.49) can be expressed as

$$\int_{w_1}^{w_2} e^{i\pi w^2/2}\,dw = \int_{w_1}^{w_2} \cos\left(\pi\,w^2/2\right)dw + i\int_{w_1}^{w_2}\sin\left(\pi\,w^2/2\right)\,dw, \qquad (9.51)$$

which in terms of Fresnel integrals takes the form

$$\int_{w_1}^{w_2} e^{i\pi w^2/2}\,dw = \left\{ \left[\int_{0}^{w_2}\cos\left(\pi w^2/2\right)dw - \int_{0}^{w_1}\cos\left(\pi w^2/2\right)dw \right] \right.$$
$$\left. + i\left[\int_{0}^{w_2}\sin\left(\pi\,w^2/2\right)dw - \int_{0}^{w_1}\sin\left(\pi\,w^2/2\right)dw \right] \right\}. \qquad (9.52)$$

Defining

$$C(w_2) = \int_{0}^{w_2}\cos\left(\pi\,w^2/2\right)dw, \quad C(w_1) = \int_{0}^{w_1}\cos\left(\pi w^2/2\right)dw, \qquad (9.53)$$

$$S(w_2) = \int_{0}^{w_2}\sin\left(\pi w^2/2\right)dw, \quad S(w_1)\int_{0}^{w_1}\sin\left(\pi w^2/2\right)dw, \qquad (9.54)$$

Now, Equations 9.49 and 9.50 can be expressed in terms of $C(w_2)$, $C(w_1)$, $S(w_2)$, $S(w_1)$ as follows:

$$E_P = \frac{E_o e^{i\left[k(r_0+\rho_0)-\omega t\right]}}{r_0\,\rho_0}\,W\sqrt{\frac{L\lambda}{2}}\left\{\left[C(w_2)-C(w_1)\right]+i\left[S(w_2)-S(w_1)\right]\right\}. \qquad (9.55)$$

The intensity I_p of the diffraction at P is proportional to $|E_P|^2$, that is

$$I_p \alpha |E_P|^2,$$

or

$$I_p \alpha |E_o^P|^2\left\{\left[C(w_2)-C(w_1)\right]^2+\left[S(w_2)-S(w_1)\right]^2\right\}. \qquad (9.56)$$

We can transition this expression into an equality by incorporating a constant on the r.h.s., which may be absorbed within $|E_o^P|^2$. Thus, we may write the relation as

$$I_P = I_o^P\left\{\left[C(w_2)-C(w_1)\right]^2+\left[S(w_2)-S(w_1)\right]^2\right\}, \qquad (9.57)$$

where

$$I_o^P = \left|E_o^P\right|^2 = \frac{\left|E_o\right|^2}{\left(r_o\, \rho_o\right)^2} W^2 \frac{L\lambda}{2}. \tag{9.58}$$

Realizing that $C(-w) = -C(w)$ and $S(-w) = -S(w)$, the quantity $\{[C(w_2) - C(w_1)]^2 + [S(w_2) - S(w_1)]^2\}$, calculated over an aperture of infinite dimensions with the origin at the center, is 2. This can be related to the intensity of an unobstructed wave front at point P. Denoting the latter by I_{UC}, we can see from (9.57) that the maximum intensity at P would be unity. Therefore, one may write the general expression for the intensity I_P at P for the wave front that has encountered the aperture as

$$I_P = I_o^P \left\{\left[C(w_2) - C(w_1)\right]^2 + \left[S(w_2) - S(w_1)\right]^2\right\}. \tag{9.59}$$

Note also that

$$I_{UC} = 2 I_o^P; \quad I_o^P = \frac{1}{2} I_{UC}. \tag{9.60}$$

Thus, Equation 9.59 becomes

$$I_P = \frac{1}{2}\left\{\left[C(w_2) - C(w_1)\right]^2 + \left[S(w_2) - S(w_1)\right]^2\right\} I_{UC}. \tag{9.61}$$

As the above equation suggests, determining Fresnel integrals $C(w_1)$, $S(w_1)$ and $C(w_2)$, $S(w_2)$ for any phasor extending between two points is sufficient to determine the intensity of diffraction at P.

Table 9.1 lists a large set of values of w, $C(w)$, $S(w)$. Although this may be useful in solving Fresnel diffraction problems, it is fairly limited and, generally, one may need to determine the integrals for a particular set of aperture dimensions.

COMMENT

It is important to draw attention to the factor I_o^P that was introduced in Equations 9.57 through 9.59. It serves as a connecting factor between the intensity I_P and the intensity of illumination on the screen caused by an unobstructed aperture. Relating I_o^P to I_{UC} via Equation 9.60 is a helpful tool toward calculating the intensity at any point P on the screen if I_{UC}, expressed by W/m^2, is given. I_{UC} is basically the intensity of the source.

TABLE 9.1
Fresnel Table Listing a Large Number of the Values of Fresnel Integrals

w	C(w)	S(w)	w	C(w)	S(w)
0.00	0.0000	0.0000	4.50	0.5261	0.4342
0.10	0.1000	0.0005	4.60	0.5673	0.5162
0.20	0.1999	0.0042	4.70	0.4914	0.5672
0.30	0.2994	0.0141	4.80	0.4338	0.4968
0.40	0.3975	0.0334	4.90	0.5002	0.4350
0.50	0.4923	0.0647	5.00	0.5637	0.4992
0.60	0.5811	0.1105	5.05	0.5450	0.5442
0.70	0.6597	0.1721	5.10	0.4998	0.5624
0.80	0.7230	0.2493	5.15	0.4553	0.5427
0.90	0.7648	0.3398	5.20	0.4389	0.4969
1.00	0.7799	0.4383	5.25	0.4610	0.4536
1.10	0.7638	0.5365	5.30	0.5078	0.4405
1.20	0.7154	0.6234	5.35	0.5490	0.4662
1.30	0.6386	0.6863	5.40	0.5573	0.5140
1.40	0.5431	0.7135	5.45	0.5269	0.5519
1.50	0.4453	0.6975	5.50	0.4784	0.5537
1.60	0.3655	0.6389	5.55	0.4456	0.5181
1.70	0.3238	0.5492	5.60	0.4517	0.4700
1.80	0.3336	0.4508	5.65	0.4926	0.4441
1.90	0.3944	0.3734	5.70	0.5385	0.4595
2.00	0.4882	0.3434	5.75	0.5551	0.5049
2.10	0.5815	0.3743	5.80	0.5298	0.5461
2.20	0.6363	0.4557	5.85	0.4819	0.5513
2.30	0.6266	0.5531	5.90	0.4486	0.5163
2.40	0.5550	0.6197	5.95	0.4566	0.4688
2.50	0.4574	0.6192	6.00	0.4995	0.4470
2.60	0.3890	0.5500	6.05	0.5424	0.4689
2.70	0.3925	0.4529	6.10	0.5495	0.5165
2.80	0.4675	0.3915	6.15	0.5146	0.5496
2.90	0.5624	0.4101	6.20	0.4676	0.5398
3.00	0.6058	0.4963	6.25	0.4493	0.4954
3.10	0.5616	0.5818	6.30	0.4760	0.4555
3.20	0.4664	0.5933	6.35	0.5240	0.4560
3.30	0.4058	0.5192	6.40	0.5496	0.4965
3.40	0.4385	0.4296	6.45	0.5292	0.5398
3.50	0.5326	0.4152	6.50	0.4160	0.5454
3.60	0.5880	0.4923	6.55	0.4520	0.5078
3.70	0.5420	0.5750	6.60	0.4690	0.4631

(Continued)

TABLE 9.1 (*Continued*)
Fresnel Table Listing a Large Number of the Values of Fresnel Integrals

w	C(w)	S(w)	w	C(w)	S(w)
3.80	0.4481	0.5656	6.65	0.5161	0.4549
3.90	0.4223	0.4752	6.70	0.5467	0.4915
4.00	0.4984	0.4204	6.75	0.5302	0.5362
4.10	0.5738	0.4758	6.80	0.4831	0.5436
4.20	0.5418	0.5633	6.85	0.4539	0.5060
4.30	0.4494	0.5540	6.90	0.4732	0.4624
4.40	0.4383	0.4622	6.95	0.5207	0.4591

EXAMPLE 9.2

As E_P is complex, using Equation 9.60, show that the value of $\{[C(w_2) - C(w_1)]^2 + [S(w_2) - S(w_1)]^2\}$ that corresponds to the intensity of diffraction at point P resulting from an unobstructed aperture is 2.

Solution

From Equations 9.55 and 9.56

$$E_P = \frac{E_0 e^{i[k(r_0+\rho_0)-\omega t]}}{r_0\,\rho_0}\, W\sqrt{\frac{L\lambda}{2}} \left\{\left[C(w_2)-C(w_1)\right]+i\left[S(w_2)-S(w_1)\right]\right\}.$$

Thus, the complex conjugate of E_P is

$$E_P^* = \frac{E_0 e^{-i[k(r_0+\rho_0)-\omega t]}}{r_0\,\rho_0}\, W\sqrt{\frac{L\lambda}{2}} \left\{\left[C(w_2)-C(w_1)\right]-i\left[S(w_2)-S(w_1)\right]\right\}.$$

The product $(E_P)(E_P^*)$ becomes

$$E_P E_P^* = \frac{E_0^2}{(r_0\,\rho_0)^2}\, W^2 \frac{L\lambda}{2} \left\{\left[C(w_2)-C(w_1)\right]^2 +\left[S(w_2)-S(w_1)\right]^2\right\}.$$

Making use of Equations 9.58 and 9.59, the above becomes

$$I_P = I_0^P \left\{\left[C(w_2)-C(w_1)\right]^2 + \left[S(w_2)-S(w_1)\right]^2\right\}$$

or

$$I_P = \left(\frac{1}{2}\right)\left\{\left[C(w_2)-C(w_1)\right]^2 + \left[S(w_2)-S(w_1)\right]^2\right\}I_{UC}.$$

Letting the value of the left-hand side be equal to I_{UP}, the value of the curled-bracketed quantity should be 2. Thus

$$\left\{\left[C(w_2)-C(w_1)\right]^2 +\left[S(w_2)-S(w_1)\right]^2\right\} = 2.$$

EXAMPLE 9.3

Consider a plane wave of wavelength 550 nm incident normally on an obstacle of a square aperture 0.500 mm on a side. Determine the intensity of the diffraction at a point lying along the straight line connecting the source and the center of the aperture 5.00 m away. (*Hint:* Results can be obtained by linear interpolation from the Fresnel table (Table 9.1) or from direct integration of integrals in Equations 9.53 and 9.54.)

Solution

From Equation 9.47, $w = z\sqrt{\frac{2}{\lambda L}}$, where from Equation 9.44, L is defined through

$$\frac{1}{L} = \left(\frac{1}{r_o} + \frac{1}{\rho_o}\right) = \left(\frac{\rho_o + r_o}{r_o \rho_o}\right),$$

that is,

$$L = \left(\frac{\rho_o r_o}{r_o + \rho_o}\right).$$

As the waves are plane, the source is effectively at infinity, and $\rho_o = \infty$. Thus, $L \approx r_o = 5.00$ m.

Now, $w_1 = z_1\sqrt{\frac{2}{\lambda L}}$, $w_2 = z_2\sqrt{\frac{2}{\lambda L}}$, $z_1 = -2.50 \times 10^{-3}$ m, $z_2 = 2.50 \times 10^{-3}$ m, L = 5.00 m.

Thus,

$$w_1 = z_1\sqrt{\frac{2}{\lambda L}} = \left(-0.250\times10^{-3}\,\text{m}\right)\sqrt{\frac{2}{\left(550\times10^{-9}\,\text{m}\right)\left(5.00\,\text{m}\right)}} = -0.213\,\text{m},$$

and

$$w_2 = z_2\sqrt{\frac{2}{\lambda L}} = \left(0.250\times10^{-3}\,\text{m}\right)\sqrt{\frac{2}{\left(550\times10^{-9}\,\text{m}\right)\left(5.00\,\text{m}\right)}} = 0.213\,\text{m}.$$

With the above values for w_1 and w_2, we have

$$C(w) = C(0.213) = 0.200, \quad S(w_2) = S(0.213) = 0.00420$$

$$C(-w) = C(-0.213) = -0.200, \quad S(-w_2) = S(-0.213) = -0.00420.$$

Using Equation 9.59, we then have

$$I_P = I_O^P\left\{\left[C(w_2)-C(w_1)\right]^2 + \left[S(w_2)-S(w_1)\right]^2\right\}$$

$$= I_O^P\left\{\left[2(0.200)\right]^2 + \left[2(0.00420)\right]^2\right\}$$

or

$$I_P = I_O^P\left\{=\left[2(0.200)\right]^2 + \left[2(0.00420)\right]^2\right\}.$$

That is,

$$I_P = \frac{1}{2}I_{UC}\left(0.1601\right) = 0.0805\,I_{UC} = 8.1\%\,I_{UC}.$$

REMARK

In solving the intensity of light in Fresnel diffraction, we utilized Table 9.1 to find C(w) and S(w) for the two given values of w: one is w_1 at A and another, w_2 at B. The solution presented above is analytic using the formula expressed in Equation 9.61. As mentioned earlier, a graphical method for such a solution via Mathematica is possible. With simple programs via Mathematica, solving and plotting numerical integrals like the Fresnel integrals can be done with ease. The following Mathematica code can be used to solve Example 9.3 and plot the graphical solution on the Cornu spiral.

You can go to Appendix E for an interestingly helpful method, Mathematica, for use in solving for numerical values of the diffraction intensity on the screen. The benefit here is that solving such a problem through Mathematica provides the corresponding plots of Cornu spiral; the line connecting the two eyes, E_1, E_2, of the spiral; and the line representing the amplitude of the resultant at the point of interest.

9.5.2 Cornu Spiral—Geometrical Solution

Another equivalent way of determining the intensity of diffraction for a rectangular slit comes through the use of a graphical schematic of Fresnel's integrals C(w) and S(w). The definition of $F(w)=e^{i\pi w^2/2}$ introduced in Equation 9.50 helps us plot $\cos(\pi w^2/2)$ and $\sin(\pi w^2/2)$ on a complex coordinate system, where the real term $\cos(\pi w^2/2)$ is displayed along the x axis and the imaginary term $\sin(\pi w^2/2)$ is on the y axis. The values of these two functions for all w generate what has been known as Cornu's spiral that has proven to be an effective tool for determining the contribution of any portion of the slit to the intensity of the diffraction at various points on the screen.

Figure 9.9 is an illustration of two Cornu spirals. In part (a) is a plot of two half-period zones of the wave front, plotted in the first quadrant of

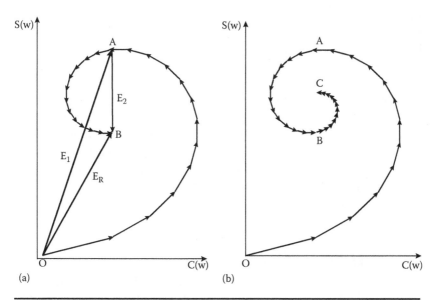

FIGURE 9.9   Cornu phasor diagram consisting of (a) two half-zones, each of which consists of 12 subzones represented by phasors, and (b) three half-period zones, each of which also consists of 12 phasors.

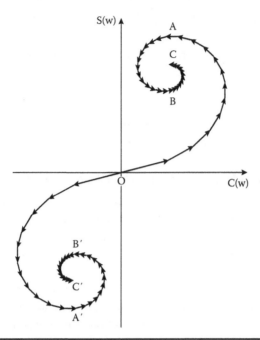

FIGURE 9.10 Cornu phasor diagram consisting of three half-zones, each of which consists of 12 subzones represented by 12 phasors. The spiral in the first quadrant is from the half-zones above the center, while the spiral in the third quadrant is contributed by the zones below the center.

the coordinate system, and in part (b) is three half-period zones sketched also in the first quadrant. In Figure 9.10, three half-period zones are displayed, three above the zero z values of the aperture measured along the slit length and three below the zero z values of the slit. In the sketch, each zone may be considered as consisting of a certain number of phasors, which if they were tremendously increased, the spiral would be a smoothly continuous curve and for z, that is, w extending from negative to positive infinity, the spiral above the zero z line would circulate in a spiral manner into so many full period zones ending with one of almost zero radius at the center of the spiral. The centers E_1 and E_2 are called the eyes of Cornu's spiral.

9.5.3 Features of Cornu Spiral

The discussion that has been presented showed a deep understanding of the physics of the problem as structured by Fresnel. The mathematics of the problem, treated by Fresnel and Kirchhoff, was geometrically supported by

Cornu in what is known as Cornu spiral. At this point, it is opportune to elaborate on several special features of Cornu's spiral:

1. Values of the variable w correspond to the length along the spiral curve. To illustrate this considering an elemental length $d\ell$, we have

$$d\ell = \sqrt{\left[dC(w)\right]^2 + \left[dS(w)\right]^2}$$

$$= \sqrt{\left[\cos^2\left(\pi,w^2\right)/2\right]^2 dw^2 + \left[\sin^2\left(\pi,w^2\right)/2\right]^2 dw^2} = dw. \qquad (9.62)$$

   Thus a value of w on the spiral can be helpful in determining the status of the diffraction at a point P directly opposite the middle of the aperture.

2. The tangent to the spiral at any point on the spiral is of a slope tan φ, where

$$\tan\varphi = \frac{dS(w)}{dC(w)} = \frac{\sin\left(\pi w^2/2\right)}{\cos\left(\pi w^2/2\right)} = \tan\left(\pi w^2/2\right). \qquad (9.63)$$

   That is,

$$\varphi = \left(\pi w^2/2\right). \qquad (9.64)$$

   This feature is also helpful since the slope of the spiral at a certain point would define the subzone of a particular half-period zone at which the point is located, and hence it provides a quantitative measure of the intensity at hand.

3. From the table, the number of values that the variable w covers is huge; they are bound by the lower limit being 0 and the upper limit being infinity. Figure 9.11 illustrates a complete Cornu spiral that one needs for calculating the diffraction at any point on the screen.

   From Equations 9.53 and 9.54, the corresponding values of the limits for C(w) and S(w) are

$$C(\infty)_{positive} = \int_0^\infty \cos\left(\pi w^2/2\right) dw = 0.5 \quad C(-\infty)_{negative} = \int_{-\infty}^0 \cos\left(\pi w^2/2\right) dw = -0.5,$$

$$S(\infty)_{positive} = \int_0^\infty \sin\left(\pi w^2/2\right) dw = 0.5 \quad S(-\infty)_{negative} = \int_{-\infty}^0 \sin\left(\pi w^2/2\right) dw = -0.5.$$

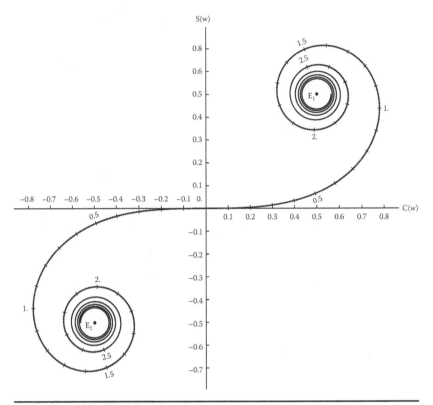

FIGURE 9.11   Cornu spiral.

Thus, from Equation 9.63

$$\frac{\mathbf{I}_P}{\mathbf{I}_{UC}} = \frac{1}{2}\left\{\left[C(w_2)-C(w_1)\right]^2 + \left[S(w_2)-S(w_1)\right]^2\right\}$$
$$= \frac{1}{2}\left\{\left[0.5-(-0.5)\right]^2 + \left[(0.5)-(-0.5)\right]^2\right\}$$
$$= 1. \tag{9.65}$$

The above result makes sense since the unobstructed wave front should give the maximum intensity at P.

4. The amplitude at a point, P, on the screen of the diffracted wave passing between two points, A, B, on an aperture is represented by the length of the line (phasor) connecting A and B on the Cornu spiral. The intensity $\mathbf{I}_p$ of the diffraction at P is related to the intensity at P of light passing through the unobstructed aperture, then is given by

$$\frac{|AB|^2}{|E_1E_2|^2} = \frac{\mathbf{I}_P}{\mathbf{I}_{UC}} \tag{9.66}$$

EXAMPLE 9.4

Redo Example 9.3 using the Cornu spiral method.

Solution

Consider the Cornu spiral in the following diagram, where the points of $w_1 = -0.213$ and $w_2 = 0.213$ are connected through line (amplitude) AB, and the line connecting E_1 and E_2 is shown.

Measuring the length of the phasor AB between the points A and B on the Cornu spiral and the phasor E_1 E_2 l and calculating the ratio between the squares of these lengths gives the following value:

$$\frac{I_P}{I_{UC}} = \frac{|AB|^2}{|E_1E_2|^2} = \frac{|5.4|^2}{|17.8|^2} = 0.092.$$

That is,

$$I_P = 9.2\% \, I_{UC}.$$

As can be noted, there is a slight difference between the answer here and that in the previous example. This is due to approximations in measuring the length of each phasor.

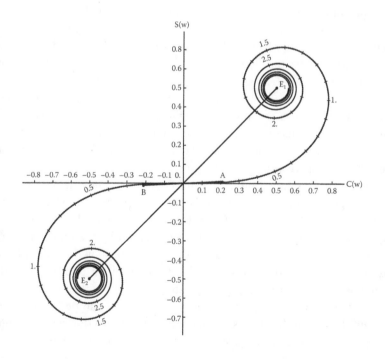

9.5.4 APPLICATIONS OF THE CORNU SPIRAL: THE STRAIGHT EDGE

Diffraction of light from an obstacle that has a knife edge was among the early diffractions observed and treated by Fresnel. Introducing Fresnel diffraction from a rectangular slit that is infinite in one direction and unobstructed along another direction represents most of what is needed for applying it to the straight-edge effect (Figure 9.12b). Long, cylindrical half-period zones may be used to analyze the diffraction on the screen, whose side view in the figure is represented by the z axis.

In Figure 9.12, (a) shows a schematic of an incident plane wave collimated on a screen via a lens placed closely behind the slit edge. In (b) is a plot of the diffraction intensity displayed on the screen. Setting up the axes in the current discussion at the edge O on the diagram, the x axis would describe the direction of propagation. This direction is perpendicular to both the edge and the screen, both in the horizontal plane; the edge (a plane obstacle with a sharp side) is in horizontal yz plane ($x = 0$), while the screen is at a distance $x = r_0$. The y axis at the edge will then be describing the width of the infinite aperture extending from $-\infty$ to ∞, while the z axis would point to the left, also extending from $-\infty$ to ∞, such that the unobstructed space extends from $-\infty$ to 0 and the obstructed part extends from 0 to ∞.

One may correlate the intensity in Figure 9.12b with the Cornu spiral of Equation 9.15 as follows. The edge of the shadow at O corresponds to point P_0 on the screen beyond which the half of the wave front that extends into infinity is contributing to the illumination reaching the screen. One can associate that with a vector, O E$'$, representing the amplitude of that wave front, and that is half the length of the vector E$'$ E, where E$'$ E represents the amplitude of the complete wave. The intensity of light at this shadow boundary point is one-fourth the intensity that we observe in the absence of the edge that literally means the space would be uncovered. We may label contribution from the uncovered space as I_{UC}.

As one moves away from Po toward P$-$, in the shadow area obstructed by the edge, the intensity of light continues to fade until it becomes zero. That corresponds to the eye E$'$ on the Cornu spiral. The amplitude of any wave reaching the screen can be represented by a vector that starts at the point of interest and ends at the eye E$'$ as shown on the plot in Figure 9.13. E$'$E represents the resultant amplitude at points where the illumination is uniform and equal to that obtained with no obstructing edge. The phasor from any point on the spiral to E$'$ can be calculated or measured through the chords connecting that point to the eye E$'$. A point like P_{++} on the screen, too distant from the edge, is expected to receive

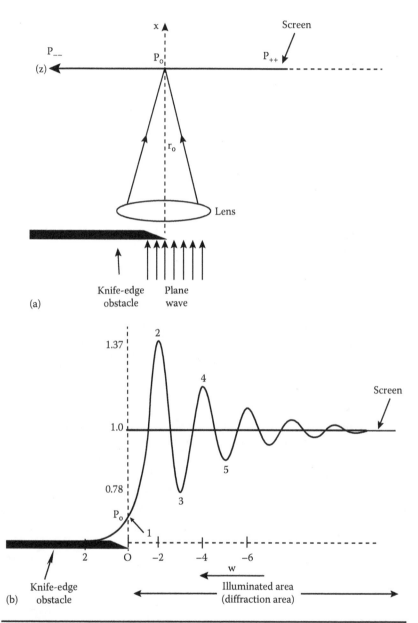

FIGURE 9.12   (a) An illustration of a straight edge, (b) a plot of the intensity for Fresnel diffraction arising from the incident waves obstructed by the edge.

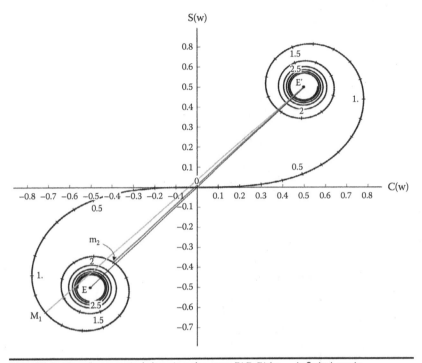

FIGURE 9.13 Cornu spiral showing the eyes E′ E; E′ (w_2 = infinity) are in the first quadrant, and E (w_2 = −infinity) is in the third quadrant. In the plot, the two lines E′M₁ and E′m₁ correspond to points 2 and 3, respectively, in Figure 9.12b.

complete illumination. The intensity at this point is proportional to the square of the length of the chord connecting the two eyes E E′. As seen from Figure 9.12, the length of EE′ is $\sqrt{2}$. This corresponds to intensity $I_o = 2$ for the unobstructed wave. Thus, if E′P is the amplitude obtained from Fresnel curve, the intensity at point P, denoted by Ip, expressed as a ratio, I_p/I_o, is

$$\frac{I_p}{I_o} = \frac{1}{2} \tag{9.67}$$

or

$$I_p = \frac{1}{2}I_o. \tag{9.68}$$

Having the first quadrant for displaying w from zero to infinity and the third quadrant for w from zero to negative infinity means that the obstructed area would refer to points in the first quadrant and the uncovered area would refer to the third quadrant. For more elaboration, adopting E′ for which w = ∞ ($C(w) = 0.5$, $S(w) = 0.5$) as the reference position for a phasor connecting E′ to another point, w_1, means that the length of this phasor E′w_1 represents the amplitude of the resultant contribution at point w_1. In other words, any point on the lower half of the spiral can act as the tip of the vector line extending from the tail at E′ to that very point. For a selection of a variety of lines starting at E′ and ending at different points on the lower spiral, one goes through a succession of maxima and minima. For example, the line from E′ to M_1, whose w (eye balling) on the spiral is −1.2, represents the amplitude of the first max of diffraction, and the line from m_1, whose w (also eye balling) on the spiral is −1.9 to E′, represents the amplitude of the resultant at the first minimum, and so on for any other points on the spiral.

From the general relation, the resultant intensity of diffraction I_P at a point, P, on the screen resulting from wave fronts encountered between two points, w_1 and w_2, is the square of the phasor connecting the two points. Following the argument that led to Equation 9.67, this ratio $\dfrac{I_P}{I_o}$ is expressed below:

$$\frac{I_P}{I_o} = \frac{1}{2}\left\{\left[C(w2) - C(w1)\right]^2 + \left[S(w2) - S(w1)\right]^2\right\} \qquad (9.69)$$

As for the knife edge, the coordinates of the reference point E′ is $z_2 = \infty$ for w that plays the role of $w_2(=\infty)$, with $C(w_2) = 0.5$, and with $S(w_2) = 0.5$, the above can be expressed for any point P of Fresnel coordinate w_1 as

$$\frac{I_P}{I_o} = \frac{1}{2}\left\{\left[0.5 - C(w1)\right]^2 + \left[0.5 - S(w1)\right]^2\right\}. \qquad (9.70)$$

Figure 9.12b illustrates four of these maxima and minima with a labeling on each point that identifies the tip of the corresponding chord (phasor) being at a position of maximum or minimum intensity (Figure 9.12). Phasors corresponding to the first maximum and first minimum in Figure 9.12 are shown in Figure 9.13 as E′M_1 and E′m_1, respectively.

EXAMPLE 9.5

Consider a plane light wave of 550.0 nm incident on a straight edge 5.00 m away from the source. Calculate (a) the relative intensity of diffraction on a screen at points for which $w_1 = -0.40$. (b) Determine the corresponding z point and correlate them to the position of the knife edge.

Solution

(a) From

$$\frac{I_P}{I_{UC}} = \frac{1}{4}\left\{\left[0.5 - C(w1)\right]^2 + \left[0.5 - S(w1)\right]^2\right\}$$

$$= \frac{1}{4}\left\{\left[0.5 - C(-0.40)\right]^2 + \left[0.5 - S(-0.40)\right]^2\right\},$$

we have

$$\frac{I_P}{I_{UC}} = \frac{1}{4}\left\{\left[0.5 - (-0.3975)\right]^2 + \left[0.5 - (-0.0334)\right]^2\right\}$$

$$= \frac{1}{4}\left\{0.8055 + 0.2845\right\} = 0.2725,$$

that is,

$$I_P = 27.25\% \, I_{UC}.$$

(b) From Equation 9.47, $w = z\sqrt{\dfrac{2}{\lambda L}}$, where from Equation 9.44, L is defined through

$$\frac{1}{L} = \left(\frac{1}{r_o} + \frac{1}{\rho_o}\right) = \left(\frac{\rho_o + r_o}{r_o \rho_o}\right),$$

that is,

$$L = \left(\frac{\rho_o r_o}{r_o + \rho_o}\right).$$

As the waves are plane, the source is effectively at infinity, and $\rho_o = \infty$. Thus, $L \approx r_o = 5.00$ m.

(c) Now,

(d)

$$w = z\sqrt{\frac{2}{\lambda L}}, \quad z = w\sqrt{\frac{\lambda L}{2}} = 11.73 \times 10^{-4} \text{ m} = 1.17 \text{ mm}.$$

PROBLEMS

9.1   Given an annular aperture that passes seven zones, 3–9, determine
the resultant and state your analysis.

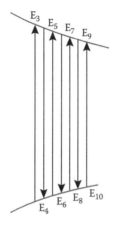

9.2   Consider a Fresnel zone plate illuminated by plane waves of 550.0 nm.
For a central Fresnel zone of radius $R_1 = 3.52 \times 10^{-3}$ m, find
(a)   The focal length of the given zone plate
(b)   The radius of the central Fresnel zone if the focal length is
required to be (i) 1.056 mm, (ii) 35.2 mm

9.3   Consider a Fresnel zone plate illuminated by plane waves of
550.0 nm. If the first five odd zones were exposed, determine the
amplitude and intensity at point P on the plate axis a distance r_o
away from the plate.

9.4   Re-solve the previous problem if in the Fresnel zone plate illuminated by plane waves of 550.0 nm, the first five even zones were
exposed; determine the amplitude and intensity at point P on the
plate axis a distance r_o away from the plate.

9.5   Redo the previous problem assuming that all but the first half-period zone are blocked.

9.6   Consider a circular aperture that passes three half-period zones. Knowing that the intensity of an uncovered plane wave at point P on the screen along the axis of symmetry is 25 W/m^2, determine the intensity of the diffraction, I_P.

9.7   In a Fresnel experiment, a plane wave of 589.6 nm is focused by a lens on a rectangular slit of 1.00 mm wide. (a) Using Equations 9.44 and 9.47, develop a relation between L and the variable w. (b) Knowing that w = 3.00 for the obtained diffraction, determine its position at a point P along the slit axis of symmetry.

9.8   A Fresnel setup consists of a slit of width W = 0.600 mm placed in the middle of a source and a screen that are separated by 0.80 m. If the source is emitting a monochromatic light of 540.0 nm, use the Fresnel table to determine the intensity of the diffraction at a point P on the screen along the line connecting the source and the center of the slit.

9.9   Consider a circular aperture with 4.00 mm diameter illuminated by a plane wave of wavelength 540.0 nm. Determine the position from the aperture of the first two maxima.

9.10   In the previous problem, find the first two minima.

9.11   Consider a plane monochromatic light of wavelength λ = 632.8 nm incident on an aperture of radius a = 3.00 mm. Find
   (a)   The radius of the central zone formed at point P on the screen along the line connecting the source and the center of the slit
   (b)   The number of Fresnel zones that would be uncovered by the aperture

9.12   Consider a straight-edge diffraction setup where a light wave of 540.0 nm is emitted by a source placed 8.0 cm behind the straight edge and a screen placed 80.0 cm in front of the straight edge. Calculate the diffraction intensity at each of the following points: (i) z_1 inside the shadow that corresponds to w_1 = 2.00 and (ii) at z_2 outside the shadow that corresponds to w_2 = −2.00.

9.13   Reconsider the previous problem, where except for the source placed at an effective infinity, all data is kept unchanged:

(a)   Calculate L for this setup (remember ρ_o is infinite).

(b)   Determine the coordinates z_1 and z_2 that correspond to $w_1 = 0.200$ and $w_2 = -0.200$, respectively. Assume that the screen is still in front of the straight edge 80.0 cm away.

9.14   For a light source placed at 80.0 cm from a straight edge, use a Cornu spiral to calculate the intensity of the (a) first maximum and (b) first minimum.

9.15   Knowing that light used in the previous problem was 540.0 nm and the source is at an effective infinity, use your results in the previous problem to calculate the separation between the first two maxima.

Optics of Multilayer Systems

Precision of thought is essential to every aspect and walk of life.

Antoine Arnauld (1612–1694)

10.1 INTRODUCTION

The subject of this chapter is the optics of thin films, deposited on a dielectric substrate. About seven decades ago, Abeles was among the early workers who applied a matrix technique to analyzing thin films in order to determine the reflectivity and transmissivity thereof. In the early 1950s, Born and Wolf reformulated the matrix method so elegantly that it attained widespread use. With the advent of modern computation technologies, at the professional and personal levels, the determination of basic film properties such as reflectivity and transmissivity of single or multiple layers using this method has become a common practice.

10.2 BASIC THEORY: DIELECTRIC LAYER

Figure 10.1 shows a sketch of a thin film of thickness h; in Figure 10.1a is a three-dimensional schematic of the film with an xyz coordinate system whose origin is at the center of the left side of the film designated as boundary a. This side represents the xy plane at z = 0, while the other side of the film, boundary b, represents the xy plane at z = h. The film has an index of refraction n_1. Figure 10.1b is a typical illustration of an electromagnetic wave (EM) of amplitude E_o incident on surface a from the left with an angle of incidence θ_o. As the figure shows, the angle of reflection is θ_r, and the angle with which the light wave emerges from the right side of the film at boundary b is θ_ℓ. The amplitude of the reflected wave is E_r, and the amplitude of the wave transmitted through the film is E_t. Figure 10.2 shows those transmitted and reflected within the film, labeled as E_{ot} and E_{tr}, respectively. The x–z plane is the plane of incidence. In the figure, the first medium, which is the medium of incidence, has an index of refraction, n_o, and the last medium to which the wave emerges has an index of refraction, n_ℓ.

Although an involved derivation of the matrix formulation for the optics of dielectric thin films is treated in many references, it may be simplified if one considers a normal incidence case. Here, the angles of incidence, reflection, and refraction are all zero. The incident and transmitted waves will propagate along the z direction, while the reflected wave will be propagating along the negative z direction (Figure 10.2). In this case, the electric and magnetic field vectors are parallel to the surface of the film.

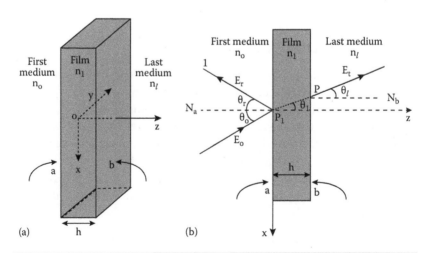

(a)

(b)

FIGURE 10.1 (a) A three-dimensional perspective of a dielectric thin film; (b) a cross-sectional side view of the film.

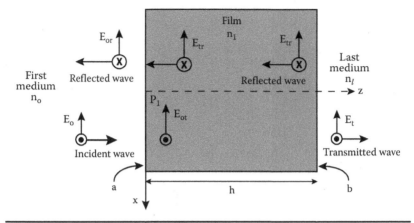

FIGURE 10.2   An exploded view of the film cross section, demonstrating an electromagnetic wave incident normal to face a. The amplitudes of the electric field in the various media are also depicted. The associated magnetic fields out of ⊙ the page and into ⊗ the page are also shown. E_o is the incident field, E_{ot} is the portion of the incident wave that propagates into the film medium, E_{tr} is the portion of E_{ot} that is reflected at the second boundary, and E_t is the transmitted wave.

The basis of the derivation is the continuity of the tangential components of the electric and magnetic fields at each of the boundaries, a and b. The magnitude and direction of the magnetic field **B** in each wave are obtained from the general form of Maxwell's equations

$$\nabla \wedge \mathbf{E} = -\frac{\partial \mathbf{B}}{\partial t},$$

which as in Equation 5.19 leads to

$$\mathbf{B} = \frac{k}{\omega}\left(\hat{\mathbf{k}} \wedge \mathbf{E}\right),$$

where $\hat{\mathbf{k}}$ is a unit vector along the propagation vector direction of **k**. Upon substitution for the values of k and ω, we have

$$\mathbf{B} = \sqrt{\varepsilon\mu}\left(\hat{\mathbf{k}} \wedge \mathbf{E}\right).$$

When dealing with magnetic materials, it is customary to work with the magnetic field intensity vector **H** instead of the magnetic field induction **B**. For "linear materials," **H** and **B** are related by $\mathbf{H} = \mathbf{B}/\mu$. Thus, the above equation becomes

$$\mathbf{H} = \sqrt{\frac{\varepsilon}{\mu}}\left(\hat{\mathbf{k}} \wedge \mathbf{E}\right) = n\left(\hat{\mathbf{k}} \wedge \mathbf{E}\right),$$

where $n = \sqrt{\dfrac{\varepsilon}{\mu}}$ is the index of refraction of the medium. The above equation applied to the first medium (usually air or vacuum), the film medium, and the last medium gives

$$\mathbf{B_o} = n_o \left(\hat{\mathbf{k}}_o \wedge \mathbf{E_o} \right) \tag{10.1}$$

$$\mathbf{B_{ot}} = n_1 \left(\hat{\mathbf{k}}_{ot} \wedge \mathbf{E_{ot}} \right) \tag{10.2}$$

$$\mathbf{B_{tr}} = n_1 \left(\hat{\mathbf{k}}_{tr} \wedge \mathbf{E_{tr}} \right) \tag{10.3}$$

$$\mathbf{B_\ell} = n_\ell \left(\hat{\mathbf{k}}_\ell \wedge \mathbf{E_\ell} \right). \tag{10.4}$$

The quantities ε_o, μ_o are the permittivity and permeability, respectively, of the first or incident medium (air); ε_ℓ, μ_ℓ are those for the last or emergence medium; and ε_1, μ_1 are the permittivity and permeability of the film. Following the same order, $\hat{\mathbf{k}}_o$ and $\hat{\mathbf{k}}_\ell$ are unit vectors along the propagation vectors of the waves in the first and last media, respectively, and $\hat{\mathbf{k}}$ is a unit vector along the propagation vector of the wave in the film.

The continuity conditions of the electric and magnetic fields *at boundary a* imply that

$$\left(E_o + E_{or} \right) = E_{ot} + E_{tr} \tag{10.5}$$

$$B_o - B_{or} = B_{ot} - B_{tr}. \tag{10.6}$$

Upon using (10.1) and (10.2), Equation 10.6 becomes

$$n_o E_o - n_o E_{or} = n_1 E_{ot} - n_1 E_{tr}. \tag{10.7}$$

The continuity conditions of the electric and magnetic fields *at boundary b* in Figure 10.2 imply that

$$\left(E_{ot} e^{ikh} + E_{tr} e^{-ikh} \right) = E_t \tag{10.8}$$

$$B_{ot} e^{ikh} - B_{tr} e^{-ikh} = B_t. \tag{10.9}$$

Upon using (10.2) and (10.3), Equation 10.9 becomes

$$n_1 E_{ot} e^{ikh} - n_1 E_{tr} e^{-ikh} = B_t. \tag{10.10}$$

Upon multiplying Equation 10.8 by n_1, it becomes

$$\left(n_1 E_{ot} e^{ikh} + n_1 E_{tr} e^{-ikh}\right) = n_1 E_t.$$
(10.11)

Adding Equations 10.10 to 10.11 gives

$$E_{ot} = \left(\frac{n_\ell E_t + B_t}{2n_1}\right) e^{-ikh}.$$
(10.12)

Subtracting Equation 10.11 from 10.10 gives

$$E_{tr} = \left(\frac{n_1 E_t - B_t}{2n_1}\right) e^{ikh}.$$
(10.13)

Substituting for E_{ot} and E_{tr} from (10.12) and (10.13) in (10.5) gives

$$\left(E_o + E_{or}\right) = \left(\frac{n_1 E_t + B_t}{2n_1}\right) e^{-ikh} + \left(\frac{n_1 E_t - B_t}{2n_1}\right) e^{ikh}.$$
(10.14)

Similarly, substituting for E_{ot} and E_{tr} from (10.12) and (10.13) in (10.7) gives

$$\left(n_o E_o - n_o E_{or}\right) = \left(\frac{n_1 E_t + B_t}{2}\right) e^{-ikh} - \left(\frac{n_1 E_t - B_t}{2}\right) e^{ikh}.$$
(10.15)

From Equations 10.5 and 10.6, we can define the left side of Equations 10.14 and 10.15 as E_a and B_a, respectively, and after some algebra and use of trigonometry on the right-hand side of both equations, they reduce to

$$E_a = \left(\cos kh\right) E_t - \left(\frac{i \sin kh}{n_1}\right) B_t$$
(10.16)

$$B_a = i\, n_1 \left(\sin kh\right) E_t + \left(\cos kh\right) B_t.$$
(10.17)

As the magnitude of the wave vector in the film is $k = 2\pi/\lambda$ and that in vacuum is $k_o = 2\pi/\lambda_o$ with λ and λ_o being the respective wavelengths and since $\lambda = \lambda_o/n_1$, then

$$k = \frac{2\pi}{\lambda} = \frac{2\pi n_1}{\lambda_o} = k_o n_1.$$
(10.18)

Equations 10.16 and 10.17 then become

$$E_a = (\cos k_o n_1 h) E_t - \left(\frac{i \sin k_o n_1 h}{n_1} \right) B_t \tag{10.19}$$

$$B_a = i n_1 (\sin k_o n_1 h) E_t + (\cos k_o n_1 h) B_t. \tag{10.20}$$

The above can be expressed in the following matrix form:

$$\begin{vmatrix} E_a \\ B_a \end{vmatrix} = \begin{bmatrix} \cos(k_o n_1 h) & \dfrac{-i}{n_1} \sin(k_o n_1 h) \\ -i n_1 \sin(k_o n_1 h) & \cos(k_o n_1 h) \end{bmatrix} \begin{vmatrix} E_t \\ B_t \end{vmatrix}. \tag{10.21}$$

In a concise matrix form, Equation 10.21 is expressed as

$$O_o = M_1 \, O_1, \tag{10.22}$$

where

$$O_o = \begin{vmatrix} E_a \\ B_a \end{vmatrix}, \quad O_1 = \begin{vmatrix} E_t \\ B_t \end{vmatrix}, \tag{10.23}$$

and the matrix for the h-thick film is

$$M_1 = \begin{bmatrix} \cos(k_o n_1 h) & \dfrac{-i}{n_1} \sin(k_o n_1 h) \\ -i n_1 \sin(k_o n_1 h) & \cos(k_o n_1 h) \end{bmatrix}. \tag{10.24}$$

Setting

$$\beta_j = k_o n_j h_j,$$

Equation 10.24 becomes

$$M_1 = \begin{bmatrix} \cos(\beta_1) & \dfrac{-i}{n_1} \sin(\beta_1) \\ -i n_1 \sin(\beta_1) & \cos(\beta_1) \end{bmatrix}.$$

For normal incidence

$$p_j = n_j \cos\theta_j = n_j; \quad \theta = 0.$$

The subscript $j = 0, 1, 2$ refers to the medium of incidence, film, and substrate, respectively. Equation 10.24 is a 2×2 matrix that is called the characteristic matrix of the film, which in a general form can conveniently be written as

$$M = \begin{bmatrix} m_{11} & m_{12} \\ m_{21} & m_{22} \end{bmatrix} \tag{10.25}$$

where the matrix elements are

$$m_{11} = \cos\beta_1, \quad m_{12} = -\frac{i}{n_1}\sin\beta_1 \quad m_{21} = -i\,n_1\sin\beta_1 \quad m_{22} = \cos\beta_1. \quad (10.26)$$

or

$$m_{11} = \cos\beta_1, \quad m_{12} = -\frac{i}{p_1}\sin\beta_1, \quad m_{21} = -i\,p_1\sin\beta_1, \quad m_{22} = \cos\beta_1 \quad (10.27)$$

As noticed, the elements of the film characteristic matrix are functions of the incident wavelength λ_o, angle of refraction, indices of refraction of medium of incidence, film, and last medium that is usually the substrate.

Equations 10.21 can be solved for E_r/E_o and E_t/E_o, giving the reflection and transmission coefficients r and t, respectively, as

$$r = \frac{E_r}{E_o} = \frac{(m_{11} + m_{12}p_\ell)n_o - (m_{21} + m_{22}n_\ell)}{(m_{11} + m_{12}p_\ell)n_o + (m_{21} + m_{22}n_\ell)} \quad (10.28)$$

$$t = \frac{E_t}{E_o} = \frac{2p_o}{(m_{11} + m_{12}p_\ell)p_o + (m_{21} + m_{22}p_\ell)}. \quad (10.29)$$

In the above,

$$p_o = n_o, \quad p_\ell = n_\ell.$$

The reflectivity R and transmissivity T of the film are

$$R = |r|^2 \quad (10.30)$$

$$T = \frac{p_\ell}{p_1}|t|^2. \quad (10.31)$$

EXAMPLE 10.1

Consider a dielectric thin film with index of refraction n = 2.0 and thickness h = 75.0 nm surrounded by air from both sides. Determine the film reflectivity and transmissivity for an incident wave of 600.0 nm. What would the value of the reflectivity and transmissivity for the film be if h = 225.0 nm, while all other quantities were kept unchanged?

Solution

First case: $h = 75$ nm; $n_1 = n = 2.0$, $n_o = n_\ell = 1.0$, $\beta_1 = \beta$.

We first determine the value in terms of h of the elements of the characteristic matrix as defined in Equation 10.26:

$$m_{11} = \cos\beta_1; \quad \beta = k_o n h = \left(\frac{2\pi}{600}\right)(2.0)(75.0) = \frac{\pi}{2}.$$

Thus

$$m_{11} = 0,$$

and

$$m_{12} = -\frac{i}{n}\sin\beta = (-i/2.0), \quad m_{21} = -in = -2i, \quad m_{22} = 0.$$

Substituting in formulae (10.28) and (10.29) gives

$$r = \frac{E_r}{E_o} = \frac{(m_{11} + m_{12}n_\ell)n_o - (m_{21} + m_{22}n_\ell)}{(m_{11} + m_{12}n_\ell)p_o + (m_{21} + m_{22}n_\ell)} = \frac{(0 - i/2) - (-2i + 0)}{(0 - i/2) + (-2i + 0)}$$

$$= \frac{1.5i}{-2.5i} = -0.60.$$

Thus,

$$R = 0.36$$

$$t = \frac{E_t}{E_o} = \frac{2n_o}{(m_{11} + m_{12}n_\ell)p_o + (m_{21} + m_{22}n_\ell)} = \frac{2}{(0 + -i/2) + (-2i + 0)}$$

$$= \frac{2}{(-i/2) + (-2i)} = -0.80i.$$

Thus,

$$T = 0.64.$$

Note that $R + T = 1.0$, that is, 100%, because the film is perfect dielectric with no loss of energy.

Second case: h = 225.0 nm.

$$m_{11} = \cos \beta_1; \quad \beta = k_0 nh = \left(\frac{2\pi}{600}\right)(2.0)(225.0) = \pi/2.$$

Thus,

$$m_{11} = 1, \quad \text{and} \quad m_{12} = -\frac{i}{n}\sin\beta = 0 \quad m_{21} = -in\sin\beta = 0, \quad m_{22} = 1.$$

The elements having the same values as those in the first case (a) should give the same reflectivity and transmissivity as in (a).
Thus,

$$R = 0.36 = 36\%.$$

Thus,

$$T = 0.64 = 64\%.$$

Note that $R + T = 1.0$, that is, 100%, because the film is perfect dielectric with no loss of energy.

Excel approach: The above results can be obtained using a general Microsoft Excel code applicable for both real and complex entries. Steps of work on such an example are detailed in Appendix G. The statement of the example is depicted in Appendix G1 with the full analytical solution as done above, while the Excel calculations in all spreadsheets are in appendix G2-sheet 1.

EXAMPLE 10.2

Redo Example 10.1 for the same type of layer that has a thickness of (a) 150.0 nm, (b) 300 nm, (c) 450 nm.

Solution

(a) The matrix elements are $m_{11} = \cos\beta_1$; $\beta = k_0 nh = (2\pi/600)(2.0)$ $(150.0) = \pi$.
Thus,

$$m_{11} = -1,$$

and

$$m_{12} = -\frac{i}{n}\sin\beta = (0) \quad m_{21} = 0, \quad m_{22} = m_{11} = -1.$$

Substituting in formulae (10.28) and (10.29) gives

$$r = \frac{E_r}{E_o} = \frac{(m_{11} + m_{12}n_\ell)n_o - (m_{21} + m_{22}n_\ell)}{(m_{11} + m_{12}n_\ell)n_o + (m_{21} + m_{22}n_\ell)}$$

$$= \frac{(-1) - (-1)}{(-1) + (-1)} = \frac{0}{-2} = 0.$$

Thus,

$$R = 0.$$

Without calculation via formulae in (10.28) and (10.29), $T = 1 - R = 1 = 100\%$.

So $R + T = 1.0$, that is, 100%, but the transmissivity has a maximum value, while $R = 0$.

(b) The matrix elements are $m_{11} = \cos\beta_1$; $\beta = k_o nh = (2\pi/600)(2.0)(30.0) = 2\pi$.

Thus,

$$m_{11} = 1, \quad \text{and} \quad m_{12} = -\frac{i}{n}\sin\beta = (0), \quad m_{21} = 0, m_{22} = 1.$$

Substituting in formulae (10.28) and (10.29) gives

$$r = \frac{E_r}{E_o} = \frac{(m_{11} + m_{12}n_\ell)n_o - (m_{21} + m_{22}n_\ell)}{(m_{11} + m_{12}n_\ell)n_o + (m_{21} + m_{22}n_\ell)}$$

$$= \frac{(1) - (1)}{(1) + (1)} = \frac{0}{2} = 0.$$

Thus,

$$R = 0.$$

Without calculation via formulae in (10.28) and (10.29), $T = 1 - R = 1 = 100\%$.

So $R + T = 1.0$, that is, 100%; the transmissivity has a maximum value, while $R = 0$.

(c) The matrix elements are $m_{11} = \cos\beta_1$; $\beta = k_o nh = (2\pi/600)(2.0)$ $(450.0) = 3\pi$.
Thus,

$$m_{11} = -1,$$

and

$$m_{12} = -\frac{i}{n}\sin\beta = (0), \quad m_{21} = 0, \quad m_{22} = m_{11} = -1$$

The elements having the same values as those in case (a) should give the same reflectivity and transmissivity as in (a).
That is,

$$R = 0; \quad T = 1 - R = 1 = 100\%.$$

Comment

From the previous cases, for a dielectric layer, the reflectivity $R = 0$ whenever nh/λ = multiples of $(1/2)$, while it has a maximum value $(\neq 1)$ when (nh/λ) is an odd multiple of $(1/4)$.

Excel approach: Again, the above example can be obtained using a general Microsoft Excel code applicable for both real and complex entries.

10.3 EXTENSION TO MULTILAYER STRUCTURES: CHARACTERISTIC MATRIX TECHNIQUE

The algebra involved in the above treatment applies to bilayer and multilayer structures with minimum difficulty. For a bilayer unit (Figure 10.3), that is, a system of two layers of thicknesses h_1 and h_2 deposited on top of each other, their characteristic matrices are

$$\mathbf{M}_1(h_1) = \begin{bmatrix} \cos(k_o n_1 h_1) & \dfrac{-i}{n_1}\sin(k_o n_1 h_1) \\ -i n_1 \sin(k_o n_1 h_1) & \cos(k_o n_1 h_1) \end{bmatrix},$$

$$\mathbf{M}_2(h_2) = \begin{bmatrix} \cos(k_o n_2 h_2) & \dfrac{-i}{n_1}\sin(k_o n_2 h_2) \\ -i n_1 \sin(k_o n_2 h_2) & \cos(k_o n_2 h_2) \end{bmatrix}.$$

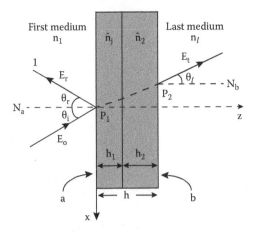

FIGURE 10.3   A side view of a bilayer in which the layers are of complex indices of refraction, n_1 and n_2.

The product of the two matrices is a matrix that represents the characteristic matrix of the bilayer. This is

$$\mathbf{M}_1(h_1)\mathbf{M}_2(h_2) = \begin{bmatrix} \cos(k_o n_1 h_1) & \dfrac{-i}{n_1}\sin(k_o n_1 h_1) \\ -i n_1 \sin(k_o n_1 h_1) & \cos(k_o n_1 h_1) \end{bmatrix}$$

$$\times \begin{bmatrix} \cos(k_o n_2 h_2) & \dfrac{-i}{n_1}\sin(k_o n_2 h_2) \\ -i n_1 \sin(k_o n_2 h_2) & \cos(k_o n_2 h_2) \end{bmatrix}. \quad (10.32)$$

The above is a 2 × 2 matrix whose elements are functions of incident wavelength, thicknesses h_1 and h_2, and indices of refraction n_1 and n_2 assumed to be constant throughout each layer. The product can be considered as the characteristic matrix $\mathbf{M}(h)$ of the bilayer having thickness $h = h_1 + h_2$. In a condensed notation, this matrix is

$$\mathbf{M}(h) = \prod_{j=1}^{2} \mathbf{M}_j(h_j); \quad j = 1, 2 \quad (10.33)$$

where the elements of the matrix $\mathbf{M}_j(h_j)$ for the jth layer are

$$m_{11}^j = \cos\beta_j, \quad m_{12}^j = -\frac{i}{n_j}\sin\beta_j, \quad m_{21}^j = -i\, n_j \sin\beta_j, \quad m_{22}^j = \cos\beta_j. \quad (10.34)$$

$$\beta_j = k_o n_j h_j. \quad (10.35)$$

For a stack of m bilayers, the stack characteristic matrix, denoted by $\mathbf{M}(mh)$, is also a 2 × 2 matrix generated from Equation 10.33 raised to the mth power (Figure 10.4). That is,

$$\mathbf{M}(mh) = \left[\mathbf{M}(h) \right]^m. \tag{10.36}$$

The treatment presented is for dielectric single, bilayer, and multilayer systems. The treatment also applies to conducting thin films as well. The only modification is that the layers' indices of refraction are complex. For the jth layer,

$$\hat{n}_j = n_j + i\kappa_j, \quad \hat{\beta}_j = k_o \hat{n}_j h_j \tag{10.37}$$

where κ is sometimes referred to as the extinction coefficient.

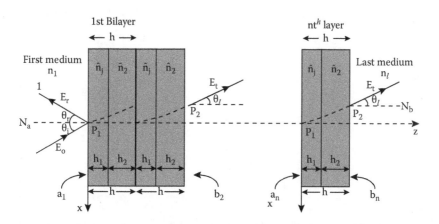

FIGURE 10.4 A side view of a multilayer structure of an arbitrary number of basic identity bilayers, the constituent of which have complex indices of refraction, n_1 and n_2.

EXAMPLE 10.3

Adding absorption to the layer in Example 10.1, let us redo the example for such a layer assuming it also has an extinction coefficient $\kappa = 0.2$.

Solution

The treatment is as presented in the solution of Example 10.1, except that the index of refraction here is complex,

$$\hat{n} = n + i\kappa,$$

where
$$n = 2.0$$
$$\kappa = 0.2$$

Using Excel, all entries are calculated following the steps described in Ga for Example 10.1. For this example (Example 10.3), the Excel display and calculations are in Appendix G (Section G1-sheet 2):

$$\cos\beta = 0.0000468994710894498 - 0.157721704053542i$$
$$\sin\beta_j = 1.01236166155351 + 7.30674103964662E - 06i.$$

Thus,

$$m_{11} = m_{22} = \cos\beta = 0.0000468994710894498 - 0.157721704053542i$$
$$m_{12} = -\frac{i}{n}\sin\beta = -0.0501132967397581 - 0.501169501102779i$$
$$m_{21} = -in\sin\beta = 0.202486945792781 - 2.024721861755881i.$$

From (10.28) and (10.29)

$$r = \frac{E_r}{E_o} = \frac{(m_{11} + m_{12}n_\ell)n_o - (m_{21} + m_{22}n_\ell)}{(m_{11} + m_{12}n_\ell)n_o + (m_{21} + m_{22}n_\ell)}$$
$$t = \frac{E_t}{E_o} = \frac{2n_o}{(m_{11} + m_{12}n_\ell)n_o + (m_{21} + m_{22}n_\ell)}.$$

Using Excel to do the calculation, the reflectivity and transmissivity turn out to be

$$R = |r|^2 = 0.284 = 28.4\%$$

and

$$T = \frac{n_\ell}{n_1}|t|^2 = 0.493 = 49.3\%.$$

Comment

As noted, $R + T < 1$. That is due to absorption. The absorptivity is

$$A = 1 - (R + T) = 1 - (0.284 + 0.493) = 0.223 = 22.3\%.$$

10.4 ULTRATHIN SINGLE FILM

For a thin film that has drastically small thickness, much smaller than the wavelength λ_o, the elements of the characteristic matrix of the film in the normal incidence mode become

$$m_{11} = 1, \quad m_{12} = -i\frac{2\pi}{\lambda_o}h, \quad m_{21} = -i(\hat{n}_1)^2 \frac{2\pi}{\lambda_o}h, \quad m_{22} = 1, \quad (10.38)$$

and the film characteristic matrix can then be approximated by

$$\mathbf{M}_1(h) = \begin{bmatrix} 1 & -i\dfrac{2\pi}{\lambda_o}h \\ -i\dfrac{2\pi}{\lambda_o}(\hat{n}_1)^2 h & 1 \end{bmatrix}. \qquad (10.39)$$

The explicit calculation of the bilayer matrix $\hat{\mathbf{M}}(h)$ in Equation 10.39 is

$$\hat{\mathbf{M}}(h) = \hat{\mathbf{M}}_1(h_1)\,\hat{\mathbf{M}}_2(h_2) = \begin{bmatrix} 1 & -i\dfrac{2\pi}{\lambda}h_1 \\ -i\dfrac{2\pi}{\lambda}(\hat{n}_1)^2 h_1 & 1 \end{bmatrix}$$

$$\times \begin{bmatrix} 1 & -i\dfrac{2\pi}{\lambda}h_2 \\ -i\dfrac{2\pi}{\lambda}(\hat{n}_2)^2 h_2 & 1 \end{bmatrix}. \qquad (10.40)$$

The product that constitutes the bilayer matrix in (10.40) is

$$\hat{\mathbf{M}}(h) = \hat{\mathbf{M}}_1(h_1)\,\hat{\mathbf{M}}_2(h_2) = \begin{bmatrix} 1 - \left(\dfrac{2\pi}{\lambda}\right)^2 \hat{n}_2^2 h_1 h_2 & -i\dfrac{2\pi}{\lambda}(h_1 + h_2) \\ -i\dfrac{2\pi}{\lambda}\left((\hat{n}_1)^2 h_1 + (\hat{n}_2)^2 h_2\right) & 1 - \left(\dfrac{2\pi}{\lambda}\right)^2 \hat{n}_2^2 h_1 h_2 \end{bmatrix}.$$

$$(10.41)$$

For a stacked layer system comprised of m bilayers, the stack characteristic matrix is obtained by raising the above matrix to the power m. That is,

$$
\hat{\mathbf{M}}^N(h) = \left[\hat{\mathbf{M}}_1(h_1)\ \hat{\mathbf{M}}_2(h_2) \right]^m
$$

$$
= \begin{bmatrix} 1 - \left(\dfrac{2\pi}{\lambda}\right)^2 \hat{n}_2^2\, h_1\, h_2 & -i\,\dfrac{2\pi}{\lambda}\left(h_1 + h_2\right) \\[3mm] -i\,\dfrac{2\pi}{\lambda}\left((\hat{n}_1)^2\, h_1 + (\hat{n}_2)^2\, h_2\right) & 1 - \left(\dfrac{2\pi}{\lambda}\right)^2 \hat{n}_2^2\, h_1\, h_2 \end{bmatrix}^m . \tag{10.42}
$$

10.5 ANALYTIC FORMULAE FOR REFLECTIVITY AND TRANSMISSIVITY OF ABSORBING FILMS

Reflectance and transmittance of an absorbing thin film (Figure 10.5) were thoroughly discussed by Born and Wolf for TE on the film with an angle of incidence θ_1. Having in mind that a detailed derivation of R and T is accessible in the cited reference, acquaintance with the following quantities is important.

The reflection coefficient for the interface 1–2 is defined as

$$
r_{12} = \frac{n_1 - \left(n_2 + i\,\kappa_2\right)}{n_1 + \left(n_2 + i\,\kappa_2\right)}, \tag{10.43}
$$

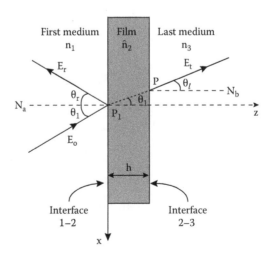

FIGURE 10.5 Schematic of a thin film.

from which the reflection amplitude ρ_{12} and associated phase change at the interface 1–2 are

$$\rho_{12}^2 = \frac{\left(n_1 - n_2\right)^2 + \kappa_2^2}{\left(n_1 + n_2\right)^2 + \kappa_2^2}, \quad \tan \varphi_{12} = \frac{2\,\kappa_2 n_1}{n_2^2 + \kappa_2^2} \tag{10.44}$$

Here, φ_{12} is the phase change at interface 1–2.

The transmission coefficient at interface 1–2 is

$$t_{12} = \frac{2\,n_1}{n_1 + \left(n_2 + i\,\kappa_2\right)}, \tag{10.45}$$

from which the transmission amplitude and associated phase change at interface 1–2 are given by

$$\tau_{12}^2 = \frac{4n_1^2}{\left(n_1 + n_2\right)^2 + \kappa_2^2}, \tag{10.46}$$

where τ_{12} is the transmission amplitude at the 1–2 interface and τ_{12}^2 is the transmission intensity at the interface.

The corresponding quantities at interface 2–3 are

$$\rho_{23}^2 = \frac{\left(n_3 - n_2\right)^2 + \kappa_2^2}{\left(n_3 + n_2\right)^2 + \kappa_2^2}, \quad \tan \varphi_{23} = \frac{2\,\kappa_2 n_3}{n_2^2 + \kappa_2^2 - n_3^2} \tag{10.47}$$

$$\tau_{23}^2 = \frac{4\left(n_2^2 + \kappa_2^2\right)}{\left(n_2 + n_3\right)^2 + \kappa_2^2}. \tag{10.48}$$

With the definition of the quantities

$$\eta = k_o\, h = \left(2\pi/\lambda\right)h, \quad \beta = k_o\, n_2\, h = \left(2\pi/\lambda\right)n\, h, \tag{10.49}$$

the reflectivity $R = |r|^2$ and transmissivity $T = \dfrac{n_3}{n_1}|t|^2$ take the following form

$$R = \frac{\rho_{12}^2 e^{2\kappa_2\eta} + \rho_{23}^2\, e^{-2\kappa_2\eta} + 2\rho_{12}\,\rho_{23}\cos\!\left(\varphi_{23} - \varphi_{12} + 2n_2\eta\right)}{e^{2\kappa_2\eta} + \rho_{12}^2\rho_{23}^2\, e^{-2\kappa_2\eta} + 2\rho_{12}\,\rho_{23}\cos\!\left(\varphi_{12} + \varphi_{23} + 2n_2\eta\right)}, \tag{10.50}$$

$$T = \frac{n_3}{n_1}\, \frac{\tau_{12}^2\, \tau_{23}^2\, e^{-2\kappa_2\eta}}{1 + \rho_{12}^2\rho_{23}^2\, e^{-4\kappa_2\eta} + 2\rho_{12}\,\rho_{23}\, e^{-2\kappa_2\eta}\cos\!\left(\varphi_{12} + \varphi_{23} + 2n_2\eta\right)}. \tag{10.51}$$

In the above treatment, the substrate is dielectric. Except for the phase changes φ_{12} and φ_{23}, introduced by Felske and Roy, the above terms are all as defined by Born and Wolf. The corrections to the phase changes at the air–film and film–substrate interfaces were subtracting π from the bracketed term in the numerator of Equation 10.50 and adding π to the bracketed angle term in the denominators of Equations 10.50 and 10.51. Otherwise, the plots displayed on page 757 in Born and Wolf, would not be obtained for the given n and κ. The arguments for these corrections are detailed by Felske and Roy in their article. The corrections, although just limited to φ_{12} and φ_{23}, are critical in securing proper calculation of the values for **R** and **T** of the film, as those reported by Born and Wolf.

For calculating the optical properties of an absorbing thin film, the corrections introduced by Felske and Roy for the phase changes φ_{12} and φ_{23} are incorporated, and all terms pertaining to a particular film in Equations 10.50 and 10.51 are defined on Excel, giving the exact plots cited in the well-known reference, *Principles of Optics* by M. Born and E. Wolf.

These quantities, for the case of normal incidence, were calculated for an absorbing thin film of the following indices of refraction: n = 3.5, κ = 0.1; n = 3.5, κ = 0.2; these were contrasted against the dielectric case for which n = 3.5, κ = 0.0. Calculations of these cases using C++ code are depicted in Figures 10.6 and 10.7.

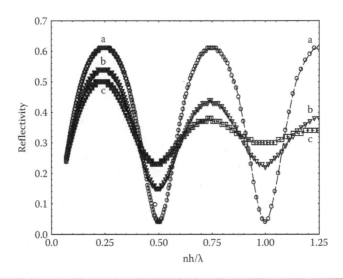

FIGURE 10.6   The reflectivity of a set of thin films, a, b, c, all of the same index of refraction but of extinction coefficients κ = 0.0, 0.1, 0.2, respectively.

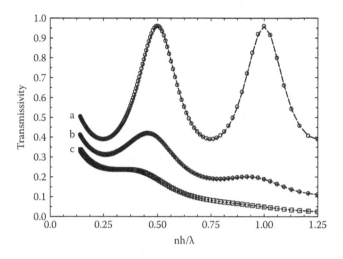

FIGURE 10.7 The transmissivity of a set of thin films, a, b, c, all of the same index of refraction but of extinction coefficients $\kappa = 0.0, 0.1, 0.2$, respectively.

Note that calculations of the right-hand side are practically laborious without resorting to programmed codes. Sophisticated coded computer programs like Fortran, C++, MATLAB®, and Excel developed in the last five decades have made this task easy.

As powerful as the reflectivity and transmissivity expressions in Equations 10.50 and 10.51 are, they are limited to be used for one-layer film. For a stack of absorbing thin films, the matrix-based CMT has proven powerful to facilitate the calculation of the optical properties for single or multilayer systems. The CMT technique is, theoretically speaking, unlimited in application to any composite material, thickness, and number of layers.

PROBLEMS

10.1   A narrow monochromatic beam of light of wavelength λ is incident normally on a film of thickness h and index of refraction n_1. If the film is in air,

(a)   Determine the normal reflectance of the beam in terms of λ, n_1, and h.

(b)   Show that for $h \ll \lambda$, the reflectance is proportional to h^2.

10.2   Letting, $\varphi = k_0 \, n_1 \, h$, apply your result in part (a) of the previous problem to the following values:

(a) $\varphi = \pi/2$, that is, $n_1 h = \lambda/4$; (b) $\varphi = \pi$, that is, $n_1 h = \lambda/2$, (c) $\varphi = 3\pi/2$, that is, $n_1 h = 3\lambda/4$, and (d) $\varphi = 2\pi$, that is, $n_1 h = \lambda$

10.3   Consider a narrow monochromatic beam of light of wavelength $\lambda = 540.0$ nm incident normal to a glass plate of 180.0 nm and index of refraction n = 1.50. For the plate with no backing substrate, that is, $n_\ell = n_{air} = 1.0$,

(a)   Determine the reflectivity of the plate.

(b)   Comment on your answer.

(c)   Verify your answer through independent calculation of the transmissivity of the plate.

10.4   Redo the previous problem keeping all but the plate to become of thickness h = 90.0 nm.

(a)   Determine the reflectivity of the plate.

(b)   Comment on your answer.

10.5   Consider a thin layer of a material of index of refraction 1.40 and thickness 98.0 nm deposited on a glass substrate of an index of refraction n = 1.60.

(a)   Write down the matrix representation of this film, expressing all elements in their explicit numerical values for $\lambda = 650.0$ nm.

(b)   Use matrix optics to calculate the reflectivity of the layer for the wave, 650.0 nm wavelength, incident normally on the layer.

10.6   A quarter wave plate of a transparent film of an index of refraction n_1 and thickness h is deposited on glass of index of refraction 1.44. For a light wave of 598.6 nm incident normal to the surface, determine

(a)   The thickness of the film desired for a minimum reflection

(b)   The transmittance of the film

10.7   Given a dielectric plate with thickness h = 200.0 nm and index of refraction n = 1.50 that is constant in the wavelength range

350.0–850.0 nm. Calculate the reflectivity R vs wavelength λ of this plate in the described wavelength range, and plot R vs λ for the described wavelength range.

10.8   Consider a dielectric plate (κ = 0.0) with a variable thickness of constant index of refraction n = 2.0. Determine the reflectivity R of this plate vs its optical thickness [m(nh = λ/4)], where m is an integer equal to 0, 1, 2, 3, 4, 5. Depict the plot against another case for a dielectric film of n = 3.0, κ = 0.0.

10.9   A metallic film (n = 3.0, κ = 0.1) is deposited on glass (n = 1.50). For a monochromatic light incident normal to the film surface, determine the normal reflectance vs the optical thickness [m(nh = λ/4)] of this film, and depict it along with the reflectivity of a film of n = 3.5, κ = 0.1.

10.10   Redo the previous problem for the transmissivity of the two cases.

10.11   A 15.0 nm silver layer (n = 3.30, κ = 0.59) is deposited on glass (n = 1.50). Treating this as an ultrathin film and for light wavelength λ = 540.0 nm, determine
(a)   The normal reflectance
(b)   The normal transmittance
(c)   The normal absorptance

10.12   A bilayer consists of a 15.0 nm silver layer (n = 3.30, κ = 0.59) and a 6.0 nm thin film of SiO (n = 0.12, κ = 3.20) is deposited on the silver film. The bilayer is deposited on a glass substrate (n = 1.50). Treating the bilayer as an ultrathin stack and for incident light of wavelength λ= 540.0 nm, determine
(a)   The normal reflectance of the bilayer
(b)   The normal transmittance of the bilayer

10.13   With no approximations, redo the previous problem, and assuming that the given optical constants of the constituent layers are fixed for the wavelength range 350.0–960.0 nm, use the CMT to
(a)   Calculate the normal reflectance and transmittance for light of wavelength λ= 540.0 nm.
(b)   Determine and plot the normal reflectance for this bilayer in the wavelength range 350.0–960.0 nm.

10.14   In the previous problem, assuming that the optical constants given at wavelength λ = 540.0 are fixed for the wavelength range 350.0–960.0 nm, determine the normal reflectance for this bilayer in the described wavelength range.

10.15   Redo problem 10.12 determining the transmissivity of the bilayer for sodium light λ = 598.6 nm when the layering order is reversed, SiO first and silver on the top. Ignore absorption.

Polarization

We do not claim that the portrait we are making is the whole truth, only that it is a resemblance.

Victor Hugo (1802–1885)

11.1 INTRODUCTION

Ordinary light is emitted by atoms that radiate when they experience a change in their energy state. The electric field vectors in these waves have all directions. This is the characterization of an unpolarized light, also called ordinary or natural light. Polarization, in contrast, means selective directions of oscillation of the electric field in the EM waves.

Polarization of light waves arises from the direction of oscillation of the electric field in those waves. Linear polarization, also called plane polarization, is among the early kinds recognized. Here, the electric field oscillates along one direction perpendicular to the direction of the propagation. If the wave is propagating along the z direction, the electric field would be directed along the x or y axis (Figure 11.1). The magnetic field would then be directed

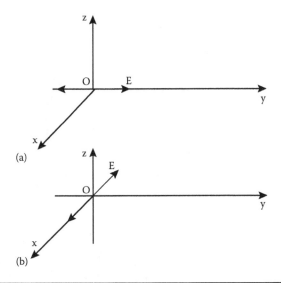

FIGURE 11.1   A light wave propagating along the positive z direction. The wave is linearly polarized in the x–y plane: (a) along the y direction and (b) along the x direction.

along the y or x axis, respectively. Our understanding of polarization of a light wave is associated with the value and direction of oscillation of the electric field. Prior to introducing ways and means of producing polarization, it is useful to understand the theoretical basis for the phenomenon. One-dimensional plane waves, propagating along some direction, say z, can satisfactorily serve the purpose.

11.2 BASIC THEORY

Polarization of light was discovered by Huygens in 1690 at a time when there was no theory of electromagnetism in existence. His discovery of polarization was related to the property of certain materials that produced what is now known as double refraction. This will be addressed later in Section 11.4.2 and 11.4.3 of the chapter. Other kinds of light polarization were discovered by Malus in 1810 and Brewster who was able to develop a connection between polarization by reflection from the boundary of a medium and the index of refraction of the medium. It was not until later in the nineteenth century when Maxwell introduced his theory of electromagnetism that a fundamental understanding of the phenomenon began to be developed.

For a wave propagating along the z direction, the electric field is oscillating in the x–y plane, producing components along each coordinate axis.

This is equivalent to saying that the wave under consideration is the super-position of two waves, one of an electric field oscillating along the x direction and another with an electric field oscillating along the y direction; both are propagating along the z direction. In reality, the wave is the super-position of many waves of differing phases, which can be treated as a single wave. In the simplest practical treatment, the wave is the superposition of just two waves, one along x and one along y with a relative phase difference of δ:

$$E_x(z, t) = E_{ox} \cos(kz - \omega t), \tag{11.1}$$

$$E_y(z, t) = E_{oy} \cos(kz - \omega t + \delta), \tag{11.2}$$

The resultant is the addition of these two component waves. That is,

$$E(z, t) = E_{ox} \cos(kz - \omega t) + E_{oy} \cos(kz - \omega t + \delta). \tag{11.3}$$

Thus, E will oscillate in the x–y plane, and the resultant is described as a plane or linearly polarized wave. The value of δ is critical to the value of the amplitude of the resultant electric field and its state of polarization.

11.3 STATES OF POLARIZATION

To simplify, we single out the relative phase by expanding the right-hand side of Equation 11.2, giving

$$E_y(z, t) = E_{oy}\{\cos(kz - \omega t)\cos\delta - \sin(kz - \omega t)\sin\delta\},$$

or

$$\frac{E_y(z, t)}{E_{oy}} = \cos(kz - \omega t)\cos\delta - \sin(kz - \omega t)\sin\delta.$$

And from Equation 11.1,

$$\frac{E_x(z, t)}{E_{ox}}\cos\delta = \cos(kz - \omega t)\cos\delta.$$

Upon subtracting the above two equations, squaring the two sides and using trigonometric relations, we get

$$\frac{E_x^2(z, t)}{E_{ox}^2} + \frac{E_y^2(z, t)}{E_{oy}^2} - 2\frac{E_x(z, t)E_y(z, t)}{E_{ox}E_{oy}} \cos\delta = \sin^2\delta. \qquad (11.4)$$

Equation 11.4 is the general form for an ellipse centered at the origin $E_x = E_y = 0$ with major and minor axes E_{ox} and E_{oy}, respectively. These axes are oblique to the coordinate axes and can be used to extract several interesting modes of polarizations, especially when $\delta = 0°$, 90°, 180°, or multiples of these values; in such a case, the algebra becomes simpler.

11.3.1 SPECIAL CASES OF POLARIZATION

11.3.1.1 Linear Polarization

11.3.1.1.1 First Case

Let

$$\delta = 0, \pm 2\pi, \pm 4\pi, \dots, \pm 2m\pi; \quad m = 0, 1, 2, \dots.$$

In this case, $\sin\delta = 0$, $\cos\delta = 1$. Equation 11.4 becomes

$$\frac{E_x^2(z, t)}{E_{ox}^2} + \frac{E_y^2(z, t)}{E_{oy}^2} - 2\frac{E_x(z, t)E_y(z, t)}{E_{ox}E_{oy}} = 0. \qquad (11.5)$$

The left-hand side of the above equation can be factored, and the equation becomes

$$\left(\frac{E_x(z, t)}{E_{ox}} - \frac{E_y(z, t)}{E_{oy}}\right)^2 = 0.$$

This implies that

$$\frac{E_x(z, t)}{E_{ox}} = \frac{E_y(z, t)}{E_{oy}}.$$

That is,

$$E_y(z, t) = \left(\frac{E_{oy}}{E_{ox}}\right) E_x(z, t).\tag{11.6}$$

Equation 11.6 shows that the relation between E_y and E_x is linear.

11.3.1.1.2 Second Case

The above argument can be followed for the case when the relative phase is δ, where

$$\delta = \pm\pi, \ \pm 3\pi, ..., \ \pm(2m+1)\pi; \quad m = 0, \ 1, \ 2, ...$$

In this case, $\sin \delta = 0$ and $\cos \delta = -1$. The ellipse in Equation 11.5 will be

$$\frac{E_x^2(z, t)}{E_{ox}^2} + \frac{E_y^2(z, t)}{E_{oy}^2} + 2\frac{E_x(z, t)E_y(z, t)}{E_{ox}E_{oy}} = 0.\tag{11.7}$$

Factoring the left-hand side gives

$$\left(\frac{E_x(z, t)}{E_{ox}} + \frac{E_y(z, t)}{E_{oy}}\right)^2 = 0,$$

from which we deduce that

$$E_y(z, t) = -\left(\frac{E_{oy}}{E_{ox}}\right) E_x(z, t).\tag{11.8}$$

Again, the above is a straight line but of a slope −1. Figure 11.2 demonstrates the two linear polarizations expressed in Equations 11.6 and 11.8. As stated earlier, the linear polarization is also called plane polarization.

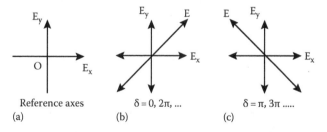

Reference axes $\delta = 0, \ 2\pi, ...$ $\delta = \pi, \ 3\pi ....$

(a)   (b)   (c)

FIGURE 11.2 Linear polarization in (a) the x–y plane of two waves that have a phase difference of (b) 0, 2π, 4π and (c) π, 3π,.... The waves are traveling along the z axis.

11.3.1.2 Elliptical and Circular Polarization

Let

$$\delta = \pm\pi/2, \ \pm 3\pi/2, \ \pm 5\pi/2...,(2m+1)\pi/2; \quad m = 0, \ \pm 1, \ \pm 2,....$$

In this case, $\sin^2 \delta = 1$ and $\cos \delta = 0$, and Equation 11.4 becomes

$$\frac{E_x^2(z, t)}{E_{ox}^2} + \frac{E_y^2(z, t)}{E_{oy}^2} = 1.$$

This is the equation of an ellipse centered at the origin $E_x = E_y = 0$ with major and minor axes along the axes E_{ox} and E_{oy}, respectively. The value of the resultant can be analyzed from the above equation either by evaluating E_x for special values of E_y or evaluating E_y for certain values of E_x as follows.

Points that can be readily seen to satisfy the equation are $E_x = 0, E_y = \pm E_{oy}$; $E_x = \pm E_{ox}, E_y = 0$; Also notice that for values $-E_{ox} < E_x < +E_{ox}$, the absolute value of $|E_y| < E_{oy}$. The initial values for E_{ox} and E_{oy} determine how one can trace the ellipse. An effective way of examining the values of E_x and E_y can be obtained by assigning values to t for a convenient value of z, for example, z = 0. In Figure 11.3a and b, two cases of elliptical polarization for which the phase difference is $\pi/2$ (case a) and $3 \pi/2$ (case b) are shown. The former is of left elliptical polarization and the latter is of right elliptical polarization. More details about such cases will follow in the following sections.

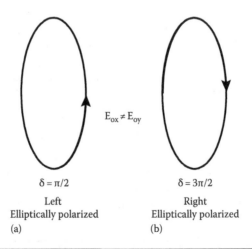

$$\delta = \pi/2 \qquad\qquad \delta = 3\pi/2$$

Left
Elliptically polarized
(a)

Right
Elliptically polarized
(b)

FIGURE 11.3 Elliptical polarization in the x–y plane of two waves of different amplitudes having a phase difference of (a) $\pi/2$ and (b) $3\pi/2$. The waves are traveling along the z axis.

In case $E_{ox} = E_{oy}$, the major and minor axes of the ellipse are equal, and the ellipse becomes a circle. That is, the polarization is circular, Figure 11.4.

COMMENT

1. In the given component waves, the relative phase was linked to the second wave. Since both forms are in the cosine form, the first one then leads the second by that phase difference, which in this case is $\pi/2$ or one-fourth of a cycle. In other words, in this example, E_x leads E_y.
2. As just noted, the circular mode of polarization is a special case of the elliptical polarization. For that matter, the linear polarization is also a special case of the elliptical polarization. In the linear polarization, one of the ellipse axes is zero.

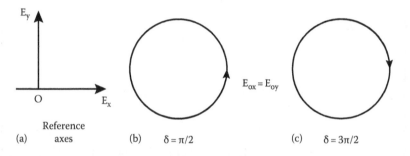

FIGURE 11.4   (a) Circular polarization in the x–y plane of two waves of equal amplitudes having a phase difference of (b) $\pi/2$ and (c) $3\pi/2$. The waves are traveling along the z axis.

EXAMPLE 11.1

Consider the following electromagnetic waves whose electric field components are

$$E_x = E_{ox} \cos(kz - \omega t + \pi/4); \quad E_y = E_{oy} \cos(kz - \omega t),$$

where $E_{ox} = E_{oy} = 1.00$ V/m. Determine the state of polarization for the resultant wave.

Solution

The simplest way of treating such a problem is to consider a snapshot of the two superimposed components and choosing a point on the z axis for the analysis. One may examine the electric field components at the chosen point for values of t that are expressed in terms of the period

T of the wave, because $\omega = 2\pi/T$, and $\omega t = (2\pi/T)t$. Choosing $z = 0$, the component waves become

$$E_x = E_{ox} \cos(\omega t + \pi/4); \quad E_y = E_{oy} \cos(\omega t)$$

At the selected t values on the left, the components become

$t = 0$: $E_x = E_{ox} \cos(0 + \pi/4) = 0.707$ V/m; $E_y = E_{oy} \cos(\omega t) = 1.000$ V/m

$t = T/8$: $E_x = E_{ox} \cos(\pi/4 + \pi/4) = 0.000$ V/m;

$E_y = E_{oy} \cos(\pi/4) = 0.707$ V/m

$t = T/4$: $E_x = E_{ox} \cos(\pi/2 + \pi/4) = -0.707$ V/m;

$E_y = E_{oy} \cos(\pi/2) = 0.000$ V/m

$t = 3T/8$: $E_x = E_{ox} \cos(3\pi/4 + \pi/4) = -1.00$ V/m;

$E_y = E_{oy} \cos(3\pi/4) = -0.707$ V/m

$t = T/2$: $E_x = E_{ox} \cos(\pi + \pi/4) = -0.707$ V/m;

$E_y = E_{oy} \cos(\pi) = -1.00$ V/m

$t = 5T/8$: $E_x = E_{ox} \cos(5\pi/4 + \pi/4) = 0.000$ V/m;

$E_y = E_{oy} \cos(5\pi/4) = -0.707$ V/m

Constructing the resultant electric field vector $\mathbf{E} = \hat{x}E_x + \hat{y}E_y$ at $z = 0$ over more than half a period of oscillation, we find that

$$E(0,t) = (0.707\hat{x} + 1.0\hat{y}), (0.000\hat{x} + 0.707\hat{y}), (-0.707\hat{x} + 0.0\hat{y}),$$

$$(-1.0\hat{x} - 0.707\hat{y}), (-0.707\hat{x} - 0.707\hat{y})$$

The wave is left elliptically polarized, since the electric field rotates counterclockwise.

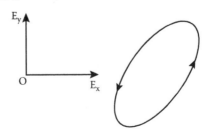

EXAMPLE 11.2

Follow the argument used in Example 11.1 to find the state of polarization of the two-component wave

$$E_x = E_0 \cos(kz - \omega t + \pi/2); \quad E_y = E_0 \cos(kz - \omega t),$$

where $E_0 = E_{oy}$ 1.00 V/m. Correlate your result with the discussion presented in Section 11.3.1, particularly the final statement at the end of the section.

Solution

Following the method used in the previous example, we analyze the values of E_x and E_y in the plane $z = 0$ at the t values selected on the left:

$t = 0:$ $E_x = E_{ox} \cos(\pi/2) = 0.000$ V/m; $E_y = E_{oy} \cos(0) = 1.000$ V/m

$t = T/8:$ $E_x = E_{ox} \cos(3\pi/4) = -0.707$ V/m; $E_y = E_{oy} \cos(\pi/4) = 0.707$ V/m

$t = T/4:$ $E_x = E_{ox} \cos(\pi) = -1.000$ V/m; $E_y = E_{oy} \cos(\pi/2) = 0.000$ V/m

$t = 3T/8:$ $E_x = E_{ox} \cos(5\pi/4) = -0.707$ V/m; $E_y = E_{oy} \cos(3\pi/4)$
$= -0.707$ V/m

$t = T/2:$ $E_x = E_{ox} \cos(3\pi/2) = -0.000$ V/m; $E_y = E_{oy} \cos(\pi) = -1.000$ V/m

$t = 5T/8:$ $E_x = E_{ox} \cos(5\pi/4 + \pi/2) = 0.707/m$; $E_y = E_{oy} \cos(5\pi/4)$
$= -0.707$ V/m

The wave is left circularly polarized.

11.4 VARIOUS PROCESSES OF POLARIZATION

Polarization has been discovered in processes like reflection or absorption. In essence, the electric field was observed to diminish partially or totally along certain directions. In such a case, the intensity of the wave decreases in a process totally different from just absorption, where the direction of the electric field is not necessarily affected. The polarizing material, usually shaped in a thin plate, is called a polarizer. The processes inside the material that induce polarization are several. Here are some of the most common:

1. Selective absorption
2. Reflection and transmission
3. Birefringence or double refraction
4. Scattering

In the following is a brief description of some of the above effects.

11.4.1 Selective Absorption

If natural light is incident on an ideal linear polarizer, only plane polarized light will be transmitted with an electric field oscillating along a specific direction perpendicular to the direction of propagation. This direction is called the transmission direction of the polarizer. This is called linear or plane polarization. For incident natural light waves of amplitude E_o, there are an infinite number of E_o field vectors oscillating in all directions. Each of these is directed at some angle θ_i with respect to the transmission direction of the polarizer, and hence, only the component $E_o \cos \theta_i$ is what will be allowed to pass through (Figure 11.5). Thus

$$E_i = E_o \cos \theta_i.$$

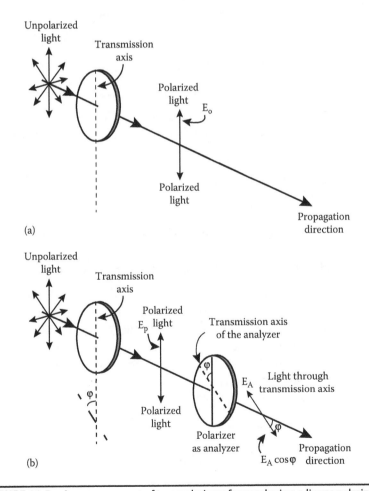

FIGURE 11.5   An arrangement of two polarizers for producing a linear polarized light and verifying Malus's law.

The intensity attributed to this component in the polarized wave is

$$\left(E_i\right)^2 = \left(E_o\right)^2 \cos^2 \theta_i.$$

The total intensity that one obtains from all components of the other electric field vectors in the wave is

$$I = \sum_1^\infty E_o^2 \cos^2 \theta_i; \quad I = 1, 2, 3, \cdots$$

As for the infinite number of E_i vectors in the incident wave, θ_i varies from $0°$–$360°$, then

$$\sum_i^\infty \cos^2 \theta_i = 1/2,$$

and

$$I = \frac{1}{2} E_o^2 = \frac{1}{2} I_o; \quad I_o = E_o^2.$$

Thus, the intensity of light passing through a linear polarizer, denoted by I_p is half the intensity of the incident natural light. That is,

$$I_p = \frac{1}{2} I_o. \tag{11.9}$$

I_p is proportional to the square of the amplitude E_p of the polarized light. That is,

$$I_p \propto E_p^2.$$

If one places another polarizer, usually called analyzer, next to the first one such that its transmission axis makes an angle φ, then the intensity of light emerging from the analyzer will be of an amplitude E_A and intensity I_A such that

$$E_A = E_p \cos \varphi$$

and

$$I_A = \left(E_A\right)^2 = E_p^2 \cos^2 \varphi = I_p \cos^2 \varphi,$$

which becomes

$$I_A = \frac{1}{2}I_o \cos^2\varphi.$$

This is known as Malus's law. Obviously, for $\varphi = 90°$, $I_A = 0$. Thus, a combination of a polarizer and an analyzer crossed by 90° pass no light through the analyzer.

EXAMPLE 11.3

Consider an arrangement of a source and two analyzers aligned on an optical bench. Let the angle between the analyzers transmission lines be 30°. For unpolarized light of intensity I_o incident upon the first polarizer of the pair,

(a) Determine the intensity I_1 of the light emerging from the first polarizer (analyzer) and that of I_2 transmitted through the second.
(b) What is the angle that reduces the intensity of the emerging light to one-fourth of its initial value?

Solution

(a) Following the previous discussion, the first polarizer transmits the beam I_1 with half of its original intensity. Thus,

$$I_1 = (1/2)I_o.$$

Using Malus's law, the second one transmits an intensity

$$I_2 = I_1 \cos^2 \varphi = (1/2)I_o \left(\cos^2 30°\right) = (3/8)I_o.$$

(b) As $I_2 = I_1 \cos^2 \varphi = (1/2) \cos^2 \varphi$, then for $I_2 = (1/4) I_o$, we set

$$(1/4)I_o = I_o (1/2)\cos^2 \varphi.$$

This makes

$$\cos^2 \varphi = (1/2),$$

that is

$$\cos\varphi = 1/\sqrt{2}.$$

Thus,

$$\varphi = 45°.$$

11.4.2 Polarization by Reflection and/or Transmission

Consider a plane wave incident on a slab of a dielectric material, and let the incident, reflected, and transmitted beam be in the plane of the page (Figure 11.6). The wave will have its electric and magnetic fields perpendicular to the direction of propagation. Looking at the electric field only, each of the three beams will have an electric field, denoted by E_p, in the plane of the page and another, E_s, perpendicular to the plane of the page. These are depicted in the figure as arrows and filled circles, respectively.

Experiments showed that light incident on the surface with an arbitrary angle of incidence gets reflected with partial polarization for both oscillations, E_p and E_s, but with E_s of dominant intensity and E_p becoming much weaker. However, the transmitted beam had more of the E_p and weaker E_s. If light emerging from the slab is passed through a series or piles of dielectric slabs, the perpendicular component of the beam continues to diminish, eliminating almost completely the perpendicular component E_s. This gets the emerging beam polarized having the parallel electric field component, E_p, only. This is the essence of polarization by transmission.

Most interesting is the fact that light incident on the surface of some materials with a particular angle of incidence, the reflected beam becomes plane polarized. Only the E_s component of the field oscillation is present (Figure 11.7). This angle of incidence is called Brewster's angle, usually denoted by φ_B. Also, it is found that for Brewster's angle of incidence, the reflected and refracted rays are mutually perpendicular.

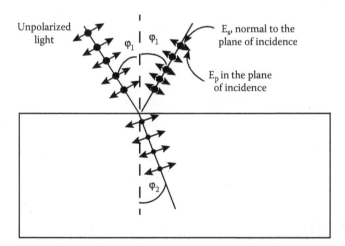

FIGURE 11.6   A schematic that demonstrates partial polarization due to reflection.

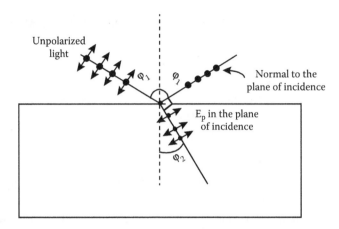

FIGURE 11.7   A schematic that demonstrates total polarization due to reflection for incidence at Brewster's angle.

That is, for $\varphi_1 = \varphi_B$

$$\varphi_2 = \pi/2 - \varphi_B.$$

Since from Snell's law of refraction

$$n_1 \sin \varphi_1 = n_2 \sin \varphi_2,$$

for Brewster's angle of incidence

$$n_1 \sin \varphi_B = n_2 \sin \left(\pi/2 - \varphi_B\right)$$

or

$$n_1 \sin \varphi_B = n_2 \cos \varphi_B.$$

Thus,

$$\tan \varphi_B = n_2; \quad n_1 = 1.00.$$

EXERCISE

Use the above discussion to show that for a light beam incident on a surface at the Brewster angle, $\varphi_1 = \varphi_B$ in Figure 11.7, the sum $\varphi_1 + \varphi_2 = \pi/2$.

EXAMPLE 11.4

Consider a beam of unpolarized light directed at one kind of quartz at an angle that equals its Brewster's angle, known to be 55°. Determine

    a. The index of refraction of this quartz
    b. The angle of refraction through the glass
    c. The angle between the glass surface and the refracted light

Solution

(a) From Brewster's law, (see Figure 11.7),

$$\tan \varphi_B = n_2.$$

Thus,

$$n_2 = \tan 55° = 1.45.$$

(b) From Snell's law,

$$n_{air} \sin \varphi_{air} = n_2 \sin \varphi_{quartz}.$$

Upon substituting for the known values, we get

$$\sin 55° = (1.45) \sin \varphi_{quartz}.$$

Thus,

$$\varphi_{quartz} = \sin^{-1}(1/1.45) = 34.4°.$$

(c) Subtracting the angle of refraction from 90° gives the angle between the refracted ray and the quartz surface.

11.4.3 POLARIZATION BY DOUBLE REFRACTION

When a beam of ordinary unpolarized light is incident upon a nonisotropic material that has the property of double refraction, for example, calcite crystal, two refracted beams inside the crystal appear instead of one. For reasoning that will be explained later, the rays are labeled as O-ray and E-ray. Among such materials are those called uniaxial that are characterized by a particular direction known as the optic axis. The material

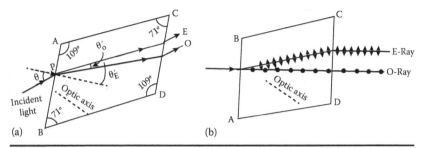

FIGURE 11.8 Double refraction in uniaxial crystal. Unpolarized light incident at (a) an arbitrary angle and (b) normal to the surface.

along this axis and perpendicular to it has two different indices of refraction. This is in contrast to refraction in isotropic materials like glass, clear plastic, and liquids, which have only one index of refraction. The effect resulting from having two indices of refraction is called double refraction or birefringence. Figure 11.8 shows that an unpolarized light incident on a calcite crystal at point P splits at the boundary inside the crystal into two rays (Figure 11.8a). The rays diverge from each other through an angle θ, and if the surfaces AB and CD are parallel, a property of the calcite cleavage process, the outgoing rays emerge from the opposite side parallel to each other. If the incident ray is normal to the surface (Figure 11.8b), the O-ray will be normal to it (Snell's law), while the E-ray in general will be refracted at some angle that is not zero. It will come out parallel to the O-ray but displaced laterally.

Snell's law still applies in the double refraction effect but only to one of the beams and not the other. The ray for which the law holds is called the ordinary (O-ray), and the other is called extraordinary (E-ray).

The discussion of the double refraction requires a clear acquaintance with the basic definition and property of a few quantities: the optic axis, the O-ray, the E-ray, and the structural geometry involving some of calcite important planes. The following definitions can be helpful:

1. *The principal section*: This is defined by the plane that combines the optic axis and the normal to one of the crystal cleavage faces. A structural property of the *principal section* is that it cuts the surfaces of a calcite crystal in a parallelogram with angles of 71° and 109°.

2. *The optic axis*: This is defined by the direction connecting two corners of the crystals and making equal angles with all faces that meet at each of those corners. As the optic axis is a direction, then any line parallel to the defined optic axis is an optic axis. Any plane that contains the optic axis is known as a principal plane.

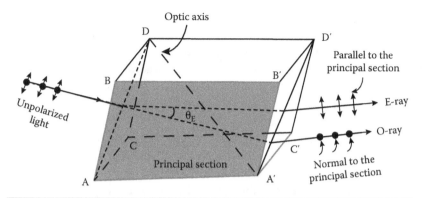

FIGURE 11.9 A side view of a calcite crystal illustrating double refraction.

3. *The principal plane of the ordinary ray*: This is the plane made by the optic axis and the ordinary ray. A special feature of the O-ray is that it always lies in the plane of incidence, but this is not necessarily true for the E-ray.
4. *The principal plane of the extraordinary ray*: This is the plane made by the optic axis and the extraordinary ray. In general, the principal planes of the two rays, the O- and E-rays, do not coincide. Also, a special case of interest that makes the optics easier to follow is when the plane of incidence is chosen to be a principal section. Figure 11.9 is a three-dimensional sketch of Figure 11.8b.

11.5 PROPAGATION OF LIGHT WAVES IN DOUBLE-REFRACTING MATERIALS

Our starting point of explaining the propagation of a light wave through a material comes from Huygens' theory that proved to be successful in explaining phenomena like reflection, refraction, interference, and diffraction. All points on the incident wave front reaching the boundary act as a line of a large number of secondary point sources, each of which emits secondary spherical wavelets propagating on all directions. At the instant of creating the secondary wavelets, they have the same velocity and phase as the primary wave. In isotropic materials, the wave front of the primary wave at any instant of time is the envelope of all secondary wavelets at that instant, and it is that envelope that progresses on behalf of the primary wave. In an anisotropic medium where double refraction develops, the success of Huygens' theory extended to explain birefringence when the properties of polarization of a doubly refracted beam in calcite are taken into consideration. The key explanation comes from a fundamental property in calcite, having an optic axis is attributed to demonstrating the presence of different indices of refraction along which the velocity of the propagating wave is not the same.

This is behind observing two linearly polarized waves that develop within the medium immediately after the primary incident wave enters the medium: an ordinary wave that propagates with a constant velocity (constant index of refraction), independent of direction, and an extraordinary wave that propagates with different velocities in different directions (directional variation of the index of refraction).

An E-wave traveling at an angle with the optic axis has a velocity c/n_1, larger or smaller than c/n_2, while its velocity along the optic axis is c/n_2. Velocities of the O- and E-waves along the direction of the optic axis are the same. It is equally important to note that the O-wave always vibrates in a direction perpendicular to the optic axis, that is, the O-wave propagates along the optic axis, while the E-wave vibrates in a principal plane that, generally speaking, is tilted by angle to the principal plane of the ordinary plane. It is a special case when the two planes coincide.

In cases where $c/n_2 > c/n_1$, the E-wave front advances along the off optic axis direction more than it does along the optic axis. An inclusive illustration of the propagation of these wave fronts is shown in Figure 11.10. As shown, the

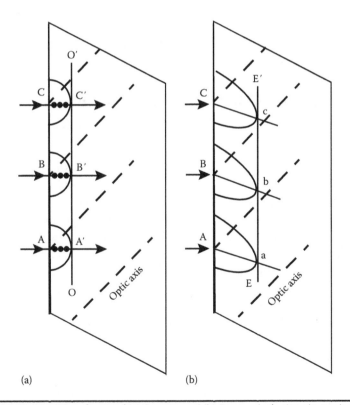

(a)   (b)

FIGURE 11.10 Wave fronts of (a) O-wave and (b) E-wave (velocity $c/n_2 > c/n_1$).

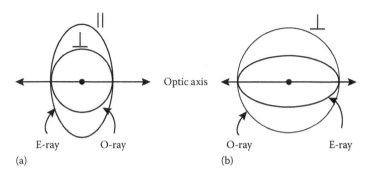

FIGURE 11.11   An illustration of (a) a negative and (b) a positive crystal. In both kinds, the O- and E-waves have the same speed along the optic axis.

O- and E-wave front surfaces are tangent to each other along the optic axis, but the E-wave front is elongated in the form of an ellipsoid rotating about the optic axis. In the figure, a higher velocity of the wavelet for the E-wave changes the generated wavelet from spherical into ellipsoidal. The envelope of the adjacent ellipsoids in this case propagates along a direction separate from the other direction of propagation of the envelope of the spherical wavelets, O-waves; thus, a double refraction is observed.

The index of refraction n_E is different for those directions of light propagation, and hence the wave has a velocity that depends on the direction of propagation. That is why such a wave is called extraordinary (E-wave) and the index of refraction, frequently denoted by n_E, is either more (Figure 11.11a) or less (Figure 11.11b) than n_O. The difference between the two indices of refraction $\Delta n = n_E - n_O$. This difference is responsible for double refraction in crystals like calcite and quartz. Crystals are classified as positive or negative based on Δn being negative as in (a) or positive as in (b).

11.6 WAVE FRONTS AND REFRACTION OF RAYS IN BIREFRINGENT MATERIALS

Following up on the discussion presented in the previous section, consider a calcite crystal, Figure 11.12, where a plane wave, represented by three arrows, is incident parallel to one of the calcite principal planes that is taken as the principal section or parallel to it. The optic axis is in the plane of the drawing and is shown by the dashed lines. A, B, and C are point sources chosen along the wave front as it hits the surface, becoming secondary sources, and thus generate spherical and ellipsoidal wave fronts. The ordinary ray in the crystal keeps traveling in the direction of the incident ray. The extraordinary ray, however, shifts in direction but remains in the plane of the sketch that for simplicity is

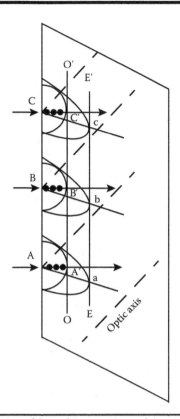

FIGURE 11.12 Formation of O- and E-waves in a calcite crystal.

taken as the one parallel to the principal section. A more comprehensive discussion can be found in *Fundamentals of Optics* by Jenkins and White and *Electromagnetic Spectrum* by C. L. Andrews that are cited in the bibliography of this text.

The advancement of the circular wave front can be followed through the advancement of the tangent common to the infinite number of the circles for the O-waves, shown by line OO', connecting A', B', C',.... The progression of the elliptical tangent can be followed through the progression of the line tangent to the infinite number of ellipses for the E-waves, and that is line E E', connecting a, b, c,.... The straight paths of the O-wave are represented by solid arrow lines, and the inclined paths of the E-wave are represented by the solid inclined lines.

Figure 11.13 shows two other cases: one where the optic axis is perpendicular to the plane of incidence and hence is perpendicular to the incident beam (Figure 11.13a) and the other case when the optic axis is parallel to the plane of incidence and hence is parallel to the incident

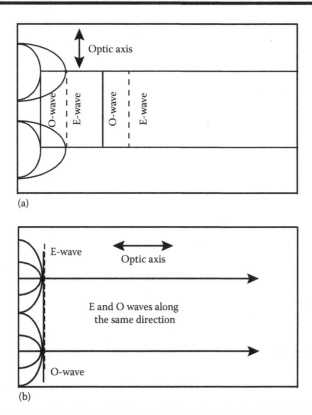

(a)

(b)

FIGURE 11.13   A projection of a side view of a calcite crystal in which (a) the optic axis is parallel to the front face and in (b) the optic axis is perpendicular parallel to the front face. In (a) the E-wave advances at v parallel, and the spherical O-wave advances at v perpendicular. In (b) both waves move through the crystal with same speed, the perpendicular speed.

beam, perpendicular to the boundary. In (a), the two rays propagate along the same path but with different speeds, while in (b), they travel with the same speed. Therefore, a phase difference between the two waves is generated, becoming bigger in (a) as they travel longer within the crystal. However, in (Figure 11.13b) the two waves travel through the crystal with equal velocities with no change in direction. Thus, they emerge from the other side in phase.

In reference to the first case (Figure 11.13a), the difference in phase between the O- and E-waves, as they emerge from the crystal opposite face, depends on the crystal's thickness. If the crystal's thickness is made such that the phase difference is $\pi/2$, light then becomes circularly polarized. The crystal is then called a quarter-wave plate. A similar notion applies to the crystal if its thickness is designed to result in a phase difference of π. The crystal is then called a half-wave plate. The phase difference and the thickness of the crystal are related through the relation

$$\Delta = \left|n_O - n_E\right| d,$$

where

Δ is the optical path difference between the two beams

d is the thickness of the crystal

n_O and n_E are the indices of refraction of the crystal in the directions of the O and E beams, respectively

Thus, the angular phase difference that corresponds to Δ is

$$\delta = \left(2\pi/\lambda\right)\left|n_O - n_E\right| d.$$

EXAMPLE 11.5

Consider a half-wave plate with its axis at an angle θ with the polarization direction of a plane polarized light produced by a polarizer. For a light beam propagating in the z direction perpendicular to the plate, determine the status of polarization of the light emerging from the plate.

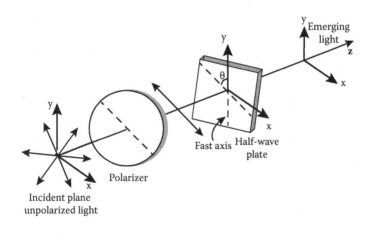

Solution

For a light wave propagating in the z direction, the electric field vector is $E = E_o \cos(kz - \omega t)$, and of two components

$$E_{ox} = E_o \cos\theta \cos(kz - \omega t), \quad E_{oy} = E_o \sin\theta \cos(kz - \omega t).$$

Analyzing the polarization at a constant z, $z = 0$, $t = 0$ would make the two components

$$E_{ox} = E_o \cos\theta, \quad E_{oy} = E_o \sin\theta.$$

The emerging light is also plane or linearly polarized, oscillating in the same plane of incidence.

If the velocities along the two components are v_1 and v_2, then the above equation after emerging from the plate would be

$$E_x = E_{ox} \cos\left[\omega(t - z/v_1)\right], \quad E_y = E_{oy} \cos\left[\omega(t - z/v_2)\right].$$

For an angular phase difference δ between the two components, they become

$$E_x = E_{ox} \cos(kz - wt), \quad E_y = E_{oy} \cos\left[(kz - \delta) - wt\right].$$

For $\delta = \pi$, we can follow the argument that led to Equation 11.7

$$\frac{E_x^2(z, t)}{E_{ox}^2} + \frac{E_y^2(z, t)}{E_{oy}^2} + 2\frac{E_x(z, t)E_y(z, t)}{E_{ox}E_{oy}} = 0,$$

which upon factoring and slight algebra gives the equation (see Equation 11.8)

$$E_y(z, t) = -\left(\frac{E_{oy}}{E_{ox}}\right)E_x(z, t).$$

The above is a line of slope equal to

$$\left(\frac{E_y\left(z,t\right)}{E_x\left(z,t\right)}\right)=-\left(\frac{E_{oy}}{E_{ox}}\right)=-\theta.$$

Thus, the direction of the vector $\mathbf{E}$ changed from one of a slope θ to another of a slope $-\theta$. This then corresponds to a rotation of the polarization plane by 2θ (Figure 11.14).

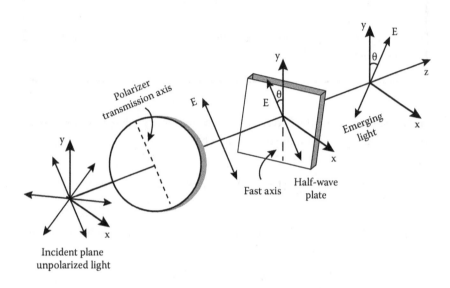

FIGURE 11.14   Rotation of the polarization plane after encountering a half-wave plate.

PROBLEMS

11.1   Consider the following EM wave with electric field components:

$$E_x = E_{ox} \cos(kz - \omega t + 3\pi/4); \quad E_y = E_{oy} \cos(kz - \omega t),$$

when $E_{ox} = E_{oy} = 1.00$ V/m. Determine the state of polarization for the resultant wave.

11.2   Describe the state of polarization of the wave with electric field components having $E_o = 1.00$ V/m for each:

$$E_x = E_o \cos(kz - \omega t); \quad E_y = E_o \sin(kz - \omega t).$$

11.3   Redo the previous problem for a case in which the sinusoidal forms of E_x and E_y are switched becoming

$$E_x = E_o \sin(kz - \omega t); \quad E_y = E_o \cos(kz - \omega t).$$

11.4   Considering the following two waves
(a) $E_x = E_o \cos(kz - \omega t); \quad E_y = -E_o \cos(kz - \omega t)$
(b) $E_x = E_o \cos(kz - \omega t); \quad E_y = E_o \cos(kz - \omega t)$
determine the state of polarization of each, and draw any special observations on any correlation between waves (a) and (b).

11.5   If the wave in part (b) in the previous problem is incident on a polarizer, what would be the angles through which the polarizer would have to be rotated if (a) maximum intensity or (b) minimum intensity is to be transmitted?

11.6   Determine the state of polarization for the resultant of the two waves:

$$\mathbf{E}_1 = \hat{\mathbf{x}} E_o \cos(kz - \omega t) + \hat{\mathbf{y}} E_o \cos(kz - \omega t);$$

$$\mathbf{E}_2 = \sqrt{2} E_o (\hat{\mathbf{x}} \cos(kz - \omega t) + \hat{\mathbf{y}} \cos(kz - \omega t)).$$

11.7   Consider two polarizers whose transmission axes are parallel. For ordinary light of 200 W/m² radiant intensity incident on the first polarizer and emerging from the second, what would be the intensity of light emerging from the second polarizer?

11.8   Consider two polarizers oriented at 30° with respect to each other. If the centers are aligned along a point source emitting an unpolarized beam of intensity I_0, find the intensity of the emerging beam if

(a) A third polarizer is placed after the second polarizer at an angle of 60° with respect to the first.

(b) A third polarizer is placed after the second polarizer at an angle of 30° with respect to the second polarizer.

11.9 Consider two polarizers that are crossed out so that no light is transmitted. If light of intensity I_0 is incident on the first polarizer, find the angle through which the second polarizer will have to be rotated to transmit light of intensity equal to $(1/2)\,I_0$.

11.10 For water in a large glass vessel, what will be

(a) The Brewster angle for water (n = 1.33) if the glass container is to be ignored?

(b) The angle of refraction into water if light is incident on the surface with Brewster's angle?

11.11 In the previous problem, for light refracted into water and reflected off the bottom glass (n = 1.45), find Brewster's angle for the water–glass interface, so that light reflected off the water-glass interface is completely polarized.

11.12 Consider a thin plate of calcite that has its optic axis parallel to the surface. Let the indices of refraction of the O- and E- waves in calcite along the optic axis for 600.0 nm wavelength be 1.658 and 1.486, respectively. Find the minimum thickness needed to produce (a) one quarter wavelength and (b) half wavelength difference for a monochromatic light of the given wavelength.

11.13 Consider a beam of light of 589.6 nm incident on a calcite crystal whose optic axis is perpendicular to the crystal plane. If the indices of refraction of the O- and E-waves in calcite along the optic axis are 1.658 and 1.486, respectively, determine

(a) The wavelength of the O- and E-waves

(b) The frequency of these waves

11.14 Consider the configuration below, where a polarizer and a half-wave plate are aligned with a source that emits unpolarized plane light wave with an intensity I_0. The transmission axis of the polarizer and the fast axis of the half-wave plate are oriented as shown in the sketch. Make use of the properties of the quarter-wave plate to find the status of polarization of the emerging beam and its intensity in terms of the incident intensity I_0.

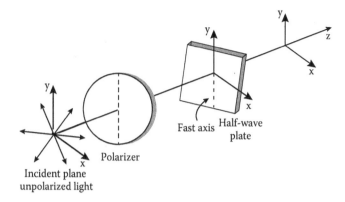

11.15 A source of 200 W/m² emits ordinary unpolarized light onto a combination of a quarter-wave plate and a linear polarizer, all aligned in (a) the polarizer first and in (b) the quarter-wave plate first. Determine the irradiance of the emergent light.

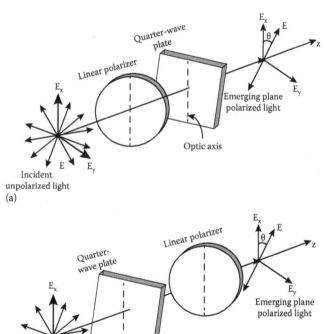

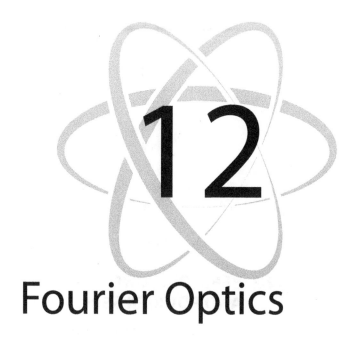

Fourier Optics

To be is to be perceived.

George Berkeley (1685–1783)

12.1 INTRODUCTION

Fourier analysis provides an elegant means of understanding and describing how optical systems process light to form images. Fourier transforms are particularly useful in analyzing optical processes such as Fraunhofer diffraction in terms of spatial frequencies. The central notion is that a periodic function $f(\theta)$ such as $f(\theta) = f(\theta + 2\pi)$, defined within the interval $-\pi < x < \pi$, can be represented by a sum of harmonic functions consisting of series of sines and cosines in the form

$$f(\theta) = \frac{a_o}{2} + \sum_{n=1}^{\infty} a_n \cos n\theta + \sum_{n=1}^{\infty} b_n \sin n\theta; \quad n = 1, 2, 3,\ldots \quad (12.1)$$

The coefficients a_n and b_n are determined from the relations

$$a_o = \frac{1}{\pi} \int_{-\pi}^{\pi} f(\theta)\, d\theta \quad (12.2)$$

$$a_n = \frac{1}{\pi} \int_{-\pi}^{\pi} f(\theta) \cos n\theta \, d\theta; \quad n = 1, 2, 3,... \tag{12.3}$$

$$b_n = \frac{1}{\pi} \int_{-\pi}^{\pi} f(\theta) \sin n\theta \, d\theta; \quad n = 1, 2, 3,... \tag{12.4}$$

In the above integrals, the function $f(\theta)$ is assumed continuous and single valued within the assigned interval that is angular $-\pi < \theta < \pi$ in this illustration but can be extended to spatial or temporal domains. Hence, the integrations, in principle, are well behaved.

12.2 PERIODIC FUNCTIONS AND FOURIER SERIES

In an attempt to connect the above discussion to light waves, we may let the angular period of 2π be linked to the sinusoidal wave functions, sin kx or cos kx, where $k = 2\pi/\lambda$, and the wavelength λ of the wave is then the spatial period, because

$$\cos kx = \cos k(x + \lambda) = \cos\left[\left(kx + (2\pi/\lambda)(\lambda)\right)\right] = \cos kx.$$

The above applies to a sine wave function as well. Equations 12.1 through 12.4 can be rewritten for a light wave function as

$$f(x) = \frac{a_o}{2} + \sum_{n=1}^{\infty} a_n \cos nkx + \sum_{n=1}^{\infty} b_n \sin nkx; \quad n = 1, 2, 3,... \tag{12.5}$$

The coefficients a_n and b_n are now given by the relations

$$a_o = \frac{1}{\ell} \int_{-\ell}^{\ell} f(x) \, dx \tag{12.6}$$

$$a_n = \frac{1}{\ell} \int_{-\ell}^{\ell} f(x) \cos nkx \, dx; \quad n = 1, 2, 3,... \tag{12.7}$$

$$b_n = \frac{1}{\ell} \int_{-\ell}^{\ell} f(x) \sin nkx \, dx; \quad n = 1, 2, 3,... \tag{12.8}$$

Expansion (12.5) is in the position-wave vector domain. Another expansion in the time–frequency domain is possible. If f(t) = f(t + T), where T is the period of the wave, then

$$f(t) = \frac{a_o}{2} + \sum_{n=1}^{\infty} a_n \cos \omega_n t + \sum_{n=1}^{\infty} b_n \sin \omega_n t; \quad n = 1, 2, 3, \ldots \qquad (12.9)$$

where a_n and b_n are given by

$$a_o = \frac{2}{T} \int_{-T/2}^{T/2} f(t) dt \qquad (12.10)$$

$$a_n = \frac{2}{T} \int_{-T/2}^{T/2} f(t) \cos \omega_n t \, dt \qquad (12.11)$$

$$b_n = \frac{2}{T} \int_{-T/2}^{T/2} f(t) \sin \omega_n t \, dt \qquad (12.12)$$

and

$$\omega_n = 2\pi n/T.$$

EXAMPLE 12.1

Expand in Fourier series the periodic function f(x) sketched below and described for one period by

$$E(x) = 0; \quad \text{for } -L \geq x \geq 0$$

$$E(x) = E_o; \quad \text{for } -0 \leq x \leq L$$

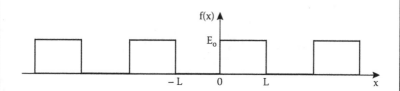

Solution

From Equation 12.1

$$f(x) = \frac{a_0}{2} + \sum_{n=1}^{\infty} a_n \cos nkx + \sum_{n=1}^{\infty} b_n \sin nkx; \quad n = 1, 2, 3, \ldots$$

The coefficients a_n and b_n are determined from the relations

$$a_0 = \frac{1}{L} \int_{-L}^{L} f(x) \, dx = \frac{1}{L} \left[\int_{-L}^{0} 0 + \int_{0}^{L} E_0 dx \right] = \frac{1}{L} \left[0 + E_0 \left| x \right|_{0}^{L} \right] = E_0$$

$$a_n = \frac{1}{L} \int_{-L}^{L} E_0 \cos nkx \, dx = \frac{1}{L} \left[0 + \int_{0}^{L} E_0 \cos nkx \, dx \right] = \frac{1}{L} \left[0 + E_0 \left(\frac{\sin nkx}{nk} \right) \Big|_{0}^{L} \right]$$

$$= \frac{1}{L} E_0 \frac{\sin nkL}{nk} = E_0 \frac{\sin nkL}{nkL}; \quad n = 1, 2, 3, \ldots$$

$$= 0 \quad \text{for } n \neq 0$$

$$b_n = L \int_{-L}^{L} f(x) \sin nkx \, dx \quad n = 1, 2, 3, \ldots$$

$$b_n = \frac{1}{L} \int_{-L}^{L} E_0 \sin nkx \, dx = \frac{1}{L} \left[0 + \int_{0}^{L} E_0 \sin nkx \, dx \right] = \frac{1}{L} \left[0 - E_0 \left(\frac{\cos nkx}{nk} \right) \Big|_{0}^{L} \right]$$

$$= -\frac{1}{L} E_0 \left(\frac{\cos nkL}{nk} - 1 \right) = E_0 \left(\frac{1}{nkL} - \frac{\cos nkL}{nkL} \right)$$

Therefore, the expansion $f(x) = \frac{a_0}{2} + \sum_{n=1}^{\infty} a_n \cos nkx + \sum_{n=1}^{\infty} b_n \sin nkx;$ becomes

$$f(x) = \frac{E_0}{2} + E_0 \sum_{n=1}^{\infty} \left(\frac{1}{nkL} - \frac{\cos nkL}{nkL} \right) \sin nkx.$$

EXAMPLE 12.2

Expand in Fourier series the periodic function f(x) described for one period by

$$E(x) = E_o; \quad \text{for} -L \geq x \geq 0$$
$$E(x) = 0; \quad \text{for } 0 \leq x \leq L.$$

A sketch of the function is depicted below.

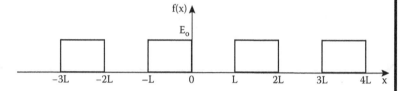

Solution

As x is spatially periodic, following Equations 12.5 through 12.8, we have

$$f(x) = \frac{a_o}{2} + \sum_{n=1}^{\infty} a_n \cos nkx + \sum_{n=1}^{\infty} b_n \sin nkx; \quad n = 1, 2, 3, \ldots.$$

The coefficients a_o, a_n, and b_n are determined as follows:

$$a_o = \frac{1}{L}\int_{-L}^{L} f(x)\,dx = \frac{1}{L}\left[\int_{-L}^{0} E_o dx + 0\right] = \frac{1}{L}\left[E_o\, x\Big|_{0}^{L} + 0\right] = E_o$$

$$a_n = \frac{1}{L}\int_{-L}^{L} E_o \cos nkx\,dx = \frac{1}{L}\left[\int_{-L}^{0} E_o \cos nkx\,dx + 0\right] = \frac{1}{L}\left[E_o\left(\frac{\sin nkx}{nk}\right)\Big|_{0}^{L}\right]$$

$$= \frac{1}{L}\,E_o\,\frac{\sin nkL}{nk} = E_o\,\frac{\sin nkL}{nkL}; \quad n = 1, 2, 3, \ldots$$

$$= 0 \quad \text{for } n \neq 0$$

$$b_n = L \int_{-L}^{L} f(x) \sin nkx \, dx; \quad m = 1, 2, 3,...$$

$$b_n = \frac{1}{L} \int_{-L}^{L} E_o \sin nkx \, dx = \frac{1}{L} \left[\int_{-L}^{0} E_o \sin nkx \, dx + 0 \right] = -\frac{1}{L} \left[E_o \left(\frac{\cos nkx}{nk} \right) \Big|_{-L}^{0} \right]$$

$$= -\frac{E_o}{L} \left[\frac{1}{nk} - \left(\frac{\cos nL}{nk} \right) \right] = \frac{E_o}{L} \left(\frac{\cos nL}{nk} - \frac{1}{nk} \right)$$

Therefore, the expansion becomes

$$f(x) = \frac{E_o}{2} + \sum_{n=1}^{\infty} \frac{E_o}{L} \left(\frac{\cos nkL}{nk} - \frac{1}{nk} \right) \sin nkx.$$

EXAMPLE 12.3

Expand in Fourier series the periodic function f(t) of period T and angular frequency $\omega_n = 2\pi n/T$. A sketch of the function is shown below and is described for one period by

$$f(t) = 0; \quad \text{for} -T/2 \geq t \geq 0$$
$$f(t) = E_o; \quad \text{for } 0 \leq t \leq T/2$$

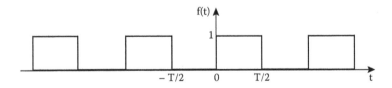

Solution

From Equation 12.9,

$$f(t) = \frac{a_o}{2} + \sum_{n=1}^{\infty} a_n \cos \omega_n t + \sum_{n=1}^{\infty} b_n \sin \omega_n t; \quad n = 1, 2, 3,...$$

where

$$a_o = \frac{2}{T}\int_{-T/2}^{T/2} f(t)\,dt = \frac{2}{T}\left[\int_{-T/2}^{0} f(t)\,dt + \int_{0}^{T/2} E_o\,dt\right] = E_o \frac{2}{T}[T/2] = E_o$$

and

$$a_n = \frac{2}{T}\int_{-T/2}^{T/2} f(t)\cos\omega_n t\,dt = \frac{2}{T}\left[\int_{-T/2}^{0} f(t)\cos\omega_n t\,dt + \int_{0}^{T/2} f(t)\ \cos\omega_n t\,dt\right]$$

$$= \frac{2}{T}\left[0 + \int_{0}^{T/2} f(t)\cos\omega_n t\,dt\right] = \frac{2}{T}\left[0 + \int_{0}^{T/2} E_o \cos\omega_n t\,dt\right] = \frac{2E_o}{T}\left(\frac{1}{\omega_n}\sin\omega_n t\right)_0^{T/2}$$

$$= \frac{2E_o}{T}\left(\frac{1}{\omega_n}\sin\omega_n T/2\right) = \frac{2E_o}{T}\left(\frac{1}{\omega_n}\sin\pi\right) = 0 \quad \text{for all } n = 1,\ 2,\ 3,\dots$$

$$b_n = \frac{2}{T}\int_{-T/2}^{T/2} f(t)\sin\omega_n t\,dt$$

$$b_n = \frac{2}{T}\int_{-T/2}^{T/2} f(t)\sin\omega_n t\,dt = \frac{2}{T}\left[\int_{-T/2}^{0} f(t)\sin\omega_n t\,dt + \int_{0}^{T/2} f(t)\sin\omega_n t\,dt\right]$$

$$= \frac{2}{T}\left[0 + \int_{0}^{T/2} E_o\ \sin\omega_n t\,dt\right] = \frac{2E_o}{T}\left(-\frac{1}{\omega_n}\cos\omega_n t\right)_0^{T/2}$$

$$= -\frac{2E_o}{T}\left[\frac{1}{\omega_n}\{\cos(\omega_n T/2)-1\}\right] = \frac{2E_o}{2\pi}(1-\cos n\pi)$$

$$= \frac{2E_o}{2\pi}(1-\cos n\pi) = \begin{cases} 0 & \text{for even } n \\ \dfrac{2E_o}{2\pi} & \text{for odd } n \end{cases}$$

As $a_n = 0$, the expansion becomes

$$f(t) = \frac{a_o}{2} + 0 + \sum_{n=1}^{\infty} b_n \sin\omega_n t;\quad n = 1,\ 2,\ 3,\dots$$

$$= \frac{E_o}{2} - \frac{E_o}{\pi}\sum_{\text{odd integer}}^{\infty}(\sin\omega_n t) = \frac{E_o}{2} - \frac{E_o}{\pi}(\sin\omega t + \sin 3\,\omega t + \sin 5\,\omega t + \dots).$$

12.3 IMPORTANT INTEGRALS

The following are some of the integrals, which are encountered in working with Fourier series:

$$\int\limits_{-\pi}^{\pi} \cos mx\, dx = \begin{cases} 0 & \text{for all } m \neq 0 \\ 2\pi & \text{for } m = 0 \end{cases} \; ; \quad m \text{ and } n \text{ are integers}$$

$$\frac{1}{2\pi} \int\limits_{-\pi}^{\pi} \sin mx \cos nx\, dx = 0$$

$$\frac{1}{2\pi} \int\limits_{-\pi}^{\pi} \sin mx \sin nx\, dx = 0; \quad \begin{cases} m \neq n \\ m = n = 0 \end{cases}$$

$$\frac{1}{2\pi} \int\limits_{-\pi}^{\pi} \sin mx \sin nx\, dx = \frac{1}{2} \quad \text{for } m = n \neq 0$$

$$\frac{1}{2\pi} \int\limits_{-\pi}^{\pi} \cos mx \cos nx\, dx = 0; \quad \{ m \neq n$$

$$\frac{1}{2\pi} \int\limits_{-\pi}^{\pi} \cos mx \cos nx\, dx = \begin{cases} \dfrac{1}{2} & m = n \neq 0 \\ 1 & \text{for } m = n = 0 \end{cases}$$

12.4 COMPLEX FORM OF FOURIER SERIES

From Euler's formula, we have

$$e^{i\theta} = \cos\theta + i\sin\theta, \quad e^{-i\theta} = \cos\theta - i\sin\theta.$$

Expressing the above in x in place of θ, the cosines and sines take the forms

$$\cos x = \frac{e^{ix} + e^{-ix}}{2}, \quad \sin x = \frac{e^{ix} - e^{-ix}}{2i}. \tag{12.13}$$

Thus, any expansion in sines and cosines can be expressed in complex form. The expansion of a periodic function $f(x)$ in exponential form may be written as

$$f(x) = \sum_{n=-\infty}^{\infty} c_n e^{inkx}, \tag{12.14}$$

where for a function of spatial period of 2λ, c_n is given by

$$c_n = \frac{1}{2\lambda} \int_{-\lambda}^{\lambda} f(x) e^{-inkx} dx. \tag{12.15}$$

If the function period is described in an angular interval of 2π, then

$$c_n = \frac{1}{2\pi} \int_{-\pi}^{\pi} f(x) e^{-inkx} dx. \tag{12.16}$$

For a periodic function $f(t)$ with a period T, and angular frequency $\omega_n = 2\pi n/T$, the expansion becomes

$$f(t) = \sum_{n=-\infty}^{\infty} c_n e^{i\omega_n t}, \tag{12.17}$$

where the coefficients c_n, evaluated for all n values between $-\infty$, ∞, are

$$c_n = \frac{1}{2\pi} \int_{-T/2}^{T/2} f(t) e^{-i\omega_n t} dt. \tag{12.18}$$

For a light wave, Equations 12.6 through 12.8 expressed in a complex form can be rewritten as

$$f(x) = \sum_{n=-\infty}^{\infty} c_n e^{ik_n x} \tag{12.19}$$

where

$$c_n = \frac{1}{\lambda} \int_{-\lambda/2}^{\lambda/2} f(x) e^{-ik_n x} dt. \tag{12.20}$$

EXAMPLE 12.4

Using the exponential form of Fourier series, expand in Fourier series
the function

$$f(t) = 0; \quad \text{for} -T/2 \geq t \geq 0$$
$$f(t) = E_o; \quad \text{for } 0 \leq t \leq T/2$$

that was treated in the previous example.
    for all n = 1, 2, 3,....

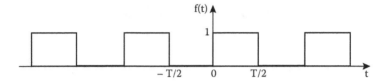

Solution

From (12.17) and (12.18), the expansion is $f(t) = \sum_{n=-\infty}^{\infty} c_n e^{i\omega_n t}$, and
from (12.19) and (12.20),

$$c_n = \frac{1}{2\pi} \int_{-T/2}^{T/2} f(t) e^{-i\omega_n t} \, dt.$$

Thus,

$$c_o = \frac{1}{2\pi} \int_0^{T/2} E_o dt = E_o \left(T/4\pi \right)$$

$$c_n = \frac{1}{2\pi} \left[\int_{-T/2}^{0} f(t) e^{-i\omega_n t} dt + \int_0^{T/2} f(t) e^{-i\omega_n t} dt \right] = \frac{1}{2\pi} \left[0 + \int_0^{T/2} f(t) e^{-i\omega_n t} dt \right]$$

$$= \frac{E_o}{2\pi} \left[\frac{e^{-i\omega_n t}}{-i\omega_n} \right]_0^{T/2} = -\frac{E_o}{2\pi} T \left[\frac{e^{-in\pi} - 1}{i2\pi n} \right]$$

$$= -\frac{E_o}{2\pi} T \left[\frac{e^{-in\pi} - 1}{i2\pi} \right] = \begin{cases} 0 & \text{for even n} \\ \dfrac{E_o}{4\pi^2 ni} T & \text{for odd n} \end{cases}$$

The expansion then becomes

$$f(t) = \sum_{n=-\infty}^{\infty} c_n e^{i\omega_n t}$$

$$= E_o(T/4\pi) + \frac{E_o}{4\pi^2 i} T\left(\frac{e^{i2\pi t/T}}{1} + \frac{e^{-i2\pi t/T}}{-1}\right) + \left(\frac{e^{i6\pi t/T}}{3} + \frac{e^{-i6\pi t/T}}{-3}\right) + \cdots$$

12.5 FOURIER TRANSFORM

Fourier transform is among several transforms that prove to be useful in addressing physical applications in physics. Fourier transforms find their way into numerous wave phenomena like sound; mechanical pulses like forces in collision events; AC electrical signals that may occur in a single pulse, whereas voltage is represented by a sinusoidal function; and a flash of light, whereas the electric and magnetic fields are sinusoidal. As Fourier series work well for periodic functions, Fourier transform is a tool that may represent a function that is not periodic over a wide range. For example, an electric signal in the form of a pulse or a light wave that has been chopped in most part but still defined in a small period can be represented by a Fourier transform. With no need for introducing a formal derivation of Fourier transforms, a function $f(x)$ that is single valued in any finite interval, and if $\int_{-\infty}^{\infty} |f(x)| dx$ exists, then the transforms are expressed in integrals that represent a continuous range of frequencies as follows:

$$f(x) = \int_{-\infty}^{\infty} \Phi(k) e^{ikx} dk \tag{12.21}$$

and $\Phi(k)$ is

$$\Phi(k) = \frac{1}{2\pi} \int_{-\infty}^{\infty} dx\, f(x) e^{-ikx} \tag{12.22}$$

As noted in Equations 12.1, 12.5, and 12.9, the sum was over n, while in the above transforms, this becomes an integral over k. The same conditions required of $f(x)$ are also required of $\Phi(k)$. The above forms are known as

Fourier integrals or Fourier transforms of one another. In (12.22), $\Phi(k)$ is the Fourier transform of f(x).

Equations 12.9 through 12.12 are in the position-wave vector domain. The parallel to Equations 12.9 through 12.12 in the frequency–time domain is the following pair:

$$f(t) = \int_{-\infty}^{\infty} d\omega \; g(\omega) e^{i\omega t} \tag{12.23}$$

$$g(\omega) = \frac{1}{2\pi} \int_{-\infty}^{\infty} dt \; f(t) e^{-i\omega t} \tag{12.24}$$

EXAMPLE 12.5

Given a step function of the form

$$f(x) = 1; \quad \text{for } 0 \le x \le \ell$$
$$f(x) = 0; \quad \text{for } -\ell \ge x \ge 0,$$

determine its Fourier transform.

Solution

From Equation 12.22,

$$\Phi(k) = \frac{1}{2\pi} \int_{-\infty}^{\infty} dx \, f(x) e^{-ikx}.$$

Thus,

$$\Phi(k) = \frac{1}{2\pi} \int_{-\infty}^{\infty} dx \, e^{-ikx} = \frac{1}{2\pi} \left[\int_{-\ell}^{0} 0 + \int_{0}^{\ell} dx e^{-ikx} \right]$$

$$= \frac{1}{2\pi} \frac{1}{-ik} \left[e^{-ik\ell} - 1 \right] = \frac{1}{2\pi} \frac{e^{-ik\ell/2}}{-ik} \left[e^{-ik\ell/2} - e^{ik\ell/2} \right]$$

$$= \frac{e^{-ik\ell/2}}{2\pi} \frac{1}{-k} \left[-2 \sin k\ell/2 \right]$$

$$= \frac{\ell e^{-ik\ell/2}}{2\pi} \frac{\sin(k\ell/2)}{k\ell/2}$$

$$= E_0 \frac{\sin(k\ell/2)}{k\ell/2} = E_0 \frac{\sin \beta}{\beta},$$

where

$$E_o = \frac{\ell\, e^{-ik\ell/2}}{2\pi}, \quad \beta = k\ell/2.$$

EXAMPLE 12.6

Redo the previous example for the periodic function for the step function

$$f(x) = 1; \quad \text{for } -\ell \le x \le \ell$$
$$f(x) = 0; \quad \text{for } -\ell \ge x \ge \ell$$

to determine its Fourier transform.

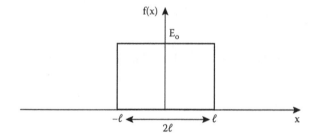

Solution

From Equation 12.22,

$$\Phi(k) = \frac{1}{2\pi}\int_{-\infty}^{\infty} dx\, f(x)\, e^{-ikx}.$$

Thus,

$$\Phi(k) = \frac{1}{2\pi}\int_{-\infty}^{\infty} dx\, e^{-ikx} = \frac{1}{2\pi}\left[\int_{-\ell}^{\ell} dx\, e^{-ikx}\right] = \frac{1}{2\pi}\frac{1}{-ik}\left[e^{-ik\ell} - e^{ik\ell}\right]$$

$$= \frac{1}{\pi k}\left[2\sin k\ell\right]$$

$$= E_o\,\frac{\sin \beta}{\beta}$$

where

$$E_0 = \frac{2\ell}{\pi}, \quad \beta = k\ell.$$

12.6 RELEVANCE OF FOURIER TRANSFORM TO DIFFRACTION

A Fraunhofer diffraction pattern of E-field corresponds exactly to the Fourier transform of the aperture function, as was noted in Section 8.3, where diffraction from a single rectangular aperture was discussed. The discussion led to Equation 8.41 that showed that amplitude of the diffraction on the screen, represented by $E_p(k_y, k_z)$, is the Fourier transform of the aperture function $A(y, z)$. Below are the two transforms:

$$E_p\left(k_y, k_z\right) = C' e^{ikr_0} \int\limits_{-\infty}^{\infty} \int\limits_{-\infty}^{\infty} A\left(y, z\right) e^{i\left(k_y y + k_z z\right)} \, dydz \qquad (12.25)$$

and

$$A\left(y, z\right) = C' e^{ikr_0} \int\limits_{-\infty}^{\infty} \int\limits_{-\infty}^{\infty} E_p\left(k_y, k_z\right) e^{-i\left(k_y y + k_z z\right)} dk_y dk_z. \qquad (12.26)$$

Of course, for a function, actually a pulse, that is expressed in terms of one variable, the one-dimensional Fourier transform applies as given in Equation 8.36. For example, for a pulse, $f(y)$, the Fourier transform is

$$E_p\left(k_y\right) = C \int\limits_{-a/2}^{a/2} A\left(y\right) e^{i\left(k_y y\right)} dy. \qquad (12.27)$$

The comparison between Equations 12.22 and 12.27 suggests that $C = 1/2\pi$. Thus, Equation 12.28 becomes

$$E_p\left(k_y\right) = \frac{1}{2\pi} \int\limits_{-a/2}^{a/2} A\left(y\right) e^{-i\left(k_y y\right)} dy. \qquad (12.28)$$

EXAMPLE 12.7

Find the Fourier transform of

$$f(x) = \begin{cases} \sin k_o x & \text{at } -\pi/2 < k_o x < \pi/2 \\ 0 & \text{at } -\pi/2 > k_o x > \pi/2. \end{cases}$$

Solution

From Equation 12.22, the Fourier transform of f(t) is

$$\Phi(k) = \frac{1}{2\pi} \int_{-\infty}^{\infty} dx\, f(x) e^{-ikx}$$

$$= \frac{1}{2\pi} \int_{-\pi/2k_o}^{\pi/2k_o} dx \sin k_o x\, e^{-ikx} = \frac{1}{2\pi k_o} \int_{-\pi/2}^{\pi/2} d(k_o x) \frac{e^{ik_o x} - e^{-ik_o x}}{2i} e^{-ikx}$$

$$= \frac{1}{2\pi k_o} \int_{-\pi/2}^{\pi/2} d(k_o x) \frac{e^{ix(k_o - k)} - e^{-ix(k_o + k)}}{2i} = \frac{1}{4\pi i}\left[\frac{e^{ix(k_o - k)}}{i(k_o - k)} + \frac{e^{-ix(k_o + k)}}{i(k_o + k)} \right]_{-\pi/2}^{\pi/2}$$

$$= \frac{1}{4\pi k_o i}\left\{ \frac{e^{i(1-k)\pi/2}}{i(k_o - k)} + \frac{e^{-i(1+k)\pi/2}}{i(k_o + k)} - \frac{e^{-i(1-k)\pi/2}}{i(k_o - k)} - \frac{e^{i(1+k)\pi/2}}{i(k_o + k)} \right\}$$

$$= \frac{1}{4\pi k_o i}\left\{ \frac{e^{-ik\pi/2}}{(k_o - k)} - \frac{e^{-ik\pi/2}}{(k_o + k)} + \frac{e^{ik\pi/2}}{(k_o - k)} - \frac{e^{ik\pi/2}}{(k_o + k)} \right\}$$

$$= \frac{1}{4\pi k_o i}\left\{ \frac{2\cos k\pi/2}{(k_o - k)} - \frac{2\cos k\pi/2}{(k_o + k)} \right\}$$

$$= \frac{\cos k\pi/2}{2\pi k_o i}\left\{ \frac{1}{(k_o - k)} - \frac{1}{(k_o + k)} \right\}$$

$$= \frac{1}{2\pi k_o i}\left(\frac{2k\cos(k\pi/2)}{(k_o^2 - k^2)} \right)$$

$$= \frac{1}{\pi k_o i}\left\{ \frac{\cos k\pi/2}{(k_o^2 - k^2)} \right\}$$

$$= \frac{e^{i\pi/2}}{\pi k_o}\left\{ \frac{\cos k\pi/2}{(k^2 - k_o^2)} \right\}.$$

PROBLEMS

12.1   For the periodic function expressed in one period

$$f(x) = \begin{cases} 1 & \text{at } 0 < x < \pi \\ 0 & \text{at } -\pi < x < 0 \end{cases}$$

find the Fourier series expansion in the interval $-\pi < x < \pi$.

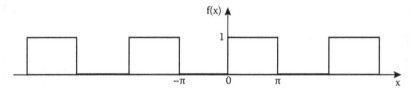

12.2   For the periodic function expressed in one period

$$f(x) = \begin{cases} -\pi/2 & \text{at } -\pi < x < 0 \\ \pi/2 & \text{at } 0 < x < \pi \end{cases}$$

find the sine–cosine Fourier series expansion in the interval $-\pi < x < \pi$.

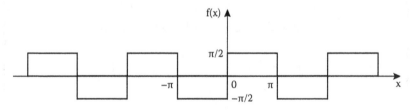

12.3   For the periodic function, sketched below and expressed in one period

$$f(x) = \begin{cases} 0 & \text{at } 0 < x < \pi \\ x & \text{at } -\pi < x < 0 \end{cases}$$

find the sine–cosine Fourier series expansion in the interval $-\pi < x < \pi$.

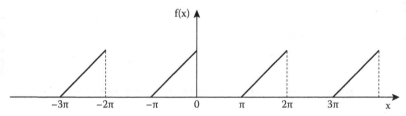

12.4   Take advantage of the solution of the previous problem in finding the sine–cosine Fourier series expansion for the periodic function expressed in one period in the interval $-\pi < x < \pi$ as described below:

$$f(x) = \begin{cases} 0 & \text{at } 0 < x < \pi \\ -x & \text{at } -\pi < x < 0 \end{cases}$$

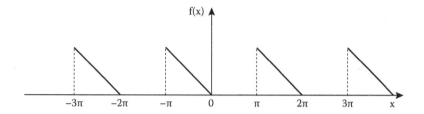

12.5   Take advantage of the solution of the two previous problems in finding the sine–cosine Fourier series expansion for the periodic function expressed in one period in the interval $-\pi < x < \pi$ as described below:

$$f(x) = \begin{cases} -x & \text{at } 0 < x < \pi \\ x & \text{at } -\pi < x < 0 \end{cases}$$

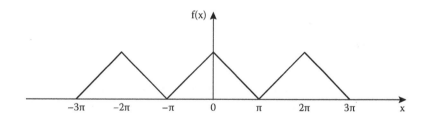

12.6   Expand in Fourier series the periodic function $f(x)$ described for one period by

$$E(x) = E_o; \quad \text{for } -\pi \geq x \geq 0$$
$$E(x) = 0; \quad \text{for } 0 \leq x \leq \pi.$$

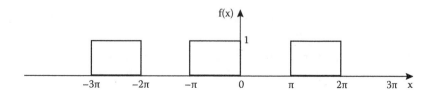

12.7 Redo the previous problem using the exponential form of Fourier series

$$E(x) = E_o; \quad \text{for} -\pi \geq x \geq 0$$
$$E(x) = 0; \quad \text{for } 0 \leq x \leq \pi.$$

12.8 For the periodic function expressing a sawtooth wave

$$f(x) = x; \quad -L < x < L,$$

find the Fourier series expansion.

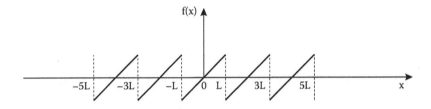

12.9 Redo the previous problem for the function f(x) = x defined in the period $-\pi < x < \pi$.

12.10 For the periodic function,

$$f(x) = \begin{cases} 1 & \text{at} -\pi < x < 0 \\ 0 & \text{at } 0 < x < \pi \end{cases}$$

use the complex form of Fourier series to find the expansion in the interval $-\pi < x < \pi$.

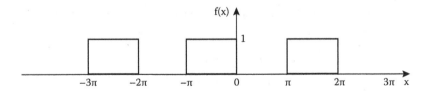

12.11 Using the exponential form of Fourier series, find the series expansion of f(x) = sin x.

12.12 Consider a periodic rectangular function f(x) of a spatial periodicity that represents the wavelength λ of a wave, the central cycle of which is a constant value A in $-\lambda/4 < x < \lambda/4$. The periodic function can be described by

$$f(x) = A, \quad \text{for}\,(s - \lambda/4) \le x \le (s + \lambda/4),$$

where s is a positive or negative integer; and zero for all other value of x. Find the Fourier sine–cosine series using the exponential form.

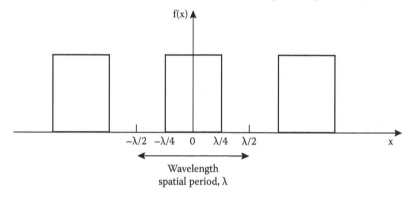

12.13 Consider the periodic function

$$f(x) = E_o; \quad \text{for}\ 0 \le x \le \ell$$
$$f(x) = 0; \quad \text{for} -\ell \le x \le 0,$$

and compute its Fourier transform.

12.14 Given a wave that is confined to two cycles only, and blocked otherwise, such that one can describe it by the following

$$f(t) = A e^{i\omega_o t} \quad \text{for} -T < t < T, \quad \text{and} \quad f(t) = 0 \text{ otherwise,}$$

find its Fourier transform.

12.15 Find the Fourier transform of

$$f(x) = \begin{cases} \cos k_o x & \text{at} -\pi/2 < k_o x < \pi/2 \\ 0 & \text{at} -\pi/2 > k_o x > \pi/2 \end{cases}$$

Photonics

The only thing I know is that I know nothing.

Socrates (470–399 BC)

13.1 INTRODUCTION

Photonics is the science of light photon generation, detection, and manipulation for any purpose. The term photonics derives from the word photon, which, in turn, derives from the Greek word meaning light. According to *Encyclopedia Britannica*, the word photon was first proposed by Gilbert N. Lewis (1875–1946; Nobel Prize, Chemistry, 1946) in 1926. He proposed it in a letter to the publication *Nature* as a new kind of atom that "carries" radiation. It should be pointed out that Gilbert was an accomplished physical chemist who helped develop the theory of covalent bonds.

Photonics, as a formal discipline, essentially, came into being in the early 1960s with the invention of light amplification by the stimulated emission of radiation (laser). It matured rapidly and by the 1980s was a staple in the development of all forms of technology involving light in the visible, infrared, and ultraviolet wavelength ranges. Major drivers of the development were military as well as medical and commercial applications ranging from basic communications to sophisticated computational systems.

In fact, the concepts behind modern photonics actually started in the nineteenth century as physicists began to explore the behavior of the radiation properties of matter in the form of electrically excited gases and heated solids. In particular, scientists began attempting to apply classical concepts of statistical and macroscopic mechanics of oscillating systems (simple harmonic motion) put forth by Isaac Newton (1642–1727) and Robert Hooke (1635–1703) to the unresolved observations of the time. Michael Faraday (1791–1867), James Clerk Maxwell (1831–1879), and Ludwig Boltzmann (1844–1906) further developed and applied these classical concepts to a remarkable degree and were able to make profound contributions to the description and understanding of the radiation properties of matter by the turn of the nineteenth century. There were at least two major successes: Boltzmann's kinetic theory of gases and the Maxwell electromagnetic wave theory of radiation. However, there were some glaring failures that ultimately led to the quantum theory of matter and motion. Four major observations that seemed to defy efforts at resolution through the application of classical theory of waves and matter: black-body radiation and absorption, the discrete spectrum of excited hydrogen and other gases, discrete absorption of kinetic energy associated with the flow of electrical current in heated mercury vapor, and the photoelectric effect. These will be considered in more detail in the following sections. Max Planck (1858–1947; Nobel Prize, Physics, 1918) pondered the idea that a system undergoing simple harmonic motion can only have energies that are integral multiples of some basic energy quantity associated with the system. In 1905 Albert Einstein (1879–1955; Nobel Prize, Physics, 1921) was able to explain the photoelectric effect using the concept of the photon representing light as a packet of energy. This came about after Max Planck and Niels Bohr (1885–1962; Nobel Prize, Physics, 1922) were able to explain black-body radiation and the spectrum of excited hydrogen gas by introducing the concept of energy quantization. The spectrum of hydrogen along with the black-body radiation phenomenon had been a mystery to classical physicists until the theory of quantum physics was born. Albert Einstein also applied the quanta concept of light to explain the photoelectric effect first observed by Heinrich Hertz in the late nineteenth century (1887) when he noticed that ultraviolet light incident upon polished metal spheres of a spark gap made it easier for the spark to occur. J.J. Thomson's (1856–1940; Nobel Prize, Physics, 1906) discovery of the electron suggests that the effect was possibly due to charges ejected from the surface resulting from energy absorption from the light. Philipp Lenard (1862–1947; Nobel Prize, Physics, 1905) confirmed this in 1900 by measuring the charge-to-mass ratio of the photoelectric particles, which turned out to be exactly the same as the electron-to-mass ratio (e/m) earlier determined by Thomson in 1897.

The field of photonics spans such a broad range of optical, electrical, and mechanical phenomena and applications that several textbooks would be necessary for a complete treatment. This chapter will, therefore, be limited in scope and depth in exploring the optical, electrical, and mechanical phenomena that are pertinent to modern photonics. In addition, several technologies that are fundamental to the advancement of modern photonics will be presented. This will include LASER, fiber optics, and other solid-state technologies involving photonic radiation. In some cases, the discussion is intended to provide the reader with a general understanding of the techniques that are used to make the material properties, utilized in the technologies, work as practical devices or systems.

13.2 CLASSICAL PHYSICS AND RADIATION: THE FOUNDATION OF MODERN PHOTONICS

13.2.1 BLACK-BODY RADIATION

Thermal radiation occurs when a massive object's temperature is above absolute zero. In the classical theory, this can be imagined as the result of the random motion of point charges near the surface of the object due to thermal agitation. Maxwell's theory shows that accelerated charges emit electromagnetic (EM) radiation perpendicular to the plane of the motion. The magnitude of the radiated field is directly dependent upon the acceleration and inversely dependent upon the distance between the field point and the charge. There is also a static field contribution that falls off as the inverse square of the same distance. Due to the number and random nature of near-surface agitations, it is expected that the radiation is composed of a broad spectrum of possible wavelengths. Furthermore, it would be expected that the rate of energy emission when integrated over the full spectrum will depend upon the surface temperature as well as the surface area. In 1879, Josef Stefan (1835–1893) developed an empirical equation that states

$$P_R = e\sigma T^4,$$

where P_R is the total energy emitted (or absorbed) averaged over all spectral frequencies (or wavelengths) per second per square meter from a surface at absolute temperature T (K). It has dimensions of W/m^2, e is a dimensionless constant called the "emissivity" and ranges between 0 and 1 depending upon the nature of the radiating surface, and σ is the Stefan–Boltzmann constant (5.6705×10^{-8} $W/m^2\text{-}K^4$). The energy emitted is supplied by the thermal agitation of the surface absorption and ends up as the thermal kinetic energy of the atoms. In 1884, Ludwig Boltzmann developed the theoretical foundation for this equation using the classical theory of thermodynamics. His model consisted of a cylindrical cavity with perfectly reflecting walls and a movable piston. The cavity was filled with thermal radiation at temperature T. He basically treated the radiation as a gas and applied the ideal gas law theory of the Carnot cycle from which he obtained a relationship between the work done by the pressure of the radiation and its temperature, T. The radiation pressure is just the average force per unit area exerted by the radiation, on the walls and piston, known from Maxwell's theory to be given by the intensity/speed of light. Boltzmann also laid the foundation for modern statistical mechanics with his probability treatment of entropy and molecular velocity distribution in developing the kinetic theory of ideal gases. Kirchhoff, in 1895, using classical thermodynamic theory, proved that the adsorption efficiency, a, and the emissivity of a radiator are the same and introduced the simple theorem, e = a. Therefore, e = 1 represents a perfect emitter or absorber and is called a "black body." In other words, a black body, which is the most efficient absorber, is also the most efficient radiator. Of course, the equation provides no information about the observed spectral distribution of the energy. The Stefan–Boltzmann law strongly suggests that the spectral distribution of the energy radiated by a body is also dependent upon the temperature. Figure 13.1 shows typical shapes of the black-body spectral energy distribution curve.

The first accurate measurements of the spectral distribution of the radiation, $I(\lambda)$, are associated with Paschen, Lummer, and Pringsheim in the 1890s using an instrument similar to a prism spectrometer but involved a lens and prisms fabricated from materials transparent to long-wavelength thermal (infrared) radiation. It is clear that the wavelengths at the peak of each curve (λ_p) have an inverse relationship to the temperature, that is, as the temperature

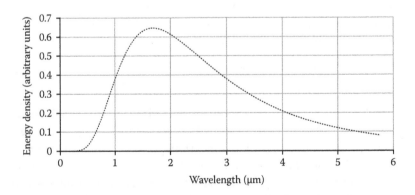

FIGURE 13.1 Typical black-body spectral distribution curve. The temperature will determine the dominant color, wavelength, or frequency of the radiation. Today, lighting is often characterized by the black-body temperature.

increases, the wavelength decreases. This was first explained quantitatively by Wien and will be considered later. Of course, it was common knowledge from "everyday" experience that objects emit more heat as their temperature increases and the "color" changes from dull red to "white hot" as the temperature goes up. Any blacksmith of the day was keenly aware of this.

Developing the theoretical foundation was a major project for physicists throughout the latter part of the nineteenth and early twentieth centuries. The approach taken was to consider the walls of an enclosed cavity with a small hole so that all incident radiation on the hole is absorbed into the cavity, so the hole acts like an ideal absorber or black body with an emissivity of 1. For the complimentary situation, any radiation in the cavity due to the thermal agitation of its walls would also be absorbed by the hole, and the hole would behave like a black-body radiator to the outside. Therefore, the cavity is a black-body radiator, and the spectral distribution is determined only by the absolute temperature of the interior walls of the cavity. Since the hole is small compared to the surface area of the cavity, very little of the energy was reflected back through the hole so the emissivity, e, of the hole was very nearly 1 and the hole would mimic a black-body radiator, producing the same spectral distribution. Although Boltzmann was effective in using a similar model to derive Stefan's empirical equation, he did not provide the spectral distribution of the energy. Building upon the work of Boltzmann, Wilhelm Wien (1864–1928) in 1893 used the fact that expansion and compression of the "Boltzmann" cavity filled with radiation imparted a Doppler shift to the frequency (wavelength) of the radiation due to the motion of the piston. Doppler first proposed the concept of motion-induced frequency shift for waves in 1846. From this Wien was able to derive a general "functional form" for the radiation energy density, $\Gamma_T(\lambda)$, at temperature T that gave a partial description of how the energy is spectrally distributed as shown in Equation 13.1. He used the data collected by Otto Lummer (1860–1925) and Ernst Pringsheim (1859–1917) to establish this relationship. All the data that was collected falls on this curve, and the product of the peak wavelength and the temperature is a constant regardless of the black-body temperature as expressed in Equation 13.2 (Wien's displacement law):

$$\Gamma_T(\lambda) = \Omega(\lambda T) / \lambda^5 \tag{13.1}$$

$$\lambda_{max} T = \text{constant} = 2.898 \times 10^{-3} \text{ m-K.} \tag{13.2}$$

Wien's efforts established that the energy density is dependent on the wavelength and temperature and that it decreased for both short and long

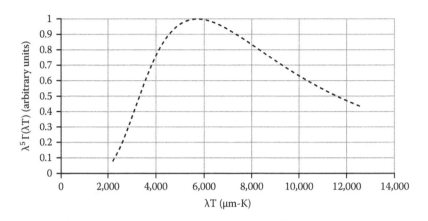

FIGURE 13.2   Wien's displacement law is based upon data by Lummer and Pringsheim. The data was taken for three black-body temperatures: 1646, 1449, and 1259 K; the points all fell close to the theoretical curve for short wavelengths. All temperatures produced a near-perfect match at the peak.

wavelengths, as shown in Figure 13.2. Although the model works well for short wavelengths it fails, quantitatively, for the longer wavelengths.

Lord Rayleigh (John W. Strutt) (1842–1919) and Sir James H. Jeans (1877–1946) considered a cavity at temperature T, with perfectly reflecting walls, filled with radiation. The EM waves result from the oscillations of the charges at the surface of cavity wall. Due to the physical restrictions of the cavity, EM standing waves are set up therein. Because each wall must be a nodal point for the standing waves, the path-length between walls must be an integral number, n, of half wavelengths for any specific frequency. For any n, there will be a set number of nodes and antinodes. Using this concept, Rayleigh and Jeans in 1905 developed an expression for the standing wave density in terms of the radiation frequency (f) given in Equation 13.3:

$$F(f)df = 8\pi c^{-3}f^2 df. \tag{13.3}$$

The energy per unit volume in the cavity is the average energy per standing wave multiplied by the number of standing waves per unit volume. Classically, the equipartition of energy theorem given by Boltzmann requires that the average kinetic energy per degree of freedom associated with any entity in thermal equilibrium with a collection of other such entities is $k_B T/2$. In the case of associating the radiation in the cavity with the linear vibration of a classical spring harmonic oscillator, in the cavity wall, with only one degree of freedom, we must remind ourselves that the average total energy of vibration of the spring is twice the average kinetic energy. This arises because the spring has both kinetic energy and potential energy, whereas in the ideal gas model, there is no potential energy; therefore, in associating the total energy to the standing waves, the average energy of each standing wave will be k_BT. Therefore, we write the energy density as the product of the standing wave density and the average total energy, k_BT:

$$\Gamma_T(f)df = 8\pi k_B T c^{-3} f^2 df. \tag{13.4}$$

Since the left-hand side of Equation 13.4 is energy density, it is a positive quantity; therefore, if one writes it in terms of the wavelength, it must be written as $\Gamma_T(\lambda)d\lambda = -\Gamma_T(f)df$ to offset the minus sign on the right-hand side of the equation. Remember that an increase in f is a decrease in λ, so a small positive change in f produces a small negative change in λ, but the energy contained in the interval is unchanged. Since $f = c/\lambda$ and $df = -cd\lambda/\lambda^2$, then

$$\Gamma_T(\lambda)d\lambda = \Gamma_T(f)c\lambda^{-2}d\lambda,$$

and one may rewrite (13.4) as

$$\Gamma_T(\lambda)d\lambda = 8\pi k_B T\lambda^{-4}d\lambda \tag{13.5}$$

or

$$\Gamma_T(\lambda)d\lambda = 8\pi k_B(\lambda T)/\lambda^5 d\lambda. \tag{13.6}$$

The latter is consistent with the Wien equation (13.2) but increases monotonically to infinity as the wavelength gets shorter. This is grossly inadequate and the best that classical thinking can produce. This became known as the *Ultraviolet Catastrophe* because it is a dramatic failure of classical physics applied to black-body radiation.

The discrepancy between the classical theory and experiment was resolved in 1901 by Max Planck (1858–1947) when he introduced a postulate that started the evolution of physics into a new paradigm. It may be stated as follows:

Any physical entity whose single "coordinate" executes simple harmonic oscillations can possess only total energies, E, which satisfy the following relationship:

$$E = nhf, \quad \text{with } n = 0, 1, 2, 3\ldots, \tag{13.7}$$

where f is the frequency of oscillation and h is a universal constant.

Planck, like many physicists of the time, was frustrated over the inability of classical physics to effectively deal with black-body radiation. He knew that he had to think "out of the box" but he did not work in a vacuum. He was keenly aware of the previous theoretical and experimental work done by all of the top physicists of the time, including Boltzmann and Maxwell who independently developed the probability distribution functions for the energy of a classical system of identical particles in a state of thermal equilibrium with each other and their container. The average energy per particle is determined by the probability distribution function whose form must be specified by the temperature T. Planck considered a black-body cavity at some temperature, filled with radiation generated by the collection of harmonically agitated charges within the wall of the cavity. Like others who attacked the problem, he was aware of the existence of the standing wave patterns and that they could occur only for a discrete set of frequencies. According to Boltzmann, the probability distribution function for such a classical system is given by

$$P(E) = A \exp(-E/k_B T); \tag{13.8}$$

A and k_B are constants ($k_B = 1.3806 \times 10^{-34}$ J/K and is called the Boltzmann constant).

Using (13.8) and allowing the energy to be related directly to the frequency and not the amplitude of the standing waves, Planck made the paradigm shift that was necessary to develop the successful theory of black-body radiation. He showed that the average energy per standing wave, using Equations 13.7 and 13.8, is

$$\langle E \rangle = hf \left[\exp(hf/k_B T) - 1 \right]^{-1}, \tag{13.9}$$

where h is now called Planck's constant and is 6.626069×10^{-34} J-s.

Then evaluating the energy density as others had attempted before him was a relatively simple task, and he obtained

$$\Gamma_T(\lambda) = 8\pi hc\lambda^{-5} \left[\exp\{hc/(\lambda k_B T)\} - 1 \right]^{-1}. \tag{13.10}$$

Figure 13.3 shows a plot of the Planck function as well as the Rayleigh–Jeans. The Planck function fits the experimental data perfectly, whereas the term "Ultraviolet Catastrophe" is seen to be highly accurate for the Rayleigh–Jeans function.

The Planck theory has been experimentally verified many times and is now considered a fundamental law of nature.

Of course, the Planck theory should produce the same results as Wien's law and the Rayleigh–Jeans law in the appropriate wavelength limits. To show this is left as an exercise for the student in the end of chapter problems.

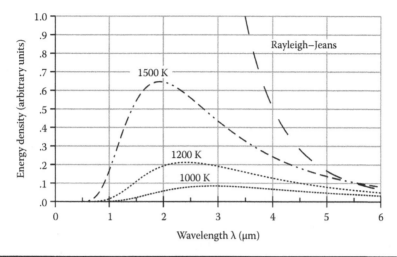

FIGURE 13.3 Black-body radiation energy density as a function of wavelength for the Planck and Rayleigh–Jeans functions. This makes very clear the failure of the classical theory in describing black-body radiation.

EXAMPLE 13.1

Assume a 100 W incandescent light bulb is operated with a maximum filament temperature of about 2900°C. (a) Assuming the filament has an emissivity of 0.40, determine the radiant power per square meter produced by the lamp. (b) What is the approximate surface area of the tungsten filament (melting point 3422°C)? (c) Determine the peak wavelength of the radiation in micrometers.

Solution

(a) Using the equation: $P_R = e\sigma T^4$ with e = 0.40, get P_R = 0.4 (5.57 × 10^{-8} W/m²-K⁴)(3173 K)⁴ = 2.25 × 10^6 W/m².

(b) Assuming all the power is converted into radiant energy,

$$\text{Area} = \text{Power}/P_R = 1.01 \times 10^7 = 100 \text{ W}/2.25 \times 10^6 \text{ W/m}^2$$

$$= 4.42 \times 10^{-5} \text{ m}^2 = 44.2 \text{ mm}^2.$$

(c) From Wien's displacement law, λ_{max} = (2.90 × 10^{-3} m-K)/3173 K = 9.14 × 10^{-7} m = 914 nm. This is well outside the normal range of human vision.

Conundrum

The filament will glow white hot to a human being, but the peak temperature is outside the normal human vision range, so why would one see the tungsten glowing white hot? Why does the glass envelope of the bulb get warm? Is the filament area realistic? Why is the filament wire in actual incandescent bulbs formed as a long, tight solenoid-shaped coil?

13.2.2 Photoelectric Effect

Heinrich Hertz (1857–1894) experimented with EM waves by producing sparks in an air gap on the secondary of a high-voltage transformer and attempting to receive the wave via another coil with a similar air gap located some distance away as depicted in Figure 13.4.

Hertz observed that in the presence of light shone onto the spherical conductors of the receiver loop's spark gap, the spark was more intense and was easier to generate. He found that the ultraviolet component of light caused this. Although he did not understand this phenomenon, he carefully noted it and published it, along with his observations concerning the then new EM waves proposed by Maxwell. It was called the Hertz effect

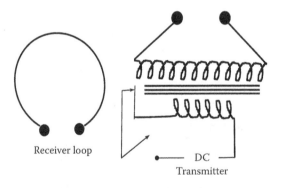

Receiver loop

DC Transmitter

FIGURE 13.4 Schematic layout of Hertz's EM wave apparatus. When direct current potential is applied, the conductor periodically switches a DC source into the primary coil of the transmitter and generates a spark across the gap. The receiver loop will reproduce a weak spark through the propagation of the EM waves if the dimension of the receiving loop is resonant with the transmitters dominant frequency. (The receiver spark gap is very small.)

for several years. Others including Aleksandr Stoletov (1839–1896), Agusto Righi (1850–1920), and Wilhelm Hallwachs (1859–1922) investigated the effect extensively. In particular, from 1888 until 1891, Stoletov performed experiments and did detailed analysis on the effect; he discovered the direct proportionality between the magnitude of the photocurrent and the intensity of the light and its dependence on gas pressure (Stoletov's law). In 1899, J.J. Thomson also studied the effect in an evacuated glass tube with a metal electrode and found that the current was caused by electrons, which he had discovered in 1897. He also reported that the current depended upon the intensity and frequency of the incident light. In 1902, Philipp Lenard did a series of experiments into the Hertz effect and also observed the fact that the magnitude of the electrical current depended upon the intensity of incident light. He further noted that the maximum kinetic energy that an electron obtained depended upon the frequency of the incident radiation, that is, it took a higher potential to stop the current generated by ultraviolet light than it did to stop the current generated by a blue light. He also found that at longer wavelengths, very little current was observed, no matter what the intensity. This energy dependence upon the frequency directly contradicted the accepted wave theory of EM radiation, which required that the energy should be dependent upon the amplitude of the incident radiation, that is, the intensity. A schematic of a typical apparatus for investigating the photoelectric effect is shown in Figure 13.5.

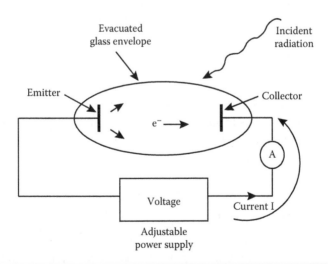

FIGURE 13.5 Photoelectric effect: schematic of an apparatus for measuring the photocurrent due to incident radiation.

If the collector is at positive potential relative to the emitter, the electron current is as shown, which is opposite to the conventional current. If the collector is negative, just enough to make the electron current zero, then the conventional current will be zero and the maximum kinetic energy of the electron is known. The potential at which this occurs is called the stopping potential V_m and the kinetic energy is given by eV_m. This potential is dependent upon the frequency of the incident radiation and independent of its intensity. This violates the classical radiation theory. Thomson's experiments showed that metals contained electrons, although it was unknown at the time as to whether or not they were bound to their atoms or free to move about the metal. Whatever the case, for an electron to be ejected, as in the photoelectric effect, required that energy be absorbed from the incident radiation, presumably through some interaction between the electric field of the EM wave and the electric charge. The oscillating field of the wave will set the electrons into oscillatory motion with an amplitude that is proportional to that of the electric field. Introductory physics shows us that the average kinetic energy of any vibrating body is proportional to the square of the amplitude of the oscillations. In addition, classical EM wave theory shows that the intensity of the EM radiation is proportional to the square of the wave amplitude, and conversely, the oscillatory amplitude of the wave is proportional to the square root of its intensity. Therefore, the average kinetic energy of the oscillating charge that absorbs the radiation energy is proportional to the square of its vibration amplitude, which is proportional to the square of the amplitude of the electric field, which is proportional to the square root of the radiation intensity. In summary, the classical theory tells one that the average kinetic energy of the ejected charge is proportional to the square root of the intensity of the radiation. Hence, higher-intensity light should give the ejected electrons higher kinetic energy. In fact, experiments indicate conclusively that the electron kinetic energy is independent of the intensity. Another difficulty with the application of classical theory is that it requires a significant amount of time for the electrons to absorb enough kinetic energy to be ejected from the metal. This is illustrated in the following example.

EXAMPLE 13.2

Consider the metal silver, which is an excellent conductor. Determine the approximate time necessary for an electron to absorb the energy, $eV_m \approx 1 \times 10^{-19}$ J (a bit less than one electron-volt [eV]) from a 1.0 W, isotropic, ultraviolet light source located approximately 1 m from the silver emitter plate of the photoelectric tube. Assume the plate has an area $A_p = 1.0$ cm^2 and is fully illuminated by the light.

Solution

First, one will need to estimate the radius of the atom. We can get this from the macroscopic density (ρ) of the element, its atomic number, A, and Avogadro's number, $N_A \approx 6.022 \times 10^{23}$ mol^{-1}. For silver $\rho \approx 9.3 \times 10^3$ kg/m^3 and A ≈ 0.108 kg/mol. The mass of the atom is $A/N_A = 0.108$ kg/$6.022 \times 10^{23} \approx 1.8 \times 10^{-25}$ kg. Assume spherical symmetry for the atom and that in the solid state, they are closely packed; the density of an atom is approximately the same as that of the macroscopic solid. Therefore, $\rho_{solid} \approx \rho_{atom}$ = mass of the atom/volume or $(A/N_A)/(4\pi r^3/3)$. From this, one obtains $r^3 = 3(A/N_A)/4\pi = 0.75(1.8 \times 10^{-25})/(3.14 \times 9.3 \times 10^3) = 4.6 \times 10^{-30}$ m$^3 \rightarrow r \approx 1.7 \times 10^{-10}$ m. Now we can calculate the time required for the photoelectrons to absorb the necessary energy. This gives an approximate cross-sectional area for an atom of silver as $A_{Ag} = \pi r^2 = 3.141(1.7 \times 10^{-10}$ m$)^2 = 8.8 \times 10^{-20}$ m^2.

For a 1.0 W isotropic source, the energy flux at a distance R from the source is given by

$$\phi_E = (\text{Energy/time})/(\text{spherical surface area centered around the source})$$

$$= (\text{Radiated power})/(4\pi R^2)$$

$$= 1.0\,W/\left(12.6(1.0\,m)^2\right) \approx 7.9\times 10^{-2}\,W/m^2 \approx 8\times 10^{-2}\,J/m^2\text{-s}.$$

The radiated power incident on the silver photo plate with area A_p is given by

$$P_{Ag} = \phi_E \times A_p = 8\times 10^{-2}\,W \times 0.0001\,m^2 = 8\times 10^{-6}\,W.$$

Given that the cross-sectional area of the atom is some fraction of the photo plate area, the power incident upon each atom is $P_{Ag} \times A_{ag}/A_p$. Therefore,

$$A_{Ag} \approx 8.8\times 10^{-20}\,m^2,$$

and the rate at which the energy that falls on the atom is

$$P_{ag} = \left(8\times 10^{-6}\,W\right)\left(8.8\times 10^{-20}/1.0\times 10^{-4}\right) = 7.0\times 10^{-21}\,J/s.$$

If one assumes that the photoelectrons ejected from the surface were bound to the atoms and somehow absorbed all of the energy, then the time, t, required to absorb 1×10^{-19} J is given by

$$t = eV_m/P_{Ag} = 1\times 10^{-19}\,J/7.0\times 10^{-21}\,J/s \approx 14\,s.$$

Lenard saw no such time delay between the onset of illumination and the emission of charges from the photo plate. His observations indicated that the charges are emitted instantly upon illumination of the plate. Subsequently, measurements made by Lawrence and Beams in 1928 using a much weaker source (orders of magnitude weaker) set the upper time limit at about a nanosecond or 10^{-9} s. It became clear that no tenable description of the photoelectric effect could be obtained using classical wave theory.

In 1905, Albert Einstein proposed a quantum theory of the photoelectric effect based upon the precepts put forth by Planck in successfully describing black-body radiation, that is, the radiation energy is proportional to the frequency, f, of the wave. He reasoned that Planck's hypothesis concerning radiant energy from a source being an integer multiple of hf implied that in the process of transitioning from one state, nhf, to the next, $(n + 1)$hf, the source would emit a burst of EM energy of magnitude hf. Einstein assumed that such a burst of emitted energy is localized in a small volume of space and remains so as it moves away from the source unlike classical waves, which spread out as they move away from the source. This "burst" is called a "quantum" of energy, and he assumed that in the photoelectric effect process, one quantum was completely absorbed by an electron in the photo emitter plate (photocathode). If this energy is enough to overcome the work necessary to get the electron to the surface of the photocathode, any excess will be seen as kinetic energy of the ejected electron. This is expressed as

$$T_m = hf - W, \tag{13.11}$$

where

T_m is the maximum kinetic energy of the electron after it escapes the surface of the photocathode

W is the work necessary to get the electron to the surface of the photocathode and depends upon its composition

W is sometimes called the "work function" of the material that composes the photocathode. It is also sometimes symbolized by Φ. f is the frequency of the radiation incident upon the photocathode. Einstein's treatment predicts that the maximum kinetic energy is a linear function of the frequency of the incident radiation. R.A. Millikan (1868–1953; Nobel Prize, Physics, 1923) verified this prediction in 1916. His experiments covered a frequency range of 6×10^{14} to 12×10^{14} Hz. The nature of his results is shown in Figure 13.6.

It became clear that the Planck–Einstein postulates were fundamental to solving the mysteries of radiation. But still, there were other results that needed attention, particularly the discrete spectrum of excited hydrogen gas.

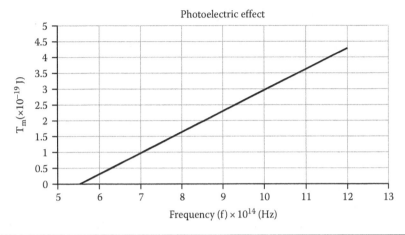

FIGURE 13.6   For the photoelectric effect, the maximum kinetic energy of ejected electrons, T_m, is shown to have a linear dependence upon the frequency of the incident radiation. The slope of the line provides an accurate value for Planck's constant, and the frequency intercept ($T_m = 0$) provides a good value for W/h and hence W, the "work function" for the material that makes up the photocathode. From the graph, slope = h $\approx 6.50 \times 10^{-34}$ J-s and W = h × (f-intercept) = W $\approx 3.6 \times 10^{-19}$ J ≈ 2.2 eV.

13.2.3 Hydrogen Spectrum

Sir Isaac Newton is generally credited with being the founding father of spectroscopy as an analytical technique in understanding the behavior of light emitted by radiating objects. Newton famously determined that solar light consisted of a series of colors from red to violet that could be observed when the white light of the sun passed through a prism. However, there were others predating Newton who were also aware of the various colors of the sun. It was Newton who showed that the colors were a constituent of white light and that they could be recombined to reform white light. He also studied the light from flames and the stars. His observations were in part responsible for his corpuscular theory of light, which was eventually replaced by the wave theory. Newton provided the results of his experiments and theoretical reflections in his book *Opticks* published in 1704. W.H. Wollaston (1766–1828) in 1802 improved upon Newton's apparatus by fabricating a system that included lenses for focusing the spectrum of the sun onto a screen. With this, he was able to see that the sun's spectrum was not spread uniformly but had dark lines where colors were missing. Real improvement in spectroscopy came when Joseph von Fraunhofer (1787–1826) replaced the prism with a diffraction grating as the dispersive element for separating the various colors of light. This improved the spectral resolution immensely and enabled Fraunhofer to accurately determine wavelengths and establish a standard that could be used in any laboratory to quantify the emissions from any light source. In fact, Fraunhofer made accurate measurements of the solar spectrum, observing and quantifying the dark bands that are still known as the Fraunhofer lines. In 1853, A.J. Angstrom (1814–1874) published his observations and theories on the spectra of gases. He postulated that an incandescent gas emits the same range of wavelengths that it is able to absorb. J.B.L. Foucault (1819–1868) had already demonstrated this in 1849, indicating that the source temperature was the important factor. Angstrom further measured and reported on the visible spectrum of incandescent hydrogen and delineated what we now call the Balmer series of hydrogen. Also, in 1855 David Alter (1807–1881) reported on his observation of the Balmer lines for hydrogen. In the 1860s, Gustav Kirchhoff (1824–1887) and Robert Bunsen (1811–1899) were able to show that individual elements had their individual spectral signatures. They showed that spectroscopy could be used in trace chemical analysis and discovered several elements that were previously unknown. Bunsen and Kirchhoff through their meticulous work established the technique of analytical spectroscopy that is the foundation of modern of spectroscopy.

In 1885 Johann Balmer (1825–1898), using the observations of Angstrom, was able to predict and quantify the first nine lines of the hydrogen spectrum in terms of a series of integers given by the following equation for the spectral line wavelengths:

$$\lambda = B n^2 / (n^2 - 4)(nm) \quad n = 3, 4, 5, 6 \text{ (visible)}, \quad (13.12)$$

where B (Balmer constant) = 364.56 nm. The accuracy is better than 1 part in 1000. Several of the prominent lines are shown in Figure 13.7.

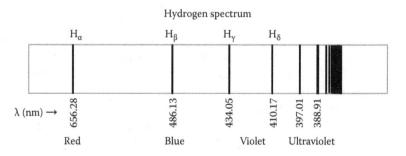

FIGURE 13.7 The hydrogen spectrum: the visible lines labeled red through violet correspond to the integers of the Balmer equation, 3, 4, 5, 6. The given wavelengths are experimental, and the Balmer equation with n = 3, 4, 5, 6, 7, 8 gives results well within 0.02 % of the indicated values.

Later in 1888, Johannes Rydberg (1854–1919), by studying the existing spectra of other elements, was able to develop a more complete equation leading to a broader description for the spectra of hydrogen and hydrogen-like atoms. Rydberg found it more convenient to work with $1/\lambda$ instead of λ and published the following equation:

$$\frac{1}{\lambda} = R\left(\frac{1}{n_1^2} - \frac{1}{n_2^2}\right)$$

(13.13)

where n_1 and $n_2 = 1, 2, 3, \ldots$ such that $n_1 < n_2$, and R = Rydberg constant $= 1.097 \times 10^7$ m^{-1} for hydrogen. It becomes clear that the Balmer formula is a special case of the Rydberg equation with $n_1 = 2$ and $n_2 = 3, 4, 5, 6$ and R = 4/B. With the Rydberg equation, other spectral series of hydrogen were predicted and eventually discovered. In either case, the equations represent empirical results with no ability to explain the underlying principles that led to their existence. Until 1913, there was no theory available in classical physics that could fundamentally explain their existence.

At this point, it is clear that any successful theory of radiation must explain all the phenomena that classical physics could not. In 1913, Niels Bohr, the Danish physicist, after studying the works of Planck and Einstein, proposed the following postulates governing the observed characteristics of the hydrogen atom:

(a)   An electron in an atom moves in a circular orbit about the nucleus under the Coulomb attractive force between the electron and the nucleus. This motion obeys the laws of classical mechanics.

(b)   Unlike classical orbital mechanics, the number of orbits available to the electron is restricted to those for which the orbital angular momentum, **L**, is an integral multiple of Planck's constants divided by 2π, that is, h/2π or $\hbar$, pronounced "h-bar": $L = n\hbar$ where $n = 1, 2, 3, \ldots$.

(c)   Despite the fact that it is accelerating, an electron in an allowed orbit does not radiate EM energy; therefore, its total energy remains constant and the orbit is stable.

(d)   If an electron, initially moving in a stable orbit of energy E_i, suddenly moves to another allowed orbit with energy E_f, it will emit (or absorb) EM radiation with some frequency, f, such that the energy of the radiation is given by $\left| E_f - E_i \right| = hf$, that is, the energy difference between the two orbital states is the product of Planck's constant h and the frequency f of the radiation emitted or absorbed.

We note that postulate (a) is a nod to the then new atomic model put forth by Ernest Rutherford (1871–1937; Nobel Prize, Chemistry, 1908) in

1911 after bombarding a very thin gold foil with a helium nucleus (alpha particle) (discovered terrestrially in 1895) and detecting the helium after it passed through the foil. He found that the angular distribution of the detected helium nucleus was consistent with a "planetary" model of the atom with a massive nucleus about which orbited the electrons. Postulate (b) introduces the notion of quantized angular momentum, which is completely different than Planck's notion of quantized energy originating from a Hooke's law force that leads to simple harmonic oscillations. However, we will see that this quantized angular momentum, indeed, leads to quantized total energy for the electron. Keep in mind that the Coulomb force of attraction between the nucleus and an electron in a hydrogen atom is an inverse-square force and that the centripetal force associated with circular motion arises from this. Postulate (c) is a reality check because according to the Maxwell theory of electromagnetic radiation, an accelerating charge should lose its energy by radiation; however, we know stable atoms of hydrogen and all the natural elements do, in fact, exist. Otherwise, we would not be here. Postulate (d) is clearly an application of the work of Planck and Einstein in describing black-body radiation and the photoelectric effect.

Given the configuration of Figure 13.8 for a one electron atom, the orbital angular momentum, relative to the orbital center, is given by:

$$L = mva. \tag{13.14}$$

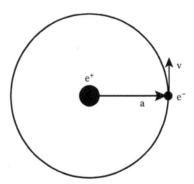

FIGURE 13.8   The Bohr hydrogen atom: A fixed atom made up of an electron, of charge e⁻ and mass m, is moving in a circular orbit, of radius "a," with speed v about a positive nucleus of charge e⁺. As one can see, this is the simplest model imaginable, and its simplicity belies its power in laying the foundation for our modern understanding of atomic structure. It represents a melding of classical and quantum physics.

Applying Bohr's second postulate (b)

$$mva = nh/2\pi = n\hbar, \quad n = 1, 2, 3,.... \tag{13.15}$$

The magnitude of the force between the nucleus and electron is given by Coulomb's law and takes the form

$$F = \left| kZe^{+}e^{-}/a^{2} \right| = kZe^{2}/a^{2}, \tag{13.16}$$

where
   K is a constant of proportionality (9.0×10^{9} N m^2/C^2 [SI])
   Z is the nuclear atomic number, which is 1 for hydrogen
   e = 1.602×10^{-19} C

Because the motion is uniform circular, the resulting centripetal force is

$$mv^{2}/a = KZe^{2}/a^{2}, \tag{13.17}$$

where m is the mass of the electron (9.1×10^{-31} kg [SI]). Using Equations 13.14 and 13.15 leads to the following:

$$KZe^{2} = mav^{2} = n^{2}\hbar^{2}/ma \quad \text{and} \quad a = n^{2}\hbar^{2}/KmZe^{2}, \quad n = 1, 2, 3,.... \tag{13.18}$$

From this, one concludes that the allowed radii are also quantized and the smallest radius allowed, often called the Bohr radius, is $a_{o} \approx 5.2 \times 10^{-11}$ m. One may also determine the (maximum) speed of an electron in the hydrogen atom as v_{m} (see exercises at the end of the chapter). We note that relativity does not enter into this work by Bohr. The student may want to ponder this, given that Bohr was well aware of the *special theory of relativity* developed by Einstein.

   It is now possible to determine the total energy of an electron in an allowed orbit of the hydrogen atom. From basic physics, we know that the potential energy (U) of a two-charge system is given by

$$U = -KZe^{2}/a$$

with the negative sign arising due to the attractive nature of the Coulomb force between unlike charges. Also, using Equation 13.17, the kinetic energy T is determined to be of the form

$$T = mv^{2}/2 = KZe^{2}/2a.$$

The total energy is given by $E = T + U = -Ke^2/2a$. Now using "a" given by Equation 13.18 allows the total energy to be expressed as

$$E = -K^2 Z^2 me^4 / 2n^2 \hbar^2, \quad n = 1, 2, 3, \ldots$$

and to emphasize the importance of the quantum number "n," the energy may be written as

$$E_n = -K^2 Z^2 me^4 / 2n^2 \hbar^2, \quad n = 1, 2, 3, \ldots, \tag{13.19}$$

and any allowed transitions from orbit, n_1, to another orbit, n_2, may be characterized by

$$\Delta E_n = E_2 - E_1 = R'\left(\frac{1}{n_1^2} - \frac{1}{n_2^2}\right), \quad n_i, n_f = 1, 2, 3, \ldots.$$

The above and the fourth postulate, (d), yields

$$|\Delta E_n| = hf = hc/\lambda \quad \text{and} \quad \text{therefore} \quad \frac{1}{\lambda} = (R'/hc)\left(\frac{1}{n_1^2} - \frac{1}{n_2^2}\right),$$

where $R' = K^2 Z^2 me^4/2\hbar^2$. The reader may want to verify that R' is related to the Rydberg constant of Equation 13.13 by $R = R'/hc$, where c is the speed of light (3×10^8 m/s) and h is Planck's constant. Note that from Equation 13.19, the lowest allowed energy state is given by $n = 1$ and that the energy is negative, meaning that the electron is bound to the atom. As "n" increases, the energy becomes less negative and the total energy approaches zero as the "n" approaches infinity. The most stable state will be that with the lowest energy, in this case $n = 1$.

The normal or ground state of hydrogen is the lowest energy, $n = 1$, so when it is somehow excited to produce a spectrum, the electron jumps to some higher state, $n > 1$, and the radiation emission occurs when it transitions back down to successively lower states until it reaches its normal ground state, producing the observed spectrum. For example, when the electrons in the excited gas transition from state $n = 7$ to $n = 1$ through successively lower states, the visible (Balmer series) spectrum is observed for transitions $\Delta n = 6 \rightarrow 2, 5 \rightarrow 2, 4 \rightarrow 2$, and $3 \rightarrow 2$.

This was a great triumph for the quantum theory, and although no attempt was made to develop similar equations for multielectron atoms, it

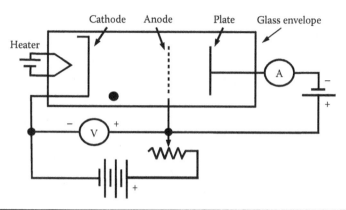

FIGURE 13.9 The Franck–Hertz apparatus used to verify that atomic energy states are quantized by showing the discrete nature of the electrical current as electrons transfer their kinetic energy to the mercury atoms that permeate the glass enclosure. The glass envelope contained a small amount of liquid mercury (Hg), which was heated to produce low-pressure Hg vapor.

seemed clear that the total energy of such electrons will also be quantized. This led to the prediction that the total energy of these complex atoms will also be quantized. This prediction was proved correct by James Franck (1882–1964; Nobel Prize, Physics, 1925) and Gustav Hertz (1887–1975; Nobel Prize, Physics, 1925) in 1914. A schematic of the apparatus is shown in Figure 13.9.

The heater forces electrons from the cathode and helps to produce the mercury vapor within the glass enclosure. As the voltage V is increased, the current rises until V reaches about 4.9 V and then it suddenly drops. This is shown in Figure 13.10.

The interpretation is that at a specific electric potential (V), the kinetic energy of the electron is absorbed by the mercury atom upon collision, and so the electron current suddenly drops causing the ammeter to read lower current. A significant fraction of the electrons with this kinetic energy lose said energy in exciting the mercury atoms. One concludes that the mercury atom can only absorb energy at this potential, which is its first excited state. The spectrum of excited mercury should show a spectral line corresponding to this energy, and Franck and Hertz were able to detect such a line. They found that the mercury showed no line at all for energies less than 4.9 eV. The second excited state was found to occur at 6.7 eV. Franck–Hertz provided a convincing proof of the validity of the general implications of the Bohr theory of the hydrogen atom and hence a validation of the new paradigm set forth by Planck in describing black-body radiation. The method of Franck–Hertz was the first general method that could be used to determine the various quantum states available for a variety of materials. Of course, one could also

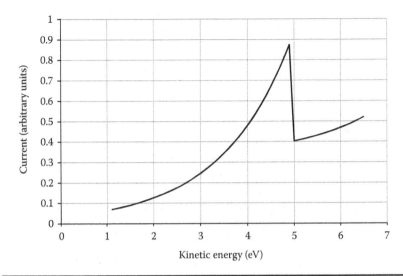

FIGURE 13.10 For the Franck–Hertz effect, this diagram shows how the kinetic energy is absorbed upon collision between an electron and a Hg atom. 1 eV = 1.602×10^{-19} J, which is the energy needed to accelerate an electron (-1.602×10^{-19} C) through a potential difference of 1 Joule per Coulomb (V). Franck and Hertz also found an energy absorption at 6.7 eV corresponding to the second excited state.

generate the spectrum and empirically construct the quantum numbers (n) as was done by Balmer and Rydberg; however, this becomes very complicated and difficult to perform as the atoms become more complex.

It should be pointed out that this was just the beginning of new era in physics and that as measurement techniques became more refined, there were details of the hydrogen spectrum not previously detected. This led to refinements to the theory that eventually led to modern quantum mechanics.

13.3 SOME NATURAL PHOTONICS

Our star, the sun, is the most massive photonic system in our local universe. It emits photons throughout the known EM spectrum (Figure 13.11) and has determined the evolutionary characteristics of all life on the planet.

The environment fostered by the sun has spurred the evolution of many sensory systems that are necessary for survival. To humans and other animals, a very useful sensory (detection) device is called the "eye" and is of paramount importance because it increases the probability of survival in predatory environments. To "see" is fundamental for most humans and depends upon the eye as a photonic system that receives, detects, and converts the energy of photons, of a particular section of the EM spectrum, into electrical impulses that are transmitted to and manipulated by the "brain" into what is termed

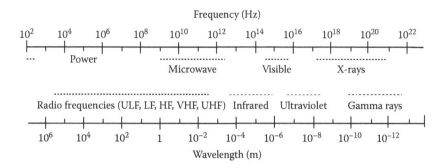

FIGURE 13.11 Electromagnetic spectrum: the visible portion is approximately 400 nm to about 800 nm (scales are logarithmic).

"vision." Independent of the survival aspect, imagine the joy of seeing a rainbow after a rain. For this, we need the sun, the eye, and the rain. (See Problem 13.1.) The eye is actually a complex synergy of biology, chemistry, and physics that forms a sophisticated light detector that, combined with the brain, results in an efficient imaging system. Evolution has taken several independent paths to eye development; however, the path taken in vertebrates is of primary interest here. (If the reader desires more information on eye development, the following reading is suggested from Chapter 21 of *Optics and Photonics, an Introduction*). The eye is depicted in Figure 13.12.

The cornea and lens are usually treated as a composite system of two lenses with an effective focal length and refractive power. However, one should understand that the cornea is a fixed focus component, which can provide close to 50% of the refractive power of the lens system. On the other hand, the lens is a variable focus device, which is somewhat malleable, and

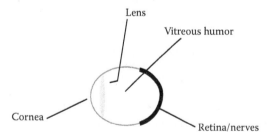

FIGURE 13.12 Schematic of the human eye: the cornea and lens represent the photon collection and imaging system of the eye. The vitreous humor represents the transmission medium and is a transparent, gel-like substance with an index of refraction similar to water. The retina is the detector array of cells that convert the photonic energy to electrical impulses transmitted to the brain for processing. It is approximately 25.4 mm from the lens in a normal eye. Not shown in the diagram is the optic nerve that transports the signals from the retina to the brain.

can be changed in shape by a collection of muscles that surround the eye in its socket. This makes for a versatile system for forming very sharp images on the retina at the back of the eye, which are transmitted via the optic nerve to the brain for processing into sight. The optical power of the lens system is defined by Equation 3.13 and is given in diopters (D). For a "normal" eye, the variable focal system allows one to clearly see objects placed from a few centimeters to very long distances in front of the cornea. Unfortunately, not all individuals have sharp vision that falls within these object distance parameters. This means that in many cases, the image is not always formed on the retina for object positions in the normal range. Either the natural position of the image is in front of the retina or behind the retina, resulting in a blurry or "unfocused" image. In the former case, the object to be viewed must be placed closer to the eye in order for the image to appear sharp and clear. This is referred to as "myopia" or "nearsightedness." In the latter case, the object must be moved farther than usual from the cornea for a clear image to form on the retina. This is referred to as "hyperopia" or "farsightedness." In either case, there will usually be a well-defined distance where an object may be in front of the cornea such that a clear image is formed on the retina. The nearest clear vision distance is called the "near point," and the most distant clear vision distance is called the "far point." This represents a refracting problem that occurs most often in older individuals unless there is an unusual medical condition or deformity affecting the vision system. For correcting this refraction problem, it is only necessary to place a corrective lens in front of and near to the cornea. See the following example.

EXAMPLE 13.3

A person who enjoys reading notices that his comfortable reading distance has, over a long time, changed from 40 to 13 cm in front of his face. This was a gradual change that occurred over several years. Obviously, corrective lenses are now required. (a) By how much has the optical power of the eyes changed from the usual reading position? (b) What is the power of the required corrective lens for reading glasses if one assumes the retina is located 25.4 mm behind the lens? (Optometrists can measure this distance very accurately.)

Solution

(a) Change in optical power from the normal reading position:
   1. One must determine the original optical power of the unaided eye at the desired reading distance. Here, the object distance is 40.0 cm and the image distance is approximately

25.4 mm. Use the thin lens equation from Chapter 3: $1/f = 1/o + 1/i = 1/0.400 \text{ m} + 1/0.0254 \text{ m} = 39.8 \text{ m}^{-1} = 39.8 \text{ D}$, $i =$ image distance, $o =$ object distance.

2. What is the optical power of the eye with the object (book) at 13 cm in front of the eye?

$$1/f = 1/0.130\,\text{m} + 1/0.0254\,\text{m} = 47.1\,\text{m}^{-1} = 47.1\,\text{D}$$

3. The optical power has increased by 7.3 D.

Conclusions: Any corrective lens must offset this increase to approach normalcy; therefore, one would expect the corrective lens power to be approximately −7.3 D.

(b) Optical power of the corrective lens:
1. One wants the corrective lens to form an upright image of the book 13 cm in front of the eye. This immediately tells one that the lens must have negative power, that is, the corrective lens must be concave in nature.
2. Use the thin lens equation from Chapter 3: $1/f = 1/o + 1/i$, where o and i are the object and image distances, respectively.
3. Because the image is on the same side of the lens as the object the image distance is negative, so $i = -13.0$ cm $= -0.130$ m and $o = 40.0$ cm $= 0.40$ m:

$$1/f = 1/.400 - 1/.13 = -5.19\,\text{m}^{-1} = -5.19\,\text{D}$$

Reality Check

Why do you think the difference between (a) and (b) is so great? See Problem 13.3. The results of part (a) require two assumptions about the eye: (1) the lens system is that of a thin lens, and (2) the image position is well defined as 2.54 cm. Do we need to know the index of refraction of the vitreous humor? Note that in part (b), no specific knowledge of the eye system is necessary. One only needs to know where the object is located to produce a clear image in the unaided eye. Also, the thin lens approximation is good in this calculation. It should be pointed out that refractive problems are not the only aberrations that occur in the eyes.

There are many other photonic systems that occur in nature that involve thin-film optics, interference and diffraction effects, and compound eyes, like those of a fly. It is suggested that Reference 1 be consulted for those readers who seek more in-depth knowledge of these topics.

13.4 HUMAN-ENGINEERED PHOTONIC SYSTEMS

There are many photonic systems produced by humanity that are taken for granted in today's society, for example, lasers, cameras, the Internet, television, smartphones, microscopes, and telescopes to mention a few. In addition, there are a few military systems that are now common knowledge but poorly understood by the public. Some of the latter are very sophisticated but are based upon an understanding of the fundamental optics principles covered in the previous chapters of this text. Several of these will be considered in the pages that follow.

13.4.1 "LIGHT PIPE"

The light pipe is very important because it is fundamental to modern communications, in the transmission of voice, images, and digital data. It is based upon the concept of total internal reflection discussed in Chapter 2, Section 2.5.2. Essentially, a light ray enters one end of a long, transparent cylindrical piece of glass or plastic at such an angle that the rays that impinge upon the walls of the cylinder do so at an angle larger (relative to a wall normal line) than the critical angle for total internal reflection, θ_c. The index of refraction of the material surrounding the cylinder must be less than that of the cylinder material in order to produce the total internal reflection. As a result, the light will "bounce" along the internal walls of the cylinder until it exits the other end. See Figure 13.13. The nice thing here is that one may bend the pipe into almost any shape and get light at the other end as long as the internal angle of

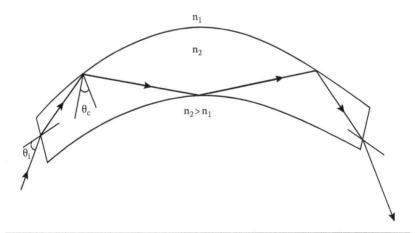

FIGURE 13.13 The light pipe is a long, thin cylindrical fiber of glass or acrylic that transmits light from one end to the other by multiple internal reflections. In practical applications, the pipe is actually a bundle of thin fibers that are clad by some other appropriate material.

incidence upon the wall remains greater than the critical angle for total internal reflection. If one imagines that the pipe is actually very long and thin, say on the order of 10 μm in diameter, it becomes very flexible and pliable. At this stage, it is called a *fiber-optic* cable. In practice, one chooses the geometry, radiation frequency, and fabrication material based upon transmission efficiency and manufacturing ability. It is important that the material of the pipe be transparent in the desired frequency range of light (radiation) in order to get maximum transmission of the energy. To this end, for long-distance fiber-optic communications, an infrared light source is often used because the wavelength is longer than that of visible light and hence less affected by scattering defects in the material. A scattering defect can be anything from a small void, an impurity, or a local change in the index of refraction or absorption properties. Basically, anything that alters the frequency, direction, or intensity of the radiation within the light pipe is an issue. In general, fiber-optic cables are more secure and can carry more information faster and over longer distances than metallic wire (copper, silver, aluminum). Unfortunately, as a practical issue, fiber-optic cables are more difficult to repair, so technicians need significant hands-on training.

In general, fiber-optic cables are constructed of a bundle of small-diameter, cylindrical fibers (light pipes) made of specialized silica glass or acrylic, each clad in a surrounding material to reduce cross talk between adjacent fibers as well as to assure cleanliness of each fiber. The index of refraction of the cladding material is lower than that of the core fiber optic so that total internal reflections occur at the interface between the two. See Figure 13.14.

Assuming the fiber cylinder of length l and diameter d_f is not severely curved, the total number of reflections that occurs is given, approximately, by $Nr = (l/d_f) \tan(\theta_r) = (l/d_f)(n_o/n_2)\theta_o$, the angle is in radians. An important property of a fiber is its numerical aperture N_A, which is defined in terms of

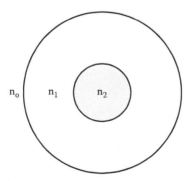

FIGURE 13.14   Cross section of an optical fiber showing the central core (shaded), the cladding, and the environment, $n_o < n_1 < n_2$.

the indices of refraction of the media as $N_A = \left(n_2^2 - n_1^2\right)^{1/2}$. This determines the incident angle within which all incident light from the source will be totally internally reflected. It determines the maximum desired dimension of any source and/or detector at each end of the fiber to most efficiently utilize the incident or detected light.

Consider a straight section of fiber optic as shown in Figure 13.14. A ray impinges on the surface at the center of the cylindrical fiber (meridian ray) with angle θ_m, the maximum incident angle, which produces total internal reflection. This defines the maximum conical half-angle of the incident rays producing total internal reflection. Any ray lying within this angle will produce total internal reflection.

Referring to Figure 13.15, it is clear that applying Snell's law at the incident interface, gives the following:

$$n_o \sin\left(\theta_m\right) = n_2 \sin\left(\theta_r\right). \tag{13.20}$$

The side wall incident (critical) angle θ_c is then refracted along the interface, so that Snell's law gives

$$n_2 \sin\left(\theta_c\right) = n_1 \sin\left(90°\right) \quad \text{or} \quad \sin\left(\theta_c\right) = n_1/n_2. \tag{13.21}$$

Furthermore, one observes that $\theta_r = 90 - \theta_c$; therefore,

$$\sin\left(\theta_r\right) = \cos\left(\theta_c\right). \tag{13.22}$$

Squaring and adding Equations 13.21 and 13.22 give

$$\left(n_1/n_2\right)^2 + \sin\left(\theta_r\right)^2 = 1 \quad \text{or} \quad \sin\left(\theta_r\right)^2 = 1 - \left(n_1/n_2\right)^2,$$

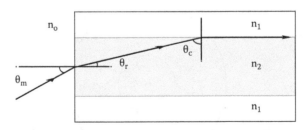

FIGURE 13.15 Refraction in a fiber optic: a section of fiber optic that shows the maximum angle θ_m of incidence that allows total internal reflection. θ_c is the critical angle for total internal reflection between the cladding and core materials of indices n_1 and n_2, respectively. Any meridian ray entering at a smaller angle will reflect internally.

and applying Equation 13.20 gets

$$n_o \sin(\theta_m) = \left(n_2^2 - n_1^2\right)^{1/2} = N_A. \tag{13.23}$$

Equation 13.23 defines the numerical aperture N_A and is also given in slightly differing form.

In practice, the small angle approximation for the incident radiation is applicable for many common systems and n_2 is only slightly greater than n_1. If one defines the relative index of refraction of the fiber such that

$$\Delta n = \left(n_2 - n_1\right)/n_2,$$

then to a good approximation the numerical aperture may be written as

$$NA \approx n_2 \left(2\Delta n\right)^{1/2}. \tag{13.24}$$

With the small angle approximation, the half-angle for the NA can be obtained from Equation 13.23 as

$$\theta_m = \left(2\Delta n\right)^{1/2} \left(n_2/n_o\right) (\text{rad}). \tag{13.25}$$

There are some fundamental and mechanical limits to the use of fiber optics such as the attenuation by scattering and absorption of the EM radiation as it passes through the fiber optic material. These fundamental issues are very complex and will not be considered in this overview (see advanced texts cited in bibliography). In addition, there are mechanical limits determined by the tensile strength and bending limits of the material and the critical angle for total internal reflection of the core material. It is easy to imagine that a sharp bend can induce an internal wall-incident ray being smaller than the critical angle and thus causing some of the radiation to be leaked out of the core fiber. Also a sharp bend can cause the material to crack, creating scattering centers for the radiation, which causes absorption and leakage, or the fiber may completely break.

13.4.2 Basic Quantum Theory

In 1924, Louis de Broglie (1892–1987; Nobel Prize, Physics, 1929) made the revolutionary proposal, contained in his doctoral thesis, that moving objects can also exhibit wave properties. de Broglie (pronounced "dibrōly") was inspired by Bohr's theory of the hydrogen atom in which each allowed orbit could be thought of as a stable standing wave state. At the time, there was no experimental evidence to suggest such a radical proposal. However, in 1927, experiments with beams of electrons exhibited interference effects similar to classical waves, and the dual nature of matter and waves was established. The principle of the "duality" of radiation and matter was now complete. This ushered in another paradigm shift in physics as important as that created by Max Planck with his theory of black-body radiation.

Consider the following: in classical mechanics, we know that the kinetic energy and momentum of a massive particle are given by

$$T = mv^2/2 \quad \text{and} \quad p = mv, \tag{13.26}$$

where m is the mass of the particle and p is its linear momentum. One observes that there is a direct relationship between the energy and the momentum of the particle:

$$T = p^2/2m \quad (\text{Classical validity only}) \tag{13.27}$$

Einstein, in the *theory of special relativity*, provides general expressions for the *total energy* and momentum of an isolated particle expressed as follows:

$$E = mc^2 \left(1 - v^2/c^2\right)^{-1/2} = \gamma\, mc^2 \tag{13.28}$$

$$p = \gamma\, mv, \tag{13.29}$$

where the definition of γ is obvious from Equation 13.28 and γm is the relativistic mass. Furthermore, Einstein shows us that the total energy may be written as

$$E^2 = \left(mc^2\right)^2 + p^2 c^2. \tag{13.30}$$

If one applies Equation 13.30 to a massless entity such as an EM wave or perhaps a photon, then

$$E = pc \tag{13.31}$$

since from Planck we know $E = hf$ where $f = c/\lambda$, then we obtain

$$p = h/\lambda, \tag{13.32}$$

where h is Planck's constant. The previous equation may also be written as

$$\lambda = h/p. \tag{13.33}$$

If one applies the de Broglie hypothesis for a classical particle, then it is possible to calculate the effective wavelength (de Broglie wavelength) of said particle. Consider the following example.

EXAMPLE 13.4

Determine the de Broglie wavelength, in meters, of a person of weight 200 lb walking at 2.5 mi/h.

Solution

The momentum of the person must be determined, so the velocity and mass of the person must be converted to the appropriate units:

$$M = (200 \text{ lb})/2.2 \text{ lb/kg}) \approx 91 \text{ kg},$$

$$v = (2.5 \text{ mi/h})(5280 \text{ ft/mi})(12 \text{ in./ft})(.0254 \text{ m/in.})(3600 \text{ s/h})^{-1}$$

$$= 0.98 \text{ m/s}$$

$$p = mv = 91(.98) \text{ kg-m/s} \approx 90 \text{ kg-m/s}$$

Applying the de Broglie hypothesis yields

$$\lambda = h/p \approx (6.63 \times 10^{-34} \text{ kg-m}^2/\text{s})/(90 \text{ kg-m/s}) \approx 7.4 \times 10^{-36} \text{ m}.$$

FOOD FOR THOUGHT

What are the practical ramifications of such a small wavelength? Is the calculation even valid since we are not considering an atomic particle?

ANSWER

As we studied black-body radiation and the quest for a solution to the problem of the Ultraviolet Catastrophe, we saw that scientists like Boltzmann, Wien, Planck, and others attempted to apply classical concepts of waves and collections of harmonic oscillators generating the radiation. This collection of particle oscillators led Planck to the final solution based on an energy density function that produced an average energy per standing wave in the radiation cavity given by

$$\varepsilon = hf/\left(e^{hf/kT} - 1\right).$$

Although Planck got the correct equation for the distribution function, he used the wrong energy for the harmonic oscillators. He used (n)hf when he should have used (n + 1/2)hf. Fortunately, he was using Maxwell–Boltzmann statistical functions to consider the collection of standing waves in the cavity, so the zero point energy (½ hf) did not come into play. It was not something he could have known at that time. The Bohr model of the atom could not explain all the features of the hydrogen spectrum much less the features of more complex atoms. It was discovered that hydrogen atoms as well as multielectron atoms all had fine structure to their spectra under varying conditions such as in the presence of a magnetic field. This eventually led to the notion of spin states for the atoms and the electrons as well as photons. Also, intensities for some lines were stronger than others. In short, the Bohr model simply could not explain how the properties of macroscopic matter derived from the structure of aggregates of atoms at the microscopic level. Remember that Dimitri Mendeleev (1834–1907) developed the periodic table of the elements based upon macroscopic observations; he discovered that there was a distinct grouping of elements with common properties. Physicists of the time knew that any successful theory must be able to explain the periodic table. With the de Broglie hypothesis and the duality of matter and energy on firm ground, a new approach to dealing with atomic structure was necessary. In 1925, Erwin Schrödinger (1897–1961; Nobel Prize, Physics, 1933) and Werner Heisenberg (1901–1976; Nobel Prize, Physics, 1932) both developed theories of quantum mechanics that encompassed the hypothesis of de Broglie. The theory by Heisenberg led to the now famous *Heisenberg uncertainty principle*, which tells us that we cannot simultaneously know the location and momentum of a particle with infinitesimal accuracy as expressed in the following equation:

$$\Delta x \Delta p \geq \hbar/2 \quad \text{Heisenberg uncertainty principle}$$

where Δx and Δp are the uncertainties in the position coordinate and momentum, respectively. This makes sense in light of the de Broglie hypothesis because a wave cannot be localized to more accuracy that a single wavelength. So one can see that the uncertainty principle and the de Broglie hypothesis are essentially the same.

Schrödinger's theory introduced the concept of a wave function $\Psi(x, t)$ associated with an object. There is no physical interpretation of Ψ other than it is associated with the de Broglie wave. It satisfies a wave equation that looks similar to the classical wave equation but with Ψ as the dependent variable whose absolute magnitude squared, evaluated for a particular location at a particular time, is proportional to the probability that an object will be located at that place at that time. Schrödinger's one-dimensional wave equation is given by

$$-\frac{\hbar^2}{2m}\frac{\partial^2\psi(x, t)}{\partial x^2} + U(x, t)\psi(x, t) = -i\hbar\frac{\partial\psi(x, t)}{\partial t}.$$

where $U(x, t)$ is a general interaction potential energy function. If an object exists somewhere in space, then the probability that it exists is 1. Since, by definition, a probability must be a positive real quantity and the wave function is in general a complex function with both real and imaginary parts, the probability density P is defined in terms of the product of Ψ with its complex conjugate Ψ^*, that is, $P = \Psi^* \Psi$ is the probability per unit distance (or volume for three dimensions) that an object will exist within that space and is always a positive real quantity. What this means mathematically is

$$\int_{-\infty}^{\infty} \Psi^*\Psi dx = 1.$$

Although more general and an improvement over the Bohr theory, it still could not account for the many experimental observations without some expansion. Ultimately, several quantum numbers for different aspects of an atom such as orbital angular momentum, spin angular momentum, and magnetic moments had to be considered to complete the theory. This resulted in the necessity for four quantum numbers, n, ℓ, m_ℓ, and m_s, in order to describe the complete state of an electron in an atom. Table 13.1 provides a description of the various quantum numbers for an atomic electron.

Wolfgang Pauli (1900–1958; Nobel Prize, Physics, 1945) in 1925, after detailed studies of many atomic spectra including helium, showed that every unobserved calculated state involved at least two electrons with identical quantum numbers. This led him to propose that a minimum of four quantum numbers were needed to specify the complete state of an atomic electron and that no two electrons can have the same set of quantum numbers. This is now called the Pauli exclusion principle, and it was a major contributor to our understanding of how electrons arrange themselves in atoms leading to a complete understanding of the periodic chart as developed by Mendeleev.

TABLE 13.1
Quantum Numbers Associated with Atomic Electrons

Name	Symbol	Allowed Values	Related Quantity
Principal	n	$1, 2, 3, \ldots$	Electron energy
Orbital	ℓ	$0, 1, 2, \ldots, n-1$	Orbital angular momentum
Magnetic	m_ℓ	$-l, \ldots, 0, \ldots, +l$	Orbital angular momentum direction (magnetic quantum number)
Spin magnetic	m_s	$-1/2, +1/2$	Electron spin direction

Note: ℓ, m_ℓ, m_s must be multiplied by $\hbar$ to get the magnitude of the associate angular momentum.

The results of experiments show that particles of odd number multiples of 1/2 spins in the same atomic system must obey the Pauli exclusion principle. These particles are referred to as *fermions* because the behavior of such systems of particles is described by a statistical distribution first determined by Enrico Fermi (1901–1954; Nobel Prize, Physics, 1938) and later by Paul A.M. Dirac (1904–1982; Nobel Prize, Physics, 1933). A system involving free electrons in a metal is a good example of a Fermionic system. On the other hand, systems of particles involving spins of 0 or integer multiples do not obey the Pauli exclusion principle and are called *bosons*. The statistical law that governs these particles was determined by S.N. Bose (1894–1974) and extended by A. Einstein. A collection of photons in a cavity is a good example of such a system.

In terms of the Schrödinger theory, if we have a collection of noninteracting identical particles, each with its own wave function Ψ, then for a system of such particles, the system wave function $\Psi(1, 2, 3, \ldots)$ may be expressed as a product of the individual wave functions:

$$\Psi(1, 2, \cdots m) = \Psi(1)\Psi(2)\cdots\Psi(m^{-1})\Psi(m). \tag{13.34}$$

Consider a system of two identical particles: one in quantum state i and the other in state j. If the two particles are exchanged, then because they are identical, it should make no difference in the probability density:

$$\left|\Psi(1, 2)\right|^2 = \left|\Psi(2, 1)\right|^2. \tag{13.35}$$

There are two symmetry possibilities for the system wave function under this particle exchange, either

$$\text{Symmetric:} \quad \Psi(2, 1) = \Psi(1, 2)$$

or

$$\text{Antisymmetric:} \quad \Psi(2, 1) = -\Psi(1, 2)$$

Both of these possibilities satisfy the equality of the probability density of Equation 13.35. The individual wave functions for each particle can be written according to Equation 13.34 as

$$\Psi_{\text{I}} = \Psi_i(1)\Psi_j(2) \quad \text{or} \quad \Psi_{\text{II}} = \Psi_i(2)\Psi_j(1)$$

Since the two particles, 1 and 2, are identical to each other, there is no way of knowing which wave function describes the system at any time. Each has the same probability of being correct at any given time. Therefore, a linear combination of the two will provide the correct description, and there are two possible ways to form that combination, one symmetric and one antisymmetric:

$$\text{Symmetric:} \quad \Psi_S = \Psi_{\text{I}} + \Psi_{\text{II}} = \Psi_i(1)\Psi_j(2) + \Psi_i(2)\Psi_j(1)$$

$$\text{Antisymmetric:} \quad \Psi_A = \Psi_{\text{I}} - \Psi_{\text{II}} = \Psi_i(1)\Psi_j(2) - \Psi_i(2)\Psi_j(1)$$

From this, it is easy to see that if $i = j$, then the antisymmetric wave function for the system is zero; therefore, such a state cannot exist within an atom. It was known by Pauli and others that certain states could be determined theoretically but could not be observed experimentally and all of these states would have required at least two electrons to have an identical set of quantum numbers; hence, he reasoned that they were excluded states because they were not observable in nature. Schrödinger was aware of this and developed his theory accordingly by introducing the notion of symmetric and antisymmetric wave functions. Clearly, no such exclusion occurs for systems described by symmetric wave functions such as bosons, which have no spin or integral spin quantum numbers, and any number of bosons may exist in a specific quantum state. It should be clear that antisymmetric wave functions are necessary to describe fermionic systems. Bosons and fermions are considered elementary particles that are the constituents of all matter.

A collection of freely moving, noninteracting, fermions is called a Fermi gas. A collection of free electrons approximate such a gas and is also called a free electron gas. Fermi was able to find solutions to Schrödinger's equation that allowed the development of a probability distribution function for electrons in a conductor, which is similar to a free electron gas. The distribution function $f(\varepsilon)$ gives the probability that an electron will occupy the energy state ε. It takes the form

$$f_F(\varepsilon) = \left\{ \exp\left[(\varepsilon - \varepsilon_F)/k_B T \right] + 1 \right\}^{-1}, \tag{13.36}$$

where ε_F is the *Fermi energy* and is equal to the maximum occupancy energy allowed in the limit as T approaches absolute zero for $\varepsilon < \varepsilon_F$. We also note that for $\varepsilon > \varepsilon_F$ in the same limit on T, $f_F = 0$, which means there is no allowed occupancy above ε_F at a temperature of absolute zero. Further investigation of f_F shows that for temperatures above absolute zero with $\varepsilon = \varepsilon_f$, f_F is 1/2. This suggests that if there are N electrons, then N/2 will occupy states for which $\varepsilon = \varepsilon_f$. Futheremore, for $\varepsilon > \varepsilon_F$ with the temperature above absolute zero, fermions from any state below ε_F will have a finite probability of occupying a state above ε_F.

13.4.3 LASER

Light Amplification by the Stimulated Emission of Radiation is a succinct description of the basic operation of a LASER. It is, in fact, a quantum effect that cannot be described classically. The concept of stimulated emissions was first proposed by Einstein in 1917. The first generation of LASER device was a cylindrical length of ruby containing controlled levels of specific impurities called "dopants." The rod was fully silvered on each end and half silvered on the other. The stimulation came from an external source of radiant energy surrounding the ruby rod. The stimulated emissions, due to atomic transitions, within the ruby were coherent and reflected back and forth between the ends, producing constructive interference within the laser, and some of the radiation was emitted through the half-silvered end in the form of a pencil beam of light. The external source surrounding the ruby rod was light or microwave radiation, depending upon the impurity dopant used in growing the ruby crystal. Eventually, gaseous and other exotic material solid-state LASERs were developed, allowing the frequency and power to be more easily controlled. These devices have become the workhorses of industry for cutting and welding applications and offer significant potential as radiation weapons for military operations.

Three types of transitions involving EM radiation are possible between two energy levels $E_1 < E_2$ in an atom. If the atom is initially in the lower state E_1, it can be raised to the higher state E_2 by the absorption of a photon of energy $\Delta E = E_2 - E_1 = hf$. This is called *stimulated absorption*. If the atom is initially in a higher state, it may undergo *spontaneous emission* by emitting a photon of frequency f. The third type, *stimulated emission*, was first proposed by Einstein. In this case an external stimulation such as an incident photon will initiate a transition from an excited state, E_2, to the lower state, E_1, causing additional photons to be emitted that are in phase with the stimulating photons. The emitted radiation is coherent. The problem is that light amplification cannot occur unless there is a metastable* state between

---

* Metastable state: An atomic or molecular state, which lasts much longer than the time associated with common excited states.

the highest excited state and the ground state because with just two states stimulated absorption will occur just as often as stimulated emission. In a three-level system, more than half the atoms must be in the metastable state for stimulated emission to dominate. Normally, most atoms in a material are inherently in the ground state, so a way has to be devised to create a "population inversion" in which the majority of the atoms are above the ground state for a sufficient time. One common way is called "optical pumping." Here, an external light source is used that produces enough photons of an appropriate frequency to excite the ground-state charges to a state that will spontaneously decay to a desired metastable state. Then other photons of a different frequency stimulate the many atoms or ions in the metastable state to decay to the ground state emitting large numbers of photons of the same frequency and phase. This is illustrated in Figure 13.16.

For example, consider a crystal of aluminum oxide (Al_2O_3) in which some of the Al^{3+} ions are replaced with Cr^{3+}. This is called a ruby and is usually identified by its red color. This represents a material that has an available metastable state. A light pump excites more than half of the chromium ions to the excited state of about 2.25 eV, and they spontaneously decay to the metastable state at about 1.79 eV. The excess energy is absorbed by the other ions in the lattice, causing the stimulated emission of radiation of wavelength of approximately 694 nm to the rapid transition from the metastable state to the ground state. The ruby rod is precisely made to have a length that is an integer multiple of a half wavelength in order to produce resonant standing waves. One end is mirrored and the other end partially mirrored to allow some of the light to exit the end. It is surrounded by a xenon flash tube used as the light pump. There is a pulse of 694 nm red light emitted from the

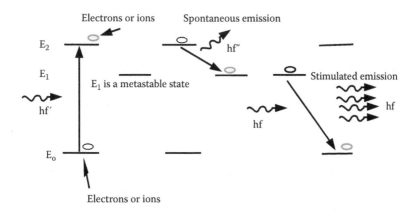

FIGURE 13.16   Atomic and/or molecular energy levels for a three-state LASER: a basic three-level process necessary for LASER action using a light pump to stimulate the lasing process. E_0 = ground state, E_2 = excited state, E_1 = metastable state.

partially mirrored end of the ruby for every flash of the light pump. The ruby laser was the first laser ever fabricated by Theodore H. Maiman (1927–2007) at Hughes Research Laboratories in 1960. Charles H. Townes (1915–2015), Alexander M. Prokhorov (1916–2002), and Nikolay G. Basov (1922–2001) were awarded the Nobel Prize in 1964 for their pioneering work that made the laser possible.

As indicated before, for a three-level lasing process, more than half the atoms must attain the metastable state; however, for a four-level lasing process, this is not the case. In this case, there is a fourth state just above the ground state and below the metastable state that is highly unstable and decays rapidly to the ground state. As a result, very few atoms reside in this state. The metastable state is above the intermediate state in energy, and the transitions from the former to the latter actually cause the laser action. Only modest light pumping is necessary to get the metastable state populated higher than the intermediate state because at any time, the latter contains so few atoms.

The common variety He–Ne laser is a multiple-level process and achieves population inversion differently. The mixture of helium (10 parts) and neon (1 part) is contained in a glass tube with a length that is an integral multiple of a half wavelength of the laser light. Each end of the tube is mirrored with one end only partially so (less than 2% transmission). The tube and mirrored ends form a resonant cavity for the light. He and Ne both have metastable states above their ground levels that are slightly different from each other, 20.61 and 20.66 eV, respectively. An electric discharge is produced in the gas by a high-voltage alternating current source. Collisions between the He and Ne with electrons produced by the discharge excite the atoms, which decay to their respective metastable states. Also, some of the excited He atoms transfer their kinetic energy via collisions to some of the ground-state Ne atoms, providing enough additional energy to excite more of the Ne atoms into their metastable states. So we see that the purpose of the He is to help create a population inversion in the Ne atoms. The Ne atoms have an intermediate state at 18.70 eV that is highly unstable and contains few atoms in that state, so the laser action takes place between its metastable state and the intermediate state. The visible light produced by the He–Ne laser has a wavelength of 632.8 nm. The description presented is a simplification limited to the most common He–Ne laser. He–Ne lasers may be fabricated that take advantage of the many other metastable states that exist because the spectra of helium and neon are complex. There exist many other metastable states that produce lasing at different wavelengths, including infrared. An important feature of this laser is that it is continuous and not pulsed like the ruby laser. A simplified energy level diagram of the process is shown in Figure 13.17.

Interestingly, in the early days of laser development, it was said that the laser was a solution looking for a problem to solve. There were many folks,

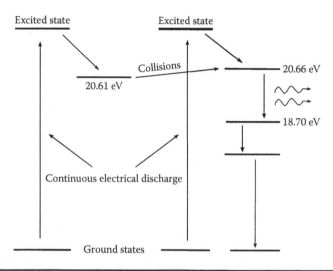

Excited state   Excited state

Collisions   20.66 eV

20.61 eV

18.70 eV

Continuous electrical discharge

Ground states

FIGURE 13.17   For a He–Ne laser, the diagram shows a simplified energy level diagram for the most common output of 632.8 nm. There are many other metastable states that may be used with different cavity lengths to get other "lasing" wavelengths.

including professionals who thought there would never be significant applications for the device. We have only touched upon a couple of the simple lasers that illustrate the principles behind the process; however, there is much more than indicated here. There are many other types of lasers, including some that utilize molecular states rather than atomic states. For example, dye lasers use molecules with energy levels so close together that they can produce laser action over a near continuum of wavelengths [6]. There are other solid-state lasers like the Nd:YAG laser that uses yttrium aluminum garnet ($Y_3Al_2 [AlO_4]_3$) crystal with about 1% neodymium (Nd) as an impurity acting similar to the chromium ions in the ruby laser. These lasers can be operated in the continuous or pulsed modes depending upon the application and how it is pumped. In the pulsed mode, they can be made to produce over 200 MW of power over very short time intervals. Also, the wavelengths may be adjusted from below 300 nm up to over 1500 nm using frequency multiplier techniques. This type of laser is ideal for use in Light Detection and Ranging (LiDAR), which in its simplest form is just distance determination by sending thousands (up to 150,000) of pulses of light each second and measuring the time for the reflected light to return and then using the known speed of light to calculate the distance between the object and the laser. This simple form allows long-range measurement of geographical or oceanic features with very high precision. When combined with absorption and phase measurements, LiDAR can determine information about the atmosphere including trace chemical content. Of course, all of this depends upon the ability to perform high-speed switching and computing. The Nd:YAG laser can be made compactly with enough power in the appropriate wavelength range to perform microsurgery on the human body. Carbon dioxide gas lasers are manufactured with power in the kilowatt range and are now common in heavy industrial operations that involve precision cutting and welding of all types of materials including steel. Very small "semiconductor lasers" are manufactured by the millions for use in communications, computers, and entertainment. They are ideal for use in fiber-optic communications involving voice, data, and video.

13.4.4 Semiconductors, p–n Junction, and Light-Emitting Diode

Some materials will sometimes behave as conductors under certain conditions and as insulators under others, for example, silicon and germanium. To explain these properties, one must look at what happens as a collection of atoms come closer and closer together and are bound into a solid. The atoms in every solid are so close together that their weakly bound valence electrons are shared, or in terms of quantum vernacular, their wave functions overlap with each other and combine in ways that form symmetric, antisymmetric, or mixed functions. This causes energy level splitting, so the more atoms in the mix, the greater the number of available levels produced by the process of forming the solid. This is understandable in that as the atoms get closer and closer together, each is exposed to strong magnetic and electric fields. In the solid state, the number of available levels is so close together that they form almost continuous bands of allowed energies. This corresponds to the *valence band* energies. Generally speaking, there will be *bandgaps* between some of the continuous bands of allowed energies. An electron in a solid cannot exist within a bandgap because these energy regions are quantum mechanically forbidden. If the higher-energy bands are filled with nearly free electrons, it is called the *conduction band*. These electrons are very mobile, and the application of an electric field will cause them to easily move. For solids that are good conductors at room temperature, the thermal (kinetic) energy provided by the heat is enough to cause the electrons to leap the forbidden gap and occupy states in the conduction band. In a good conductor, the Fermi level will be in the middle of the valence band, which will be half filled with electrons. A solid cannot be a good conductor if the valence band is completely filled with electrons, that is, all possible quantum states are fully occupied. In this case, the material is a good insulator and the Fermi level is near the top of the valence band. Bands in some solids may also overlap and are sometimes referred to as semimetals. There are those materials in which the Fermi energy is located in the middle of the bandgap between the valence and conduction bands. These are said to be *semiconductors* because they can be made to be good conductors or good insulators. The valence and conduction bands are separated by a smaller gap than that of an insulator. For example, in diamond, the energy gap is about 6 eV wide and the Fermi energy is located at the top of the valence band. At room temperature (300 K), $k_B T$ is about 0.025 eV, and so there is no way that thermal energy can supply enough energy to cause valence electrons to leap the gap and reside in its conduction band. Furthermore, since the valence band is filled, the electrons are tightly bound and not very mobile so that it would take a very strong electric field to give them enough kinetic energy to leap the 6 eV gap. Diamond is classified as an insulator. On the other hand, pure silicon (Si) has the diamond crystal structure, which is the same as the element carbon (C), and it has a bandgap energy of about 1 eV. At low temperatures, Si is similar to diamond as an insulator, but at room temperature, there is enough thermal energy to cause a small fraction of the valence electrons to leap the gap energy into its conduction band. When an electric field is applied, there will be a very small current flow. Furthermore, if a small amount of impurity such as arsenic (As) is added to Si, it can become a good conductor, and this feature is shared by all materials that behave as semiconductors. Si has four outer electrons and As has five, so when As is used as an impurity in Si, it replaces some of the Si atoms by covalent bonding with its nearest neighbors, leaving one weakly bound (.05 eV) electron, which can easily be moved about the crystal. This is Si that has been *doped* with As and the conductivity has been dramatically increased. The new energy levels associated with the As are called *donor* levels and lie just below the conduction band edge of the Si. The Fermi energy of the pure (*intrinsic*) Si has been shifted upward closer to the bottom edge of the conduction band. The Si is now called *n-type* because any electric current induced by the application of an electric field will be carried by the excess electrons introduced. A schematic of the band structure of intrinsic and arsenic-doped Si is shown in Figure 13.18a. Another useful dopant in Si is gallium (Ga) because it has only three electrons in its outer energy shell, which means that when the Ga replaces the Si in the crystal structure, there is now a missing electron at the lattice site previously occupied by the missing Si atom. So there is a *hole* in the electronic structure of the crystal lattice. If an electric field is applied across the crystal, a weakly bound electron in the outer shell of Si will migrate or hop to the vacant level of Ga and leave another electron vacancy (hole) in the nearest neighbor silicon atom, so it is as if the missing level is migrating through the Si crystal. Therefore, this material is called *p-type* because it behaves as if the hole is carrying the charge as the current flows. The Ga impurity creates *acceptor* energy levels just above the upper edge of the Si valence band. This is shown in Figure 13.18b.

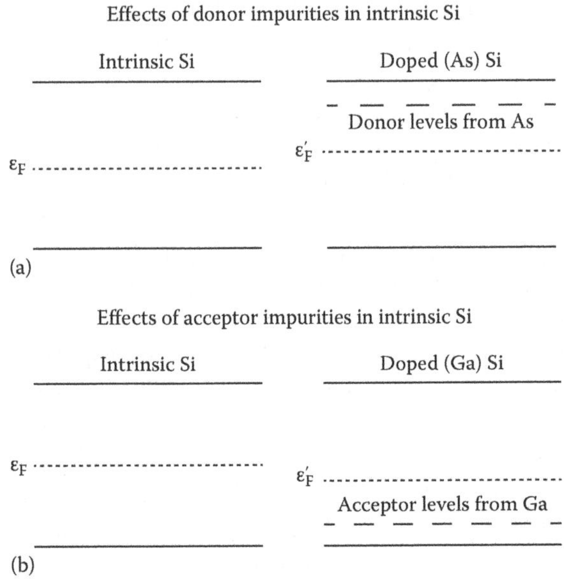

Effects of donor impurities in intrinsic Si

Intrinsic Si	Doped (As) Si

Donor levels from As

ε_F .... ε'_F ....

(a)

Effects of acceptor impurities in intrinsic Si

Intrinsic Si	Doped (Ga) Si

ε_F .... ε'_F ....

Acceptor levels from Ga

(b)

FIGURE 13.18 The effects of (a) donor and (b) acceptor impurities in pure Si are shown. These effects are common to all semiconductors, assuming the appropriate dopant is used. In each case, note that the Fermi energy is shifted from the center of the band.

When n-type and p-type regions coexist in close proximity within the same semiconductor crystal, there is a transition region in which holes and electrons recombine and form a charge free zone called the *depletion* region. The Fermi energy must be constant across the transitions region and that forces the valence and conduction bands to shift in order to maintain charge neutrality and the lowest energy for the system in equilibrium. This is depicted in Figure 13.19.

At nonzero temperature, there will be thermal agitation of the electrons into the conduction band causing a small electron current (I_T) to flow, but there will also be a small recombination of electron current (I_R) in the opposite direction as the electrons flow across the junction to recombine with the holes. The net result is that once equilibrium is reached, the two currents cancel each other and a region extending to either side of the "junction" is void of mobile charge carriers. This region is called the depletion zone.

Figure 13.20 is a simplified schematic showing three configurations of voltage bias: (a) no bias voltage, (b) reverse voltage bias, and (c) forward voltage bias. In (a), there is no net current flow, and so all is neutral; in (b), after a short time, an equilibrium condition is attained where there is a depletion of charge in a region near the junction extending on either side, and no current

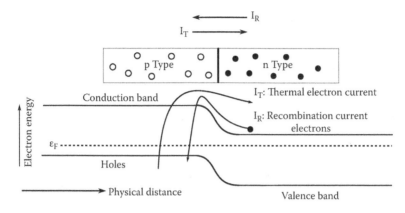

FIGURE 13.19 Band structure for P and N regions in close proximity on a common semiconductor crystal. The conduction and valance bands shift in order to maintain a minimum energy configuration and a constant Fermi energy across the physical dimension. This is the basic junction diode structure.

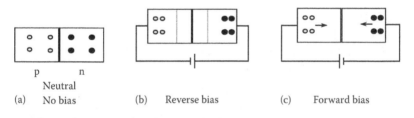

FIGURE 13.20 This diagram shows a simplified semiconductor junction diode under various biasing conditions: (a) neutral in which no current flows; (b) reverse bias in which current flows until equilibrium is reached, forming a depletion region near the junction shown by the two gray lines, then only a small leakage current flows; and (c) forward bias in which the hole and electron currents combine to give a net forward current flow.

flows other than a small leakage current; and in (c), the holes and electrons flow in opposite directions, giving a net current flow equal to the sum of the hole and electron currents.

A "light-emitting diode" (LED) is a p–n junction formed such that the energy gap between impurity states and their respective intrinsic band edges may be overcome by forward biasing and energy is produced primarily in the form of radiation. The band structure is shown in Figure 13.21. Diodes can now be engineered so that the emitted photons range from infrared to ultraviolet, including visible light. Because the radiation from a single LED is usually over a very narrow wavelength range, for most practical purposes, it can be treated as monochromatic. There are now high-power white-light LEDs that replace normal incandescent bulbs. They use less power for producing

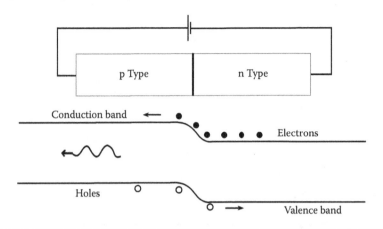

FIGURE 13.21   The band structure of an LED p–n junction is shown. As electrons in the conduction band recombine with the holes n, the valence band radiation is emitted; the frequency (wavelength) depends upon the forbidden energy gap.

comparable radiant energy than other light sources such as incandescent and fluorescent lights.

The first LEDs were introduced commercially in the early 1960s and have grown into a major industry. High-power LEDs also exist that are used as laser light pumps to create charge population inversions.

Various combinations of materials can be used to make LEDs—for example, infrared to red wavelength range, gallium arsenide (GaAs) and aluminum gallium arsenide (AlGaAs) of varying fractions of each component; green, gallium phosphide (GaP), aluminum gallium phosphide (AlGaP), aluminum gallium indium phosphide (AlGaInP), indium gallium nitride (InGaN), and gallium nitride (GaN); and blue, zinc selenide (ZnSe) and indium gallium nitride (InGaN). White light is attained by using various combinations of LED colors or using blue and/or ultraviolet LEDs with an appropriate phosphor. For the wavelengths from infrared to ultraviolet and white-light radiation, the bandgap voltages range from approximately 1.6 eV to a little over 4.0 eV, respectively. It should be pointed out that the emission of radiation in an LED is produced by the recombination of electron–hole pairs, which release energy. In semiconductors like silicon and germanium, the energy is absorbed by the crystal structure as heat. In other semiconductors like gallium arsenide, the energy is released primarily as visible photons. In the forward bias condition, the electrons and holes both move toward the depletion region, and this is where they recombine. The electrical current needed for this is small. If the bias current is made high enough, the spontaneous emission cannot keep up with the rate of arrival of the charge carriers, and a population inversion occurs in the depletion region. The spontaneous emission of the LED now becomes a light pump source for the stimulation for the emission of other photons in an avalanche effect. This is the condition for laser action to occur. If the dimensions of the junction area are carefully controlled, then light amplification can be made to occur. In a semiconductor LASER, opposite ends of the p–n junction are made parallel and partially reflective to form a resonant cavity within the depletion region for the desired radiation frequency.

Insulators tend to have bandgaps that are too wide for visible light photons to be absorbed, so they should appear transparent to such; however, we know that is not always the case. This, for the most part, is due to the scattering of the photons resulting from defects in the structures of the material. For example, highly processed quartz is transparent, but most natural quartz is cloudy and opaque due to defects and impurities. Conductors tend to be opaque because they can absorb most of the visible light photons due to the smaller forbidden energy bandgaps. Semiconductors also tend to be opaque because their forbidden energy bandgaps are in the visible light range. They tend to be transparent at the longer wavelengths approaching the infrared. For example, gallium is often used as a window for infrared detection systems, even though it is opaque to the visible spectrum. Its filtering action reduces unwanted noise in the system.

13.4.5 GENERAL MOTION OF AN ELECTRON IN AN ENERGY BAND OF AN INTRINSIC SEMICONDUCTOR

In order to have a more complete understanding about the motion of electrons in a semiconductor, it is necessary to understand the equations of motion. We will apply the de Broglie hypothesis and consider wave properties of an electron as it moves through the periodic crystal lattice. The Schrödinger equation includes an interaction potential, which in the case of a free space electron would be zero everywhere. This cannot be the case for a nearly free electron moving through the lattice, because the potential will be periodic. As an electron moves through such a crystal lattice, it interacts with the lattice as well as holes, other electrons, impurities, and any defects that may occur therein. This will involve collisions of all sorts including elastic and inelastic. If there are no impurities or defects in the crystal, then it is ideal. In any case, at nonzero absolute temperature, the lattice vibrates like a quantum harmonic oscillator with a discrete set of allowed energy states given by $E = (n + 1/2)\hbar\omega$ where $n = 0, 1, 2, 3, ...$ This tells us that the energy of the thermally induced lattice vibration is quantized very similar to the photons of EM waves. Physically, one may think of the lattice sites vibrating and producing a slightly large scattering cross section available for collisions with moving charges. The quantum of energy is called a *phonon* in direct analogy with photons for EM waves. When n is zero, the energy is the zero point energy of the particular mode of vibration of the lattice. Remember, phonons represent quantized elastic waves in the crystal lattice with which charge carriers will interact. This is a direct consequence of the allowed solutions under periodic boundary conditions of the Schrödinger equation. Felix Bloch (1905–1983; Nobel Prize, Physics, 1952) proved in 1928 that the solution to the Schrödinger equation for one electron moving under the influence of a periodic potential had to be of the form

$$\Psi(\mathbf{k}, \mathbf{r}) = u(\mathbf{k}, \mathbf{r}) \exp(i\mathbf{k} \cdot \mathbf{r}), \qquad (13.37)$$

where

  r is the position of the de Broglie wave associated with the electron
      (charge carrier)
  k is the wave vector giving the direction of the wave with magnitude of $2\pi/\lambda$
  u(k) is a function that reflects the periodicity of the crystal lattice

The exponential term is a plane wave. The wave function is just the product of a special periodic function with a plane wave. This is a direct result of the fact that u(k, r) = u(k, r + a) where a is the periodicity of the lattice in each direction. Equation 13.37 is called a Bloch function. For the simple case of cubic symmetry with periodicity a, Equation 13.37 yields the following:

$$k_x, k_y, k_z = 0, \pm 2\pi/a, \pm 4\pi/a, \pm 6\pi/a,$$

which shows that the wave number is quantized. Furthermore, the allowed energy states are given by

$$\varepsilon = \frac{\hbar^2}{2m}\left(k_x^2 + k_y^2 + k_z^2\right). \qquad (13.38)$$

Furthermore, an electron moving in a crystal is subject to external forces such as electric and magnetic fields as well as from interactions with the crystal lattice. When an electron breaches the energy gap, one may think of it being "nearly free" so that its total energy is given by Equation 13.38 plus the gap energy E_g. This is also interesting because it leads to a useful interpretation of the mass of the charge carriers given by Equation 13.38. We see that $\varepsilon_k = k^2\hbar^2/2m$; therefore, the second derivative of ε_k with respect to k is just the inverse of the mass. This means that the mass is strongly dependent upon how the energy changes in k or momentum space. This has a strong impact upon the mobility of the charge carriers, especially near the band edges.

Experimentally, it is known that the hole and electron mobilities are different, and this is explained by this dependence of the effective mass m* defined by

$$1/m^* = \hbar^2 d^2 \varepsilon_k /dk^2. \tag{13.39}$$

The physical interpretation is that the charges interact with the lattice and there is a transfer of energy back and forth between them, and this is a way to interpret the effect of the lattice and the energy bandgap upon the motion of the charges in a semiconductor. The effective mass may also depend upon the physical direction of the charge motion, in which case Equation 13.38 will necessarily change.

If an electric field is applied to a conductor (or semiconductor), we know that the charge carriers will experience a force and tend to accelerate but will reach a constant drift velocity, v_d, due to the collisions with the lattice phonons, other charges,[*] impurities, and defects in the crystal. The mobility is determined largely by the effective mass of the carriers and the average time of collisions that the mobile charges undergo. If an electric field, E, is applied to charge q, it experiences a force, F, given by

$$F = qE. \tag{13.40}$$

Since the equation of motion is given by

$$F = \frac{\partial p}{\partial t}; \quad p = mv = \hbar k,$$

then

$$qE = \frac{\partial p}{\partial t} = \hbar \frac{\partial k}{\partial t} \tag{13.41}$$

So if a collection of moving charges are in equilibrium with the lattice, the net force will be zero, and the collective acceleration is stopped by the lattice interactions with the charges set in motion by the electric field. If we consider an individual collision, it's clear that in the absence of the field, the charge's motion is random with some mean free path and average speed, v_{av}, that determines the time between collisions, Δt. In the presence of the field, during this same small time interval, the collection of charges will have moved in a direction parallel to the electric field. Therefore, for the time Δt, Equations 13.40 and 13.41 lead to $\hbar \Delta k = qE\Delta t = m\Delta v$. Physically, we interpret Δv to be the

---

[*] Charge–charge collisions are rare and less important to mobility than charge–lattice collisions.

final drift velocity of a charge imposed by the interactions between the moving charge and everything else in the material during the time interval Δt:

$$\Delta v = v_d = q\, E\Delta t/m. \tag{13.42}$$

In general, v_d will be much smaller than the random collision speed v_{av} mentioned above.

For n_q charges per unit volume in a conductor (or semiconductor) moving under the influence of an electric field, **E**, with velocity, v_d, the current density is defined as the current per unit area that passes through a surface. In its simplest form, it is symbolized by **J** and given by

$$J = n_q q v_d \left(A/m^2\right). \tag{13.43}$$

In terms of Ohm's law, **J** can be written as

$$J = \sigma_q E, \tag{13.44}$$

where σ_q is the *electrical conductivity* and is obtained from the previous three equations as

$$\sigma_q = n_q q^2 \Delta t/m. \tag{13.45}$$

The mobility μ_q of charge q, by definition, is the magnitude of the drift velocity per unit of electric field and is positive for electrons and holes; therefore, utilizing Equation 13.42, we have

$$\mu_q = v_d/E = q\,\Delta t/m \quad \text{and} \quad \sigma_q = n_q q \mu_q. \tag{13.46}$$

Because the electrical conductivity is defined for both holes and electrons, the contributions from each must be considered when working with charge motion in semiconductors:

$$\sigma_T = \sigma_e + \sigma_h$$

$$\sigma_T = n_e e \mu_e + n_h e \mu_h. \tag{13.47}$$

The hole and electron carrier concentrations n_h and n_e, respectively, are determined by the Fermi function and the density of allowed states.

13.4.6 Photon Detectors

There are many materials with properties that can be utilized as photon detectors. Semiconductors and the p–n junction have become the mainstay in much of the detection technology of the modern era. Two fundamental properties of semiconductors will be considered: photoconductance and photovoltaic.

13.4.6.1 Photoconductance Detectors

Photoconductance is the property that allows a material to change its electrical conductance (σ_ε) (resistivity [ρ] or resistance [R]). For certain semiconductors, in light of the quantum theory, it is easy to see why it occurs. Imagine a pure (intrinsic) semiconductor with a bandgap in a range that will absorb energy from radiation in the visible range. Electrons in the valence band will attain enough kinetic energy to lift them into the conduction band and thus increase the electrical conductance of the material. This reduces the macroscopic resistance to electrical current flow through the material when an electric field is applied. Of course, the bandgap will determine the longest wavelength that can be absorbed by the material. For example, cadmium sulfide (CdS) has a bandgap of about 2.46 eV, and it can absorb radiation of wavelengths shorter than about 5050 A. A useful circuit for utilizing this property is shown in Figure 13.22.

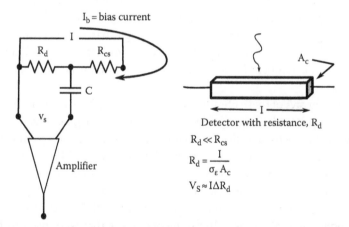

FIGURE 13.22 Simple photoconductor light detector: I_b is a constant bias current source for the detector resistance, A_c is the cross-sectional area of the detector, and I is the length of the detector. σ_ε is the free charge carrier electrical conductivity and is proportional to the free charge carrier concentration; it is a characteristic of the detector material, not its geometric factors. C is a blocking capacitor that prevents any DC currents from entering the amplifier.

The bias current is constant and can be adjusted for optimum performance of the detector. The resistance R_{cs} is large compared to the detector resistance R_d to assure that small changes in the detector resistance have no significant impact upon the bias current. The signal voltage v_s is actually a small DC voltage change caused by absorption of the radiation energy impinging upon the detector. Its value is given by Ohm's law to be

$$v_s \approx I_b \Delta R_d = -I_b R_d \Delta\sigma_\varepsilon / \sigma_\varepsilon,$$

which because of Equation 13.47 can also be written as

$$v_s \approx I_b \Delta R_d = -I_b R_d \Delta n/n, \qquad (13.48)$$

where Δn is the excess charge carrier concentration in the conduction band resulting from the incident radiation.

We have assumed that all radiation is absorbed and that all photons transfer electrons from the valence band to the conduction band. The detector material is not a black body, so it cannot absorb all the incident radiation; however, with proper thin-film, antireflection coatings that are transparent to a particular wavelength range, it is possible to get high absorption efficiency for some semiconductors. The issue of photons making electron–hole pairs is more complex in that a photon must be absorbed and it must transfer enough of its energy to an electron to force it to jump to the conduction band and then the electron and hole must be collected before they recombine or they do not contribute to the signal. We see that the active area and thickness of the detector become important to having enough charges generating current that can be collected as a signal. Response *quantum efficiency* is defined as the ratio of the number of detectable output events to the number of incident photons.

Linear arrays of these photoconductor detectors have been used to produce early (1960s) military-grade passive infrared imaging systems. In its simplest form, such a detector is used by scanning an image across the detector and using the output to drive a small linear array of LEDs whose light is scanned in phase with the field scanner, across a phosphor-coated screen to produce a monochrome image. Such a system is depicted in Figure 13.23. The systems were large and bulky but "man portable." Most were used in heavy equipment like tanks. The semiconductors that were used to fabricate the detectors had forbidden bandgaps near 0.1 eV, so they had to be cooled to liquid nitrogen temperatures of about 77 K in order to resolve temperature gradients of less than 0.05 K in the 8–14 µm wavelength range, which is a relatively clear window for atmospheric transmission. Another advantage is that the long-wavelength infrared is not scattered as much as

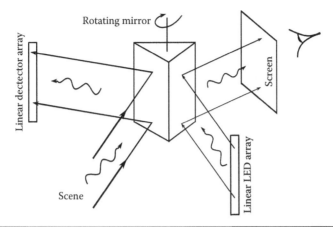

FIGURE 13.23   Schematic of a simple scanning imager: The rotating mirror is front surface and rotates such that the scene is swept across the linear detector. The output of the detector array electronics drives a similar array of light-emitting diodes whose output light is swept across the screen that enables the viewer to see the scene image. The physical geometry of the mirror can be any polygon that meets the spin rate needs for the application. In addition, optics exists in the input and output light chains to create a usable image.

visible light by dust, fog, and smoke. So this type of imaging system penetrates these kinds of environments better than a visible light–based system. Systems such as these were still in use until the early 1990s. These are called passive systems because they do not require a separate radiation source to illuminate the target area. These systems utilize the targets' own radiation generated internally or from other environmental sources absorbed and reradiated by the target.

Compared to other types of photon detectors, photoconductance-based systems are some of the easiest to fabricate into viable systems that perform imaging.

13.4.6.2 Photovoltaic Detectors

Photovoltaic cells are single- or multiple-layer stacks of p–n junctions. The most common is a single junction made from silicon. A simplified band diagram of a single junction is shown in Figure 13.24.

In (a), there is no radiation incident upon the p–n junction, so an equilibrium state exists. As the material is illuminated near the junction in the p region, additional electrons are transferred to the conduction band, resulting in a strong electric field within the junction region directed from the p region to the n region due to the enhanced charge separation. A new equilibrium condition is reached and the Fermi energy shifts up closer to the conduction band to compensate. This is reflected at the terminals as an electric

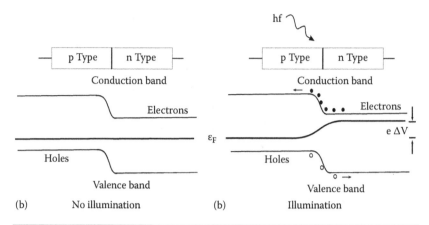

FIGURE 13.24 The band structure of a photovoltaic p–n junction: (a) depicts the case with no illumination and (b) shows the case with the junction illuminated.

potential. If an external load is attached to the terminals, the charge carriers will move to lower energy through the load. As long as the radiation is continued, there will be current flowing through the load. The characteristics can be measured with standard equipment to produce a current (I–V) plot. The equation for an ideal p–n junction is given by

$$I = I_{sat}\left\{\exp\left[e\Delta V/k_B T\right] - 1\right\}, \tag{13.49}$$

where I_{sat} is the saturation current for a p–n junction, which is the maximum current that can pass through the junction in the forward bias condition. It is limited by the diffusion properties of the electrons and holes near the junction region and has only a weak dependence on the absolute temperature T of the junction. e is the magnitude of the electronic and hole charge (1.602 × 10⁻¹⁹ C), and $e\Delta V$ is the shift in the level of the Fermi energy, which is associated with the junction voltage ΔV. An I–V plot of a photovoltaic device is shown in Figure 13.25.

As given, Equation 13.49 assumes that if the terminals of the junction are "short-circuited," there will be no current flow in the absence of illumination. However, given the nature of the devices and the environments in which they are housed and operate, there will always be background radiation incident on the active area of the junction that produces enough energy to generate some electron–hole pairs and hence a small, undesired current. There is also a slight "ohmic" leakage current when the Fermi energy level shifts. Finally, there is an efficiency factor that arises in the argument of the exponential term of Equation 13.49. Therefore, for

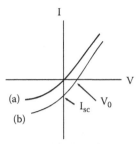

FIGURE 13.25 Ideal current–voltage (I–V) characteristics of a photovoltaic p–n junction: (a) shows the case for no illumination of the junction, while (b) shows an illuminated junction.

real p–n junctions used as detectors, Equation 13.49 must be modified to present a more accurate model of the signal current:

$$I = I_{sat}\left\{\exp\left[\alpha e\Delta V/k_B T\right]-1\right\}+I_{sc}+G_{sh}\Delta V, \tag{13.50}$$

α is an efficiency factor for photodiodes: 1 for ideal and 0.5–0.33 for real junctions.

I_{SC} is the "short circuit" current due to background radiation and is determined empirically; it will have a linear dependence upon the active area of the detector. Also, any incident flux will affect this current, so it is not independent of the external radiation flux density Φ_{qS}. G_{sh} is the effective shunt conductance arising because of an Ohm's law leakage current for nonzero ΔV. It is proportional to the cross-sectional area of the junction and is on the order of $10^{-8}\ \Omega^{-1}$.

In general, we are interested in small changes in I with respect to small changes in ΔV that occur due to external radiation flux density changes on the detectors' surface. In this case, the argument of the exponential term in Equation 13.50 becomes linear. Therefore, we can write that the signal current is given by

$$i_s \approx \left\{I_{sat}\, \alpha e/\left(k_B T\right)+g_{sc}+G_{sh}\right\}\delta V, \tag{13.51}$$

where

   δV represents the changes in ΔV

   g_{sc} is an effective short circuit conductance associated with the short circuit response to δV

It has to be determined empirically.

Because ΔV will depend upon the detector active area A_d and the flux density Φ_{qS}, from the source of incident radiation, it is reasonable to write an equation that, for small signals, reflects the signal current in terms of the incident signal radiation flux density Φ_{qS}. Therefore,

$$i_s \approx e\eta A_d \Phi_{qS}, \tag{13.52}$$

where η is the fraction of incident photons that produce meaningful units of current by collecting electron–hole pairs before they recombine. One unit of current would be the charge e divided by the maximum recombination time or the inverse of the recombination rate. η is the system quantum efficiency.

The best operating point is when there is no current if the terminals are "short-circuited," and since this never occurs in real p–n junctions that are used as photovoltaic detectors, a reverse or "back" bias circuit may be necessary, depending upon desired performance, such as in an imaging system. A simple arrangement for this is shown in Figure 13.26.

V_b can also be adjusted to increase or decrease the width of the depletion region assuring a more likely probability of photon absorption in proximity thereto. This reduces the time it takes for free charge carriers to reach the depletion region from their site of photon absorption. Increasing Vb further can generate an electric field that accelerates the carriers to sufficient kinetic energy that collisions with the atoms at the lattice sites generate additional free charge carriers. This tends to produce a multiplier effect and possibly improve the signal to noise.

One advantage of photovoltaic detector systems is that they can be used to fabricate practical mosaic staring arrays as opposed to linear scanning arrays, which tend to be the domain of photoconductive arrays for practical reasons of signal sampling. This improves the signal to noise over a linear array immensely because each picture element (pixel) spends more time on a

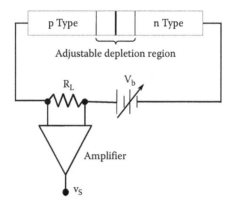

FIGURE 13.26 Photovoltaic detector circuit with back bias: The back bias V_b may be adjusted to provide optimum performance by offsetting some of the noise current but also by manipulating the width of the depletion region.

TABLE 13.2
Materials Suitable for Quantum Detectors

Detector Material	Detector Temperature (K)	Wavelength Range (μm)
Si	300	0.4–1.1
PbS	300, 193, 77	1–3.3, 1–3.6, 1–5.2
PbSe	300, 145–250, 77	1–4.5, 1–6.5, 2–7.3
InAs	300, 195, 77	1–3.7, 1.5–3.6, 1–3.3
InSb	300, 77	2.2–7.5
Ge:Au[a]	77	2–9.7
Ge:Hg	5 (liquid He)	2–14.2
Ge:Cd	5	2–24
Ge:Zn	5	5–40
GE:Cu	5	4–31
Si:As and Si:Ga	20	3–24, 2–17
$Hg_{1-x}Cd_xTe$	77 (liquid N)	5–12
$Pb_{1-x}S_xTe$	77	2–11

Source: Hecht , E., *Optics*, 4th edn., Addison Wesley, San Francisco, CA, 2002.
[a] A:B means A doped with B; He, N are helium and nitrogen

dedicated portion of the scene than is possible with a scanning linear array in a single frame sample time necessary for human vision. Many staring arrays can be made on one substrate of very strong materials like silicon, which can also contain the underlying signal processing electronics. The technology is mature enough now that it is common for N × N arrays to have millions of pixel elements to give resolution approaching that of the human eye.

These detectors are the most difficult to fabricate into viable imaging systems on a large scale but offer significant advantages in sensitivity, versatility, and reliability that outweigh any disadvantages. They can be made to operate in high-performance modes at many different wavelengths for special application. The highest-performance systems use detector materials that must be cooled to liquid nitrogen or even liquid helium temperatures (Table 13.2).

13.4.6.3 Thermal Photon Detectors

Thermal detectors do not use quantum energy transitions to function; they use heating, cooling effects, and thermal conduction. One might say that thermal conductance, heating, and cooling are as much about quantum effects as is electrical conductivity and the generation and transport of charges. Indeed, that is basically true because they are intimately related. However, thermal effects are macroscopic effects, and it is not necessary to know the microscopic details of the quantum processes that reflect into the macroscopic realm of thermal detectors.

There are many macroscopic phenomena that can be used to detect thermal changes in our environment, and several of them are promising and merit serious consideration in making imaging systems. Most tend to be designed to detect thermal changes over a narrow wavelength range, but one thing that all thermal detectors have in common is that they are capable of detecting energy over a broad spectrum of radiation because they do not depend upon quantum energy transitions. Practical high-performance systems, however, depend upon optical coatings for the efficient absorption of energy. One other thing of importance is that they all work at or near room temperature, and for that alone, they will never be as sensitive as quantum-based detectors in the infrared region. Because they require small temperature changes in the detective element, any imaging system based on thermal detectors will need to have the radiation target scene "chopped." Chopping refers to the process of blocking the scene in a controlled repetitive pattern at a rate compatible with human vision. Choppers may be electromechanical or electro-optical.

There are several physical effects that can be used to make thermal detectors: thermal expansion, the Peltier effect (thermocouple), the thermo-pneumatic effect (pressure), the bolometric effect, the pyroelectric effect, and the ferroelectric effect.

For our case study, we will use the pyroelectric effect because it is one of the least complex to analyze and will provide the transfer and response function that are common to all thermal detectors. The *pyroelectric effect* may be described as follows: The pyroelectric effect corresponds to the spontaneous change in the electric polarization (P) as a result of a temperature change. P is the dipole moment per unit volume (C/m^2). dP/dT is called the pyroelectric coefficient, p_e (C/m^2-K), and provides a measure of the flow of charge as the polarization changes. This represents a structure "phase" transition, and if one can choose a point on the transition curve at a temperature near the desired operating temperature where dP/dT is sufficiently large and fast, then the detectable current becomes a useful signal source. In the case of a pyroelectric detector element, two metal contacts are deposited on the material so as to detect the polarization change as an electrical current flow.

The following is common to all thermal detectors in that the thermal properties are all governed by the same physics. In any transfer of thermal energy to or from a massive object, one must account for the source, the heat capacity of the mass, and the conductivity of heat to and from the mass whether by physical contact or radiation.

If we let Φ (W) be our chopped source of radiant energy, C_d (J/K) the heat capacitance of the detective element, and G (W/K) the thermal conductance to the environment, we can write

$$\Phi = \Phi_o \exp(i\omega t),$$

where ω is the chopper frequency in radians per second. The fraction of the energy absorbed by the detector will be determined by the emissivity (e_d) of the detector's surface. The modulation of the radiation implies that the time variations in both the polarization and temperature changes (δ) will be of similar form:

$$\delta P = \delta P_o \exp(i\omega t) \tag{13.53}$$

and

$$\delta T = \delta T_o \exp(i\omega t), \tag{13.54}$$

where δP_o and δT_o are the maximum values of these quantities about their respective equilibrium values. From the definition of the pyroelectric coefficient, we can also write that

$$d(\delta P) = p_e \, d(\delta T) \quad \text{and} \quad d(\delta P)/dt = p_e \, d(\delta T)/dt. \tag{13.55}$$

The energy absorption by the detective element is governed by the law of conservation of energy; therefore, we can write

$$C_d \, d(\delta T)/dt + G(\delta T) = e_d \Phi. \tag{13.56}$$

Using Equations 13.53 through 13.56, it can be shown that

$$\delta T = e_d \, \Phi \left\{ 1 + \omega^2 (C_d/G)^2 \right\}^{-1/2} /G$$

or

$$\delta T = e_d \, \Phi \left\{ 1 + \omega^2 \tau_d^2 \right\}^{-1/2} /G, \tag{13.57}$$

where τ_d is called the thermal time constant. Basically, it is a measure of how long a time it takes the detector to heat up or cool down. It is analogous to the electrical time constant.

The current density associated with this temperature change in the detector is obtained from Equation 13.55, and its magnitude is given by

$$J_d = \omega p_e \delta T,$$

and assuming the energy is absorbed uniformly through the detector surface of area A_d, the electrical current is

$$I_d = A_d \omega p_e \delta T. \tag{13.58}$$

Note that the current has the same time dependence as the temperature change.

This must be converted to an electrical signal, so now we must be a little more specific about the physical configuration of the detector and how the signal is removed and how we get the signal.

Keep in mind that the detector element is essentially a dielectric with two metal contacts, so it is essentially a capacitor. Consider the electrical configuration in Figure 13.27 where the detector is rectangular with a surface area A_d and a small thickness. Remember if the detector element is designed to be part of an imaging system, the active elements must respond within the human eye frame rate, about a thirtieth of a second, so the detector is relatively small, less than a cubic millimeter.

Ohm's law tells us that the magnitude of the signal voltage is given by

$$v_s = I_d \left| Z \right| = I_d R_{eq} \left(1 + \omega^2 \tau_e^2 \right)^{-1/2}$$

$\tau_e = R_{eq} C_{eq}$ is the electrical time constant

$$V_s = \omega p_e A_d e_d \Phi R_{eq} G^{-1} \left\{ 1 + \omega^2 \left(\tau_d \right)^2 \right\}^{-1/2} \left\{ 1 + \omega^2 \tau_e^2 \right\}^{-1/2}.$$

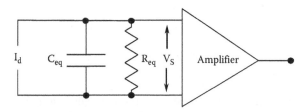

FIGURE 13.27   The electrical model of a pyroelectric thermal detector, with the detector shown as a current source. The equivalent capacitance C_{eq} represents that of the parallel combination of the detector and the amplifier. The equivalent resistance R_{eq} is the combination of the detector resistance, an external shunt resistance (not shown), and the amplifier input resistance.

The responsiveness of the system is defined by $R_s = v_s/\Phi$; therefore,

$$R_s = \omega p_e A_d e_d R_{eq} G^{-1} \left\{ 1 + \omega^2 \left(\tau_d \right)^2 \right\}^{-1/2} \left\{ 1 + \omega^2 \tau_e^2 \right\}^{-1/2}. \qquad (13.59)$$

From the response function, we observe that for chopper frequencies such that $1/\tau_d \leq \omega \leq 1/\tau_e$, the system response is relatively constant, but outside this range, the response rolls off to lower values. We note that at no time in the discussion has the wavelength or quantum energy of the radiation been a factor. These detectors are inherently broadband to EM radiation. It is often desirable to limit this by using a filter or special coatings because unwanted radiation causes noise. Pyroelectrics have one characteristic that makes them problematic for some applications: They are also piezoelectric; therefore, vibration and shock generate much electrical noise that is difficult to eliminate. They operate well at or near room temperatures, so their performance will never match photon detectors. They can be fabricated into staring mosaic arrays using the same techniques used on quantum detectors. All thermal devices have performances that are comparable to each other. As are all detectors, they are limited by fundamental noise sources in the detectors, the processing electronics and the environment. For thermal detectors, other limits involve the interaction with their environment, primarily through thermal conductivity via their physical contact with necessary processing electronics and physical structures. In addition, the Stefan–Boltzmann law imposes a loss of energy that cannot be eliminated. The radiation conduction is fundamental and provides the upper limit on all thermal detector performance at least for those that are uncooled and operate near room temperature.

There are many materials that can be used for thermal detection and imaging, and Table 13.3 presents a few of the most successful ones.

TABLE 13.3 Thermal Detector Materials		
Detector Material	**Operating Temperature (K)**	**Pyroelectric Coefficient (10^{-6} C/m²-K)**
LiNbO$_3$	300	0.03
Ba$_x$Sr$_{1-x}$TiO$_3$ (BST)	300	550–1929
KTa$_{1-x}$Nb$_x$O$_3$ (KTN)	300	2.5
LiTaO$_3$ (LT)	300	230
PbZr$_x$Ti$_{1-x}$O$_3$	300	0.09

13.4.6.4 Figures of Merit for Photon Detectors

It is necessary to develop techniques for evaluating the quality of detectors, which requires quantifying their potential performance in given applications, such as firefighting, military, medical, etc.

Several performance parameters have emerged as the useful means for evaluating the quality of a detector: noise-equivalent power (NEP), detectivity (D), normalized detectivity (D-star or D*), and responsivity (R_d). Each must be specified in reference to the detector's active area, the frequency, and bandwidth over which the noise measurements are made. They may be specified at or over a specific detection wavelength range. This depends upon the specified application:

$$R_d = v_s/\Phi \left(\text{volts/watt, V/W}\right)$$

where

v_s is the rms signal voltage at the output

Φ is the rms incident power

NEP: The root mean square (rms) value of incident power required to produce an rms signal to noise ratio of one.

Practically speaking, this means that you will not detect signals that are below the inherent electronic noise level of the detector. It has dimensions of watts (W). NEP = Φ (v_n/v_s), Φ = incident power, v_n, v_s = noise and signal voltages, respectively, NEP = v_n/R_d

D: Detectivity = $(\text{NEP})^{-1}$ $(\text{W})^{-1}$

D*: A normalization of D to take into account the responsive area of the detector and the electronic bandwidth dependence (cm-Hz$^{1/2}$/W), D* = $(A_d \, \Delta f)^{1/2}$ D, where A_d and Δf are the detector area and measurement bandwidth, respectively.

A typical quantum detector will have parameters comparable to the following [2]:

$$\text{NEP} = 1.1 \times 10^{-10}\,\text{W}, \quad D^* = 1.8 \times 10^9\,\text{cm-Hz}^{1/2}/\text{W}, \quad R_d = 3.2 \times 10^4\,\text{V/W}$$

A typical thermal detector will have parameters comparable to the following [2]:

$$\text{NEP} = 1.0 \times 10^{-8}\,\text{W}, \quad D^* = 2.0 \times 10^8\,\text{cm-Hz}^{1/2}/\text{W}, \quad R_d = 2.0\,\text{V/W}$$

PROBLEMS

13.1 Show that the Planck theory leads to Wien's displacement law in the short wavelength limit and to the Rayleigh–Jeans result in the long wavelength limit for the radiation.

13.2 Estimate the surface area of your body and use the Stefan–Boltzmann law to determine how much energy you radiate. Remember, you must use absolute temperature.

13.3 Rework Example 13.3 knowing that the index of refraction for the vitreous humor within the eye is 1.34. How does this change the results of the calculations? Explain.

13.4 Is it plausible to think that the wavelength of light changes in traveling from air to water but the frequency will remain unchanged? Explain.

13.5 Design and sketch a simple submarine periscope that uses two right-angle prisms with no coatings. What is the minimum index of refraction of the material used in the prisms if it is to function properly?

13.6 Water has an index of refraction of about 1.33. (a) Determine the critical angle for total internal reflection for visible light passing from water into air ($n_{air} \approx 1.00029$). (b) Determine the critical angle for an air–diamond interface, $n = 2.42$.

13.7 Derive Equation 13.24, assuming that $n_1 \approx n_2$.

13.8 Calculate the numerical aperture of a fiber optic with no cladding in air if its index of refraction is 1.55.

13.9 A fiber-optic cable made up of a bundle of one thousand, 10 micron fibers, is supposed to deliver a circle of light to a CCD camera detector chip that is a square of 2.50 cm dimension on each side. (a) Determine the Numerical Aperture of the fiber. (b) Determine the distance required between the end of the bundle and the center of the chip that will allow the light circle to exactly fall within the borders of the detector chip. For each fiber of the bundle, the cladding index of refraction is 1.50 and that of the core is 1.52. Refer to Figures 13.14 and 13.15.

13.10 Determine the maximum speed of an electron in the hydrogen atom. (*Hint:* Consider Equations 13.14 through 13.18.) What fraction of the speed of light is this? Is the theory of special relativity pertinent to this problem? Explain.

13.11 Using Equations 13.28 and 13.29, derive Equation 13.30 and show that the kinetic energy of Equation 13.27 is no longer valid at relativistic speeds. (*Hint:* Imagine applying a force to a particle and taking it from rest up to a final relativistic speed). Use the work–energy theorem from basic physics.

13.12 Referring to the previous problem, show that in the limit as the speed of a particle becomes much less than that of light, the relativistic equation for the kinetic energy reverts to the classical Newtonian form, Equation 13.26.

13.13 Calculate the de Broglie wavelength of a 20 keV electron. Knowing this, explain why electron microscopes have such higher resolution when compared to light-based microscopes.

13.14 An LED requires a forward bias voltage of 2.50 V. Calculate the wavelength of the dominant radiation.

13.15 From what you know about the He–Ne laser, calculate to three significant figures the wavelength of the red light emitted in nanometer, micrometer, and Angstrom units.

13.16 Derive Equation 13.57 for the pyroelectric detector.

13.17 If a thermal detector has no significant loss to the environment except through radiation, (a) using the Stefan–Boltzmann law, derive an expression for the effective thermal conductance G to the environment. (b) Determine the numerical value for a detector of surface area $A_d = 0.7$ mm^2 and emissivity $e_d = 0.90$ if it is operating at a temperature of 0°C. Be sure to show the final units.

13.18 It is known that for any detectors, the statistical fluctuations in the arrival and emission of photons lead to noise generated in the detector. For thermal detectors, it has been shown that this results in the mean square fluctuations in radiation power given by $\langle \delta\Phi^2 \rangle = 4k_b T_d^2 G \Delta f$ [2] where Δf is the bandwidth over which the noise is referenced, usually 1 Hz. (a) Use the result from Problem 13.17 to show that $\langle \delta\Phi^2 \rangle = 16 A_d k_b \sigma e_d T_d^5 \Delta f$. (b) Determine the NEP and D^* of the detector. The latter results give the theoretical limit of performance.

13.19 Using Equation 13.58, verify Equation 13.59.

Appendix A: Trigonometry

A.1 TRIGONOMETRIC FUNCTIONS

There are several basic trigonometric functions defined in terms of two of the three sides of a right-angled triangle shown in Figure A.1. These functions are

$$\sin\theta = \frac{\text{Opposite}}{\text{Hypotenus}} = \frac{a}{c}; \quad \operatorname{cosec}\theta = \frac{1}{\sin\theta} = \frac{c}{a} \qquad (A.1)$$

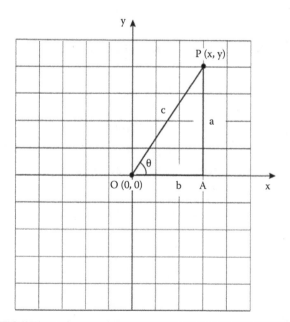

FIGURE A.1   Right angle triangle used to define trigonometric functions.

$$\cos\theta = \frac{\text{Adjacent}}{\text{Hypotenus}} = \frac{b}{c}; \quad \sec\theta = \frac{1}{\cos\theta} = \frac{c}{b} \qquad \text{(A.2)}$$

$$\tan\theta = \frac{\text{Opposite}}{\text{Adjacent}} = \frac{a}{b}; \quad \cotan\theta = \frac{1}{\tan\theta} = \frac{b}{a} \qquad \text{(A.3)}$$

$$\tan\theta = \frac{\sin\theta}{\cos\theta}; \quad \cotan\theta = \frac{\cos\theta}{\sin\theta}. \qquad \text{(A.4)}$$

An inverse trigonometric function is asked when the trigonometric function is given, and the question is to determine the function itself. Here are a few examples.

For $\sin\theta = \dfrac{\text{Opposite}}{\text{Hypotenus}} = \dfrac{a}{c}$, the inverse sine is $\theta = \sin^{-1}(a/c)$.

As a numerical example, take $\sin 30° = 0.50$. So, the inverse operation function can be expressed in the following question.

What is the angle whose sine is 0.50? This is equivalent to $\sin^{-1}(0.50) =$??
The answer is 30°. It is a question asking for the angle whose sine is 0.50.
Similarly, $\cos^{-1}(0.50) =$??
The answer is 60°.

A.2 TRIGONOMETRIC RELATIONS

$$\sin^2\theta + \cos^2\theta = 1 \qquad \text{(A.5)}$$

$$\sin 2\theta = 2\sin\theta\cos\theta \qquad \text{(A.6)}$$

$$\cos 2\theta = \cos^2\theta - \sin^2\theta \qquad \text{(A.7)}$$

$$\sec^2\theta = \tan^2\theta + 1 \qquad \text{(A.8)}$$

$$\cosec^2\theta = \cotan^2\theta + 1$$

$$\sin(\alpha \pm \beta) = \sin\alpha\,\cos\beta \pm \cos\alpha\,\sin\beta \qquad\text{(A.9)}$$

$$\cos(\alpha \pm \beta) = \cos\alpha\,\cos\beta \mp \sin\alpha\,\sin\beta \qquad\text{(A.10)}$$

$$\cos A + \cos B = 2\cos\frac{A+B}{2}\cos\frac{A-B}{2}$$

$$\cos B - \cos A = 2\sin\frac{A+B}{2}\sin\frac{A-B}{2}$$

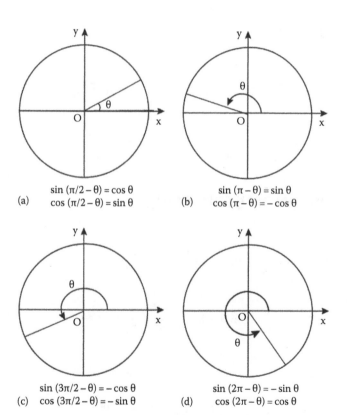

(a) $\sin(\pi/2 - \theta) = \cos\theta$
     $\cos(\pi/2 - \theta) = \sin\theta$

(b) $\sin(\pi - \theta) = \sin\theta$
     $\cos(\pi - \theta) = -\cos\theta$

(c) $\sin(3\pi/2 - \theta) = -\cos\theta$
     $\cos(3\pi/2 - \theta) = -\sin\theta$

(d) $\sin(2\pi - \theta) = -\sin\theta$
     $\cos(2\pi - \theta) = \cos\theta$

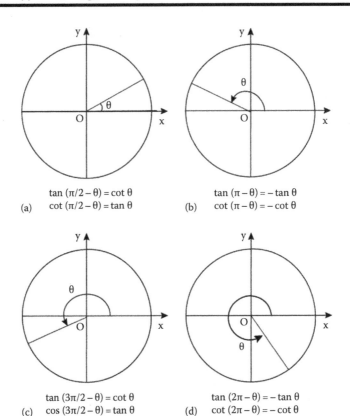

$$\tan (\pi/2 - \theta) = \cot \theta$$
(a) $\cot (\pi/2 - \theta) = \tan \theta$

$$\tan (\pi - \theta) = -\tan \theta$$
(b) $\cot (\pi - \theta) = -\cot \theta$

$$\tan (3\pi/2 - \theta) = \cot \theta$$
(c) $\cos (3\pi/2 - \theta) = \tan \theta$

$$\tan (2\pi - \theta) = -\tan \theta$$
(d) $\cot (2\pi - \theta) = -\cot \theta$

A.3 TRIGONOMETRIC SERIES

$$\cos \theta = 1 - \frac{\theta^2}{2!} + \frac{\theta^4}{4!} - \frac{\theta^6}{6!} + \cdots \tag{A.11}$$

$$\sin \theta = \theta - \frac{\theta^3}{3!} + \frac{\theta^5}{5!} - \frac{\theta^7}{7!} + \cdots \tag{A.12}$$

Also,

$$e^\theta = 1 + \theta + \frac{\theta^2}{2!} + \frac{\theta^3}{3!} + \frac{\theta^4}{4!} + \frac{\theta^5}{5!} - \frac{\theta^6}{6!} + \cdots \tag{A.13}$$

$$e^{i\theta} = 1 + i\theta - \frac{\theta^2}{2!} - \frac{i\theta^3}{3!} + \frac{\theta^4}{4!} + \frac{i\theta^5}{5!} - \frac{\theta^6}{6!} + \cdots \tag{A.14}$$

Sorting the terms in Equation A.14 such that the real terms are in one set and the imaginary ones in another, the above becomes

$$e^{i\theta} = \left(1 - \frac{\theta^2}{2!} + \frac{\theta^4}{4!} - \frac{\theta^6}{6!} + \cdots\right) + i\left(\theta - \frac{\theta^3}{3!} + \frac{\theta^5}{5!} - \frac{\theta^7}{7!}\right)$$

Thus,

$$e^{i\theta} = \cos\theta + i\sin\theta \qquad (A.15)$$

and

$$e^{-i\theta} = \cos\theta - i\sin\theta. \qquad (A.16)$$

A.3.1 EXPONENTIALS

$$\sum_{kn=0}^{\infty} \frac{1}{n!} = e = 2.718$$

$$e^x = \sum_{n=0}^{\infty} \frac{x^n}{n!} = 1 + x + \frac{x^2}{2!} + \frac{x^3}{3!} + \frac{x^4}{4!} + \cdots$$

$$e^{-x^2} = 1 - x^2 + \frac{x^4}{2!} - \frac{x^6}{3!} + \cdots$$

A.3.2 FOURIER SERIES

A periodic function, $f(x)$ of period 2ℓ, integrable over the interval $(-\ell, \ell)$ is expressed in terms of the following geometric series:

$$f(x) = \frac{a_0}{2} + \sum_{n=1}^{\infty}\left(a_n \cos\frac{n\pi x}{\ell} + b_n \sin\frac{n\pi x}{\ell}\right)$$

where

$$a_n = \int_{-\ell}^{\ell} f(t)\cos\frac{n\pi t}{\ell}dt, \quad b_n = \int_{-\ell}^{\ell} f(t)\sin\frac{n\pi t}{\ell}dt$$

A.3.3 BINOMIAL SERIES

$$(1+x)^{-1} = 1 - x + x^2 - x^3 + \cdots = \sum_{n=1}^{\infty} (-1)^{n-1} x^{n-1}$$

$$(1+x)^{-2} = 1 - 2x + 3x^2 - 4x^3 + \cdots = \sum_{n=1}^{\infty} (-1)^{n-1} nx^{n-1}$$

$$(1+x)^{1/2} = 1 + \frac{1}{2}x - \frac{1.1}{2.4}x^2 + \frac{1.1.3}{2.4.6}x^3 - \frac{1.1.3.5}{2.4.6.8}x^4 + \cdots$$

$$(1+x)^{-1/2} = 1 - \frac{1}{2}x + \frac{1.3}{2.4}x^2 - \frac{1.3.5}{2.4.6}x^3 + \cdots$$

A.3.4 GEOMETRIC SERIES

This is defined as a progression of numbers, each of which is obtained from multiplying the previous number by a fixed number or a constant. Here are three examples:

$$1, 5, 5^2, 5^3, 5^4, \ldots$$

$$1/3, 3/3, 9/3, 27/3, \ldots$$

$$a, ar, ar^2, ar^3, ar^4, \ldots$$

The sum of n terms in the third series is

$$S_n = \frac{a(1-r^n)}{1-r}.$$

In the above example, if n = ∞, the sum of the series becomes

$$S = \lim_{n \to \infty} S_n = \frac{a}{1-r}.$$

Appendix B: Complex Numbers

B.1 COMPLEX PLANE

It has become rather frequent in our computations to come across quantities that involve the square root of a negative number, implying that it is an unreal value. A typical case where this occurs is in the solution for the quadratic equation

$$az^2 + bz + c = 0,$$

where a, b, and c are constants. The solution through the quadratic formula is

$$z = \frac{-b \pm \sqrt{b^2 - 4ac}}{2a}.$$

The quantity $b^2 - 4ac$ is called the discriminant, denoted by d. In case d turns out to be negative, the square root of a negative number is part of the treatment. The handling of such a number in a compound mathematical calculation became possible by defining the square root of negative unity, usually denoted by i. Thus,

$$i = \sqrt{-1}.$$

This leads to classifying numbers into two kinds, real referred to as "Re" and imaginary referred to as "Im." The space that embraces both kinds is treated as

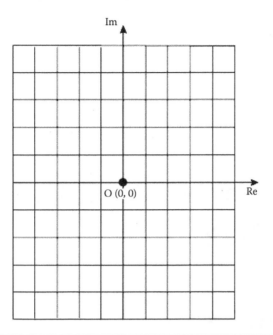

FIGURE B.1   Two dimensional complex plane showing Real and Imaginary axes.

two dimensional, one horizontal ($\equiv$ x axis) for the real numbers and another vertical ($\equiv$ y axis) for the imaginary numbers (Figure B.1). The plane formed by the two axes is called the complex plane.

Figure B.2 demonstrates how points like P_1 (3, 0), P_2 (0, –2i), P_3 (3, 2i), P_4 (3, –2i), P_5 (–3, 4i) are represented on the complex plane.

B.2 COMPLEX NUMBERS

Any number that consists of real and imaginary parts is considered a complex number and is usually expressed in the form

$$\tilde{z} = x + iy, \tag{B.1}$$

where
  x is the real part of $\tilde{z}$
  y is the imaginary part of $\tilde{z}$

The plane x-y is called a complex plane. The absolute value of the complex number is $|\tilde{z}| = \sqrt{x^2 + y^2}$. For example, for the points $P_1, P_2, ..., P_5$, they can be represented by the complex numbers

$$\tilde{z}_1 = 3 + 0i, \quad \tilde{z}_2 = 0 - 2i, \quad \tilde{z}_3 = 3 + 2i, \quad \tilde{z}_4 = 3 - 2i, \quad \tilde{z}_5 = -3 + 4i.$$

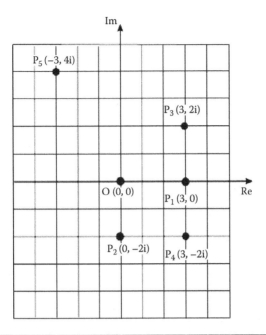

FIGURE B.2   Complex numbers represented as points on a complex plane.

The line connecting any of these points with the origin is the length of the vector of Cartesian components x and y. This length represents the absolute magnitude of the complex number, that is,

$$|\tilde{z}| = \sqrt{x^2 + y^2}. \tag{B.2}$$

The angle θ in the right-angled triangle sketched in the complex plane represents the phase of the complex number $\tilde{z}(= x + iy)$. This is defined as

$$\tan\theta = \frac{y}{x}. \tag{B.3}$$

The complex number $\tilde{z}(= x + iy)$ can then be written in a polar form as

$$\tilde{z} = re^{i\theta} = r\left(\cos\theta + i\sin\theta\right);$$

θ is expressed in radians. The absolute value of the number is $|\tilde{z}| = \sqrt{x^2 + y^2} = \sqrt{r^2 \cos^2\theta + r^2 \sin^2\theta} = r.$

The equation $\tilde{z} = re^{i\theta}$ describes in the complex plane a circle of radius r.

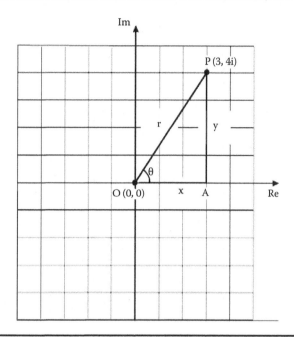

FIGURE B.3   Absolute value r and angle θ (polar form) of a complex number P(3, 4i).

The absolute values of the lines extending from the origin to the points P_1, P_2, P_3, P_4, P_5 described earlier (Figure B.2) are

$$\left|\tilde{z}_1\right| = \sqrt{3^2 + 0^2} = 3, \quad \left|\tilde{z}_2\right| = \sqrt{0^2 + (-2)^2} = 3, \quad \left|\tilde{z}_3\right| = \sqrt{3^2 + (-2)^2} = \sqrt{13},$$

$$\left|\tilde{z}_4\right| = \sqrt{3^2 + (-2)^2} = \sqrt{13}, \quad \left|\tilde{z}_5\right| = \sqrt{3^2 + (4)^2} = 5.$$

For the point P in Figure B.3, the length of OP is the absolute value of the vector OP whose components are 3 and 4i. That is, the complex number describing OP is $\tilde{z} = 3 + 4i$.

B.3 COMPLEX CONJUGATE

For any number, $\tilde{z} = x + iy$, (Equation B.1), the complex conjugate, denoted by $\tilde{z}^*$, is defined as

$$\tilde{z}^* = x - iy. \tag{B.4}$$

The complex conjugate of a complex number is complex. However, the product of a complex number and its complex conjugate is always real. For $\tilde{z} = x + iy$, the complex number is $\tilde{z} = x - iy$. The product $\tilde{z}\tilde{z}^*$ is

$$\tilde{z}\tilde{z}^* = (x + iy)(x - iy) = x^2 + y^2. \tag{B.5}$$

For $\tilde{z} = 3 + 4i$, $\tilde{z}^* = 3 - 4i$, the product $\tilde{z}\tilde{z}^*$ is

$$\tilde{z}\tilde{z}^* = (3 + 4i)(3 - 4i) = 9 + 16 = 25.$$

B.4 COMPLEX NUMBERS AND TRIGONOMETRIC FUNCTIONS

From the Taylor series expansion of $\cos\theta$, $\sin\theta$, and e^x, we can write the following series:

$$\cos\theta = 1 - \frac{\theta^2}{2!} + \frac{\theta^4}{4!} - \frac{\theta^6}{6!} + \cdots \tag{B.6}$$

$$\sin\theta = \theta - \frac{\theta^3}{3!} + \frac{\theta^5}{5!} - \frac{\theta^7}{7!} + \cdots \tag{B.7}$$

and for the exponential,

$$e^x = 1 + x + \frac{x^2}{2!} + \frac{x^3}{3!} + \frac{x^4}{4!} + \frac{x^5}{5!} - \frac{x^6}{6!} + \cdots$$

Replacing x with $i\theta$, the above becomes

$$e^{i\theta} = 1 + i\theta - \frac{\theta^2}{2!} - \frac{i\theta^3}{3!} + \frac{\theta^4}{4!} + \frac{i\theta^5}{5!} - \frac{\theta^6}{6!} + \cdots \tag{B.8}$$

or

$$e^{i\theta} = \left(1 - \frac{\theta^2}{2!} + \frac{\theta^4}{4!} - \frac{\theta^6}{6!} + \cdots\right) + i\left(\theta - \frac{\theta^3}{3!} + \frac{\theta^5}{5!} - \frac{\theta^7}{7!} + \cdots\right).$$

That is,

$$e^{i\theta} = \cos\theta + i\sin\theta. \tag{B.9}$$

As $|e^{i\theta}| = (\cos^2\theta + \sin^2\theta)^{1/2} = 1$, the complex number $e^{i\theta}$ lies in the complex plane on the unit circle around the origin of coordinates. The complex conjugate of $e^{i\theta}$ is $e^{-i\theta}$. This is written by replacing i in Equation B.9 by $-i$, giving the same unit circle around the origin of coordinates.

$$e^{-i\theta} = \cos\theta - i\sin\theta \tag{B.10}$$

It is also important to note that every complex number, expressed as a function of θ, is periodic with period 2π.

B.5 EXPONENTIAL EXPRESSIONS FOR SINUSOIDAL FUNCTIONS (SINE AND COSINE FUNCTIONS)

From (B.9) and (B.10), we have

$$e^{i\theta} = \cos\theta + i\sin\theta$$

$$e^{-i\theta} = \cos\theta - i\sin\theta.$$

Upon adding the above two equations and subtracting them, we get first

$$\cos\theta = \frac{1}{2}\left(e^{i\theta} + e^{-i\theta}\right) \tag{B.11}$$

and second

$$\sin\theta = \frac{1}{2i}\left(e^{i\theta} - e^{-i\theta}\right). \tag{B.12}$$

Appendix C: Mathematical Operators—Cartesian and Spherical Coordinates

C.1 UNIT VECTORS IN THREE DIMENSIONS

Let us consider a unit vector $\hat{\mathbf{r}}$ along $\mathbf{r}$. In Figure C.1, the component along the positive z axis of the unit vector $\hat{\mathbf{r}}$ is $|\hat{\mathbf{r}}|\cos\theta = \cos\theta$. Therefore, its component in the x-y plane, that is, along the $\hat{\mathbf{r}}'$ direction is $|\hat{\mathbf{r}}|\sin\theta = \sin\theta$. Since $\hat{\mathbf{r}}'$ makes an angle φ with the x axis, the component $\hat{\mathbf{r}}\sin\theta$ along the x direction is

$$\left(|\hat{\mathbf{r}}|\sin\theta\right)\cos\varphi = \sin\theta\cos\varphi.$$

That is,

$$\hat{\mathbf{r}}_x = x = \left(|\hat{\mathbf{r}}|\sin\theta\right)\cos\varphi = \sin\theta\cos\varphi.$$

The component of $\hat{\mathbf{r}}$ along the y axis then is $\hat{\mathbf{r}}_y = y = \left(|\hat{\mathbf{r}}|\sin\theta\right)\sin\varphi = \sin\theta\sin\varphi$. The component of $\hat{\mathbf{r}}$ along the z axis then is $\hat{\mathbf{r}}_z = \left(|\hat{\mathbf{r}}|\cos\theta\right) = \cos\theta$.

The component along the negative z axis of $\hat{\theta}$ is $-|\hat{\theta}|\cos\alpha = \cos\alpha$. As $\alpha = \pi/2 - \theta$, $\hat{\theta}_z = -\sin\theta$. The component in the x-y plane, that is, along r' is

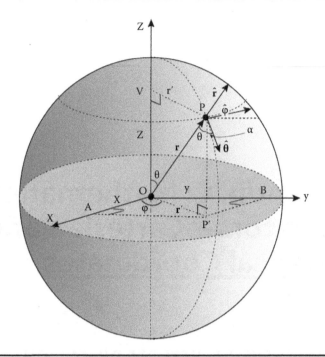

FIGURE C.1 Illustration of the unit vectors $\hat{r}$, $\hat{\theta}$, $\hat{\varphi}$ on Cartesian coordinates.

$\hat{\theta} \cos \theta = \cos \theta$. Since the **r**′direction makes an angle φ with the x axis, the component along the x direction of the $\hat{\theta} \cos \theta$ vector is

$$\hat{\theta}_x = \left(\cos \theta \right) \cos \varphi = \cos \theta \cos \varphi.$$

Along the y direction, the component of $\hat{\theta} \cos \theta$ vector is

$$\hat{\theta}_y = \left(\cos \theta \right) \sin \varphi = \cos \theta \sin \varphi.$$

And as shown earlier, $\hat{\theta}_z = -\sin \theta$.

The $\hat{\varphi}$ unit vector is tangent to the circle whose radius is r'. Therefore, the $\hat{\varphi}$ unit vector makes an angle $\pi/2 - \varphi$ with the negative direction of the x axis. Therefore, the component of $\hat{\varphi}$ along the negative x axis is $\hat{\varphi} \cos\left(\pi/2 - \varphi \right)$, or

$$\hat{\varphi}_x = -\left| \hat{\varphi} \right| \sin \varphi = -\sin \varphi.$$

This means that the component of $\hat{\varphi}$ along the y axis is

$$\hat{\varphi}_y = \cos \varphi.$$

However, its component along the z direction is

$$\hat{\varphi}_z = 0.$$

From the above discussion, the unit vectors in a spherical polar coordinate system, expressed in terms of the Cartesian unit vectors, are

$$\hat{\mathbf{r}} = \sin\theta\cos\varphi\,\hat{\mathbf{x}} + \sin\theta\sin\varphi\,\hat{\mathbf{y}} + \cos\theta\,\hat{\mathbf{z}} \qquad\text{(C.1)}$$

$$\hat{\theta} = \cos\theta\cos\varphi\,\hat{\mathbf{x}} + \cos\theta\sin\varphi\,\hat{\mathbf{y}} - \sin\theta\,\hat{\mathbf{z}} \qquad\text{(C.2)}$$

$$\hat{\varphi} = -\sin\varphi\,\hat{\mathbf{x}} + \cos\varphi\,\hat{\mathbf{y}}. \qquad\text{(C.3)}$$

Using the above set of equations, one can derive the set of Cartesian unit vectors $\hat{\mathbf{i}}$, $\hat{\mathbf{j}}$, $\hat{\mathbf{k}}$ in terms of the spherical unit vectors $\hat{\mathbf{r}}$, $\hat{\theta}$, $\hat{\varphi}$. These are

$$\hat{\mathbf{i}} = \sin\theta\cos\varphi\,\hat{\mathbf{r}} + \cos\theta\cos\varphi\,\hat{\theta} - \sin\varphi\,\hat{\varphi} \qquad\text{(C.4)}$$

$$\hat{\mathbf{j}} = \sin\theta\sin\varphi\,\hat{\mathbf{r}} + \cos\theta\sin\varphi\,\hat{\theta} + \cos\varphi\,\hat{\varphi} \qquad\text{(C.5)}$$

$$\hat{\mathbf{k}} = \cos\theta\,\hat{\mathbf{r}} - \sin\theta\,\hat{\theta}. \qquad\text{(C.6)}$$

C.2 PARTIAL DERIVATIVES

Let us consider the sphere depicted in Figure C.1 with radius r and a point P on its surface. The Cartesian coordinates x, y, and z of P expressed in spherical polar coordinates r, θ, and φ are

$$x = r\sin\theta\cos\varphi \qquad\text{(C.7)}$$

$$y = r\sin\theta\sin\varphi \qquad\text{(C.8)}$$

$$z = r\cos\theta, \qquad\text{(C.9)}$$

where θ is the polar angle that the position vector **r** makes with the z axis and φ is the azimuthal angle that the projection of **r** on the x-y plane makes with the x axis.

The partial derivatives of x, y, and z with respect to (r, θ, and φ) are

$$\frac{\partial x}{\partial r} = \sin\theta\cos\varphi, \quad \frac{\partial x}{\partial\theta} = r\cos\theta\cos\varphi, \quad \frac{\partial x}{\partial\varphi} = -r\sin\theta\sin\varphi \quad \text{(C.10)}$$

$$\frac{\partial y}{\partial r} = \sin\theta\sin\varphi, \quad \frac{\partial y}{\partial\theta} = r\cos\theta\sin\varphi, \quad \frac{\partial y}{\partial\varphi} = -r\sin\theta \quad \text{(C.11)}$$

$$\frac{\partial z}{\partial r} = \cos\theta, \quad \frac{\partial z}{\partial\theta} = -r\sin\theta, \quad \frac{\partial z}{\partial\varphi} = 0. \quad \text{(C.12)}$$

Expressing each of the differentials dx, dy, dz independently in the partial derivatives derived in (C.10), (C.11), and (C.12) gives

$$dx = \frac{\partial x}{\partial r}dr + \frac{\partial x}{\partial\theta}d\theta + \frac{\partial x}{\partial\varphi}d\varphi, \ dy = \frac{\partial y}{\partial r}dr + \frac{\partial y}{\partial\theta}d\theta + \frac{\partial y}{\partial\varphi}d\varphi, \ dz = \frac{\partial z}{\partial r}dr + \frac{\partial z}{\partial\theta}d\theta + \frac{\partial z}{\partial\varphi}d\varphi.$$
$$\text{(C.13)}$$

Substituting for the partial derivatives derived in Equation C.13 generates the following set:

$$dx = \frac{\partial x}{\partial r}dr + \frac{\partial x}{\partial\theta}d\theta + \frac{\partial x}{\partial\varphi}d\varphi$$
$$= \left(\sin\theta\cos\varphi\right)dr + \left(r\cos\theta\cos\varphi\right)d\theta + \left(-r\sin\theta\sin\varphi\right)d\varphi \quad \text{(C.14)}$$

$$dy = \frac{\partial y}{\partial r}dr + \frac{\partial y}{\partial\theta}d\theta + \frac{\partial y}{\partial\varphi}d\varphi$$
$$= \left(\sin\theta\sin\varphi\right)dr + \left(r\cos\theta\sin\varphi\right)d\theta + \left(r\sin\theta\cos\varphi\right)d\varphi \quad \text{(C.15)}$$

$$dz = \frac{\partial z}{\partial r}dr + \frac{\partial z}{\partial\theta}d\theta + \frac{\partial z}{\partial\varphi}d\varphi = \cos\theta\,dr - r\sin\theta\,d\theta. \quad \text{(C.16)}$$

Multiplying (C.14) by sin θ cos φ, (C.15) by sin θ sin φ, and (C.16) by cos θ gives

$$\sin\theta\cos\varphi\,dx =$$
$$\left(\sin^2\theta\cos^2\varphi\right)dr + \left(r\sin\theta\cos\theta\cos^2\varphi\right)d\theta + \left(-r\sin^2\theta\sin\varphi\cos\varphi\right)d\varphi$$
$$\text{(C.17)}$$

$\sin\theta\sin\varphi\,dy =$

$$\left(\sin^2\theta\sin^2\varphi\right)dr + \left(r\cos\theta\sin\theta\sin^2\varphi\right)d\theta + \left(r\sin^2\theta\sin\varphi\cos\varphi\right)d\varphi$$

(C.18)

$$\cos\theta\,dz = \cos^2\theta\,dr - r\sin\theta\cos\theta\,d\theta,$$ (C.19)

respectively.

Adding (C.17), (C.18), and (C.19) gives dr for the right-hand side, and the result becomes

$$dr = \sin\theta\cos\varphi\,dx + \sin\theta\sin\varphi\,dy + \cos\theta\,dz.$$ (C.20)

In another set of manipulations, let us multiply Equation C.14 by $(1/r)\cos\theta$ $\cos\varphi$, Equation C.15 by $(1/r)\cos\theta\sin\varphi$, and Equation C.16 by $(-1/r)\sin\theta$ and add them. This gives

$$d\theta = (1/r)\left(\cos\theta\cos\varphi\right)dx + \left(\cos\theta\sin\varphi\right)dy - \sin\theta\,dz.$$ (C.21)

Also, let us multiply Equation C.17 by $\left(-\dfrac{\sin\varphi}{r\sin\theta}\right)$ and Equation C.18 by $\dfrac{\cos\varphi}{r\sin\theta}$ and add them. This gives

$$d\varphi = \frac{1}{r\sin\theta}\left(-\sin\varphi\,dx + \cos\varphi\,dy\right).$$ (C.22)

From (C.20), we have

$$\frac{\partial r}{\partial x} = \sin\theta\cos\varphi, \quad \frac{\partial r}{\partial y} = \sin\theta\sin\varphi, \quad \frac{\partial r}{\partial z} = \cos\theta.$$ (C.23)

From (C.21), we have

$$\frac{\partial\theta}{\partial x} = \frac{\cos\theta\cos\varphi}{r}, \quad \frac{\partial\theta}{\partial y} = \frac{\cos\theta\sin\varphi}{r}, \quad \frac{\partial\theta}{\partial z} = -\frac{\sin\theta}{r}.$$ (C.24)

From (C.22), we have

$$\frac{\partial\phi}{\partial x} = -\frac{\sin\varphi}{r\sin\theta}, \quad \frac{\partial\phi}{\partial y} = \frac{\cos\varphi}{r\sin\theta}, \quad \frac{\partial\varphi}{\partial z} = 0.$$ (C.25)

C.3 THE DEL OPERATOR

From the chain rule,

$$\frac{\partial}{\partial x} = \frac{\partial}{\partial r}\frac{\partial r}{\partial x} + \frac{\partial}{\partial \theta}\frac{\partial \theta}{\partial x} + \frac{\partial}{\partial \varphi}\frac{\partial \varphi}{\partial x}, \tag{C.26}$$

$$\frac{\partial}{\partial y} = \frac{\partial}{\partial r}\frac{\partial r}{\partial y} + \frac{\partial}{\partial \theta}\frac{\partial \theta}{\partial y} + \frac{\partial}{\partial \varphi}\frac{\partial \varphi}{\partial y}, \tag{C.27}$$

$$\frac{\partial}{\partial z} = \frac{\partial}{\partial r}\frac{\partial r}{\partial z} + \frac{\partial}{\partial \theta}\frac{\partial \theta}{\partial z} + \frac{\partial}{\partial \varphi}\frac{\partial \varphi}{\partial z}. \tag{C.28}$$

Substituting for $\frac{\partial r}{\partial x}, \frac{\partial \theta}{\partial x}$, and $\frac{\partial \varphi}{\partial x}$ from (C.14), for $\frac{\partial r}{\partial y}, \frac{\partial \theta}{\partial y}$, and $\frac{\partial \varphi}{\partial y}$ from (C.24), and for $\frac{\partial r}{\partial z}, \frac{\partial \theta}{\partial z}$, and $\frac{\partial \varphi}{\partial z}$ from (C.25) in Equation C.26 through C.28, we get

$$\frac{\partial}{\partial x} = \sin\theta\cos\phi\,\frac{\partial}{\partial r} + \frac{\cos\theta\cos\varphi}{r}\frac{\partial}{\partial \theta} - \frac{\sin\varphi}{r\sin\theta}\frac{\partial}{\partial \varphi} \tag{C.29}$$

$$\frac{\partial}{\partial y} = \sin\theta\sin\phi\,\frac{\partial}{\partial r} + \frac{\sin\varphi\cos\theta}{r}\frac{\partial}{\partial \theta} + \frac{\cos\varphi}{r\sin\theta}\frac{\partial}{\partial \varphi} \tag{C.30}$$

$$\frac{\partial}{\partial z} = \cos\theta\,\frac{\partial}{\partial r} - \frac{\sin\theta}{r}\frac{\partial}{\partial \theta}. \tag{C.31}$$

One more set of transformations that expresses the Cartesian unit vectors $\hat{\mathbf{i}}$, $\hat{\mathbf{j}}$, $\hat{\mathbf{k}}$ in terms of the spherical coordinate vectors $\hat{\mathbf{r}}$, $\hat{\boldsymbol{\theta}}$, $\hat{\boldsymbol{\varphi}}$ is needed. From the diagram below, one can see that

$$\hat{\mathbf{i}} = \left(\sin\theta\cos\varphi\right)\hat{\mathbf{r}} + \left(\cos\theta\cos\varphi\right)\hat{\boldsymbol{\theta}} - \left(\sin\varphi\right)\hat{\boldsymbol{\varphi}} \tag{C.32}$$

$$\hat{\mathbf{j}} = \left(\sin\theta\sin\varphi\right)\hat{\mathbf{r}} + \left(\cos\theta\sin\varphi\right)\hat{\boldsymbol{\theta}} + \left(\cos\varphi\right)\hat{\boldsymbol{\varphi}} \tag{C.33}$$

$$\hat{\mathbf{k}} = \left(\cos\theta\right)\hat{\mathbf{r}} - \left(\sin\theta\right)\hat{\boldsymbol{\theta}}. \tag{C.34}$$

Substituting for $\dfrac{\partial}{\partial x}, \dfrac{\partial}{\partial y}$, and $\dfrac{\partial}{\partial z}$ from Equations C.23 through C.25 and for $\hat{\mathbf{i}}$, $\hat{\mathbf{j}}$, and $\hat{\mathbf{k}}$ from Equations C.26 through C.28 in the del Cartesian form,

$$\nabla = \hat{\mathbf{i}}\frac{\partial}{\partial x} + \hat{\mathbf{j}}\frac{\partial}{\partial y} + \hat{\mathbf{k}}\frac{\partial}{\partial z} \tag{C.35}$$

$$\nabla = \left(\sin\theta\cos\varphi\,\hat{\mathbf{r}} + \cos\theta\cos\varphi\,\hat{\boldsymbol{\theta}} - \sin\varphi\,\hat{\boldsymbol{\varphi}}\right)$$

$$\times\left(\frac{\partial}{\partial r}\sin\theta\cos\varphi + \frac{\partial}{\partial\theta}\frac{\cos\theta\cos\varphi}{r} - \frac{\partial}{\partial\varphi}\frac{\sin\varphi}{r\sin\theta}\right)$$

$$+ (\sin\theta\sin\varphi\,\hat{\mathbf{r}} + \cos\theta\sin\varphi\,\hat{\boldsymbol{\theta}} + \cos\varphi\,\hat{\boldsymbol{\varphi}})$$

$$\times\left(\frac{\partial}{\partial r}\sin\theta\sin\varphi + \frac{\partial}{\partial\theta}\frac{\cos\theta\sin\varphi}{r} + \frac{\partial}{\partial\varphi}\frac{\cos\varphi}{r\sin\theta}\right)$$

$$+ (\cos\theta\,\hat{\mathbf{r}} - \sin\theta\,\hat{\mathbf{r}} - \sin\theta\,\hat{\boldsymbol{\theta}})\left(\frac{\partial}{\partial r}\cos\theta - \frac{\partial}{\partial\theta}\frac{\sin\theta}{r}\right).$$

The above after some algebra results in

$$\nabla = \hat{\mathbf{r}}\frac{\partial}{\partial r} + \hat{\boldsymbol{\theta}}\left(\frac{1}{r}\frac{\partial}{\partial\theta}\right) + \hat{\boldsymbol{\varphi}}\left(\frac{1}{r\sin\theta}\frac{\partial}{\partial\varphi}\right). \tag{C.36}$$

C.4 THE LAPLACIAN OPERATOR

Squaring each of Equations C.20 through C.22 and adding them gives

$$\nabla^2 = \vec{\nabla}\cdot\vec{\nabla} = \left(\hat{\mathbf{i}}\frac{\partial}{\partial x} + \hat{\mathbf{j}}\frac{\partial}{\partial y} + \hat{\mathbf{k}}\frac{\partial}{\partial z}\right)\cdot\left(\hat{\mathbf{i}}\frac{\partial}{\partial x} + \hat{\mathbf{j}}\frac{\partial}{\partial y} + \hat{\mathbf{k}}\frac{\partial}{\partial z}\right),$$

which in Cartesian is

$$\nabla^2 = \frac{\partial^2}{\partial x^2} + \frac{\partial^2}{\partial y^2} + \frac{\partial^2}{\partial z^2}, \tag{C.37}$$

and in spherical polar coordinates, it becomes

$$\nabla^2 = \left(\frac{\partial^2}{\partial r^2} + \frac{2\partial}{r\,\partial r}\right) + \frac{1}{r^2\sin\theta}\frac{\partial}{\partial\theta}\left(\sin\theta\frac{\partial}{\partial\theta}\right) + \frac{1}{r^2\sin^2\theta}\frac{\partial^2}{\partial\varphi^2}, \tag{C.38}$$

or

$$\nabla^2 = \frac{1}{r^2}\frac{\partial}{\partial r}\left(r^2\frac{\partial}{\partial r}\right) + \frac{1}{r^2\sin\theta}\frac{\partial}{\partial\theta}\left(\sin\theta\frac{\partial}{\partial\theta}\right) + \frac{1}{r^2\sin^2\theta}\frac{\partial^2}{\partial\varphi^2}. \qquad \text{(C.39)}$$

C.5 AREA AND VOLUME ELEMENTS

Consider Figure C.2 where a circle of an arbitrary radius is shown. Consider a point P of Cartesian coordinates (x, y). The plane polar coordinates (r, θ) of P are related to the Cartesian x, y coordinates as follows:

$$x = r\cos\theta, \quad y = r\sin\theta$$

The radial distance r from the origin is $r = \sqrt{x^2 + y^2}$.

An element of area (dx)(dy) at the point P is dA = (dr)(r dθ) = r dr dθ.

Figure C.3 is a three-dimensional representation of Cartesian and spherical coordinates of a point P, whose Cartesian coordinates are x, y, z and spherical coordinates are r, θ, φ. Here, θ is called the polar angle that r makes with the z axis, and φ is the azimuthal angle that the projection of r on the x–y plane makes with the x axis. If an elemental radial change dr in r occurs

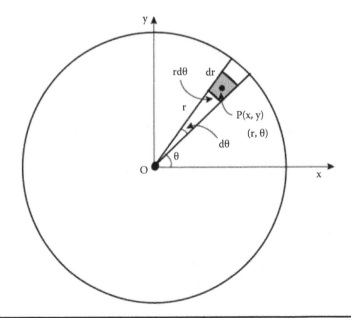

FIGURE C.2   An element of area at point P(x, y) within a circle of arbitrary radius.

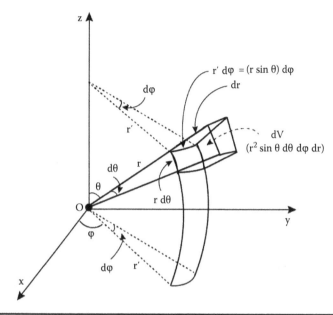

FIGURE C.3 Three dimensional representation of point P in Cartesian (x, y, z) and spherical (r, θ, φ) coordinates.

that corresponds to an elemental change dθ in the polar angle and another dφ in the azimuthal angle, a parallelepiped is constructed. For such changes, the corresponding changes in the Cartesian coordinates are dx, dy, dz. The constructed parallelepiped has a base of the elemental area dA and elemental volume dV, where

$$dA = \left(r \sin \theta \, d\varphi\right)\left(r \, d\theta\right) = r^2 \sin \theta \, d\theta \, d\varphi,$$

$$dV = \left(r \sin \theta \, d\varphi\right)\left(r \, d\theta\right) dr = r^2 \sin \theta \, d\theta \, d\varphi \, dr.$$

Appendix D: Matrices

D.1 INTRODUCTION: MATRIX FORM

The concept and role of matrices (singular: matrix) became of immense use in solving linear equations in a set where each equation involves two or more variables, x, y, z, etc. The old way consisted of successive operations that each involved expressing one variable in terms of the others, reducing the number of equations each time by one until, eventually one has an equation in one variable that would be determined. Substitution of the determined variable into the reduced set of equations helps repeating the successive reduction of the number of equations by one leading at the end to determining the rest of the variables. In a large number of equations, this process is tedious. Arranging the coefficients of the variables in each equation in a row and going through a reduction of these rows in a systematic way proved very helpful and time saving. Below is a typical 2 × 2 arrangement matrix, M_1:

$$M_1 = \begin{pmatrix} a & b \\ c & d \end{pmatrix}.$$

The 2 × 2 is a reference to the number of rows and columns. Below is another one of three rows and two columns:

$$M_2 = \begin{pmatrix} a & b \\ c & d \\ e & f \end{pmatrix}.$$

D.2 DETERMINANT OF A MATRIX

As shown, matrix M_1 is square (number of rows equals the number of columns), while M_2 is not. Matrices have no numerical value. However, there is a special importance to another notion called "determinant," abbreviated as (det), which exists for matrices that are square. The elements of the determinant of a square matrix are the same as those of the matrix but usually placed between two vertical lines as shown below, displayed in a way similar to that of the matrix.

Thus, $\det M_1 = \begin{vmatrix} a & b \\ c & d \end{vmatrix}$.

The determinant has a value obtained from multiplying the subtraction of the product of the diagonals as illustrated below:

$$\det M_1 = \begin{vmatrix} a & b \\ c & d \end{vmatrix} = ad - bc.$$

For a 3×3 matrix, $M = \begin{pmatrix} a & b & c \\ d & e & f \\ g & h & i \end{pmatrix}$, the determinant is

$$\det M = \begin{vmatrix} a & b & c \\ d & e & f \\ g & h & i \end{vmatrix} = a \begin{pmatrix} e & f \\ h & i \end{pmatrix} - b \begin{pmatrix} d & f \\ g & i \end{pmatrix} + c \begin{pmatrix} d & e \\ g & h \end{pmatrix}.$$

D.3 PROPERTIES AND OPERATION OF MATRICES

1. *Product of two matrices*: Let us explain that through the following examples.

EXAMPLE 1

Let us calculate the product of the following two matrices, $M_1\, M_2$, where

$$M_1 = \begin{pmatrix} a_1 & b_1 \\ c_1 & d_1 \end{pmatrix}, \quad M_2 = \begin{pmatrix} a_2 & b_2 \\ c_2 & d_2 \end{pmatrix}$$

$$M_1 M_2 = \begin{pmatrix} a_1 & b_1 \\ c_1 & d_1 \end{pmatrix} \begin{pmatrix} a_2 & b_2 \\ c_2 & d_2 \end{pmatrix} = \begin{pmatrix} a_1 a_2 + b_1 c_2 & a_1 b_2 + b_1 d_2 \\ c_1 a_2 + d_1 c_2 & c_1 b_2 + d_1 d_2 \end{pmatrix} = M.$$

The product M is a 2 × 2 matrix whose elements are obtained through a "row times column" multiplication.

It is important that in the multiplied two matrices, the number of columns in the first matrix should be equal to the number of rows in the second one. In the above product M_1 had two rows, same as in M_2.

EXAMPLE 2

Consider the two matrices $M_1 = \begin{pmatrix} 1 & 2 \\ 3 & 4 \end{pmatrix}$, $M_2 = \begin{pmatrix} 5 & 6 \\ 7 & 8 \end{pmatrix}$. Determine their product $M = M_1 \, M_2$:

$$M = M_1 M_2 \begin{pmatrix} 1 & 2 \\ 3 & 4 \end{pmatrix} \begin{pmatrix} 5 & 6 \\ 7 & 8 \end{pmatrix} = \begin{pmatrix} 1 \times 5 + 2 \times 7 & 1 \times 6 + 2 \times 8 \\ 3 \times 5 + 4 \times 7 & 3 \times 6 + 4 \times 8 \end{pmatrix}.$$

Thus, $M = \begin{pmatrix} 19 & 22 \\ 43 & 50 \end{pmatrix}$.

---

2. *Product of two matrices*: In general, matrix multiplication is not commutative, that is, $M_1 M_2 \neq M_2 M_1$.
3. *Multiplication of matrices is associative*: That is, $(M_1 + M_2)M_3 = M_1M_3 + M_2M_3$ and $M_3(M_1 + M_2) = M_3M_1 + M_3M_2$.
4. *Identity matrices*: A diagonal matrix with all elements equal to one in the principal diagonal is called the identity matrix 1.

D.4 TRANSPOSE OF A MATRIX

The transpose M^T of a matrix M is obtained by interchanging the rows and columns of M. Thus, for a matrix

$$M = \begin{pmatrix} 19 & 22 \\ 43 & 50 \end{pmatrix},$$

its transpose is $M^T = \begin{pmatrix} 19 & 43 \\ 22 & 50 \end{pmatrix}.$

D.5 INVERSE MATRIX

How to find the inverse of a matrix? As mentioned earlier, a matrix is a display of numbers; it does not have a numerical value. Only square matrices can have a determinant and can have inverses. However, some square matrices do not have an inverse. If A is any matrix, then A^{-1} is the inverse matrix of A.

To find A^{-1} for a given matrix A, first find the determinant of the matrix. Suppose we have $A = \begin{pmatrix} a & b \\ c & d \end{pmatrix}$, $\det A = \begin{vmatrix} a & b \\ c & d \end{vmatrix} = ad - bc$.

Second, find the cofactors C_{ij} of all elements; i represents a row in a matrix and j represents a column in a matrix. Then write the matrix C whose elements are C_{ij}; transpose it (interchange rows and columns), as matrix C^T; and divide by det A. So

$$A^{-1} = \frac{1}{\det A} C^T.$$

EXAMPLE

Find the inverse of the given matrix $A = \begin{pmatrix} -2 & 0 & 1 \\ 1 & -1 & 2 \\ 3 & 1 & 0 \end{pmatrix}$.

1. Find the determinant, $\det A = -2 \begin{vmatrix} -1 & 2 \\ 1 & 0 \end{vmatrix} - 0 \begin{vmatrix} 1 & 2 \\ 3 & 0 \end{vmatrix} + 1 \begin{vmatrix} 1 & -1 \\ 3 & 1 \end{vmatrix}$

$= -2 \times -2 - 0 + 1 * 4 = 8.$

2. Find the cofactors of each element: $-2 \times -2 - 0 + 1*4 = 8.$

First row: $+ \begin{vmatrix} -1 & 2 \\ 1 & 0 \end{vmatrix} = -2,$ $- \begin{vmatrix} 1 & 2 \\ 3 & 0 \end{vmatrix} = 6,$ $+ \begin{vmatrix} 1 & -1 \\ 3 & 1 \end{vmatrix} = 4.$

Second row: $- \begin{vmatrix} 0 & 1 \\ 1 & 0 \end{vmatrix} = 1,$ $+ \begin{vmatrix} -2 & 1 \\ 3 & 0 \end{vmatrix} = -3,$ $- \begin{vmatrix} -2 & 0 \\ 3 & 1 \end{vmatrix} = 2.$

Third row: $+ \begin{vmatrix} 0 & 1 \\ -1 & 2 \end{vmatrix} = 1,$ $- \begin{vmatrix} -2 & 1 \\ 1 & 2 \end{vmatrix} = 5,$ $+ \begin{vmatrix} -2 & 0 \\ 1 & -1 \end{vmatrix} = 2.$

Then

$$C = \begin{pmatrix} -2 & 6 & 4 \\ 1 & -3 & 2 \\ 1 & 5 & 2 \end{pmatrix}, \quad C^T = \begin{pmatrix} -2 & 1 & 1 \\ 6 & -3 & 5 \\ 4 & 2 & 2 \end{pmatrix},$$

so $A^{-1} = \left(\dfrac{1}{8}\right) \begin{pmatrix} -2 & 1 & 1 \\ 6 & -3 & 5 \\ 4 & 2 & 2 \end{pmatrix}$.

Appendix E: Physical Constants

E.1 SELECTION OF SOME PHYSICAL CONSTANTS

Physical Quantity	Symbol	Value
Speed of light	c	3.00×10^8 m/s
Planck's constant	h	6.63×10^{-34} J-s
		4.14×10^{-15} eV-s
Reduced Planck's constant	$\hbar$	1.05×10^{-34} J-s
		6.58×10^{-16} eV-s
Electron mass	m_e	9.11×10^{-31} kg
		0.511 MeV/c^2
Electron charge	e	1.60×10^{-19} C
Proton mass	M_p	1.67×10^{-27} kg
		938 MeV/c^2
Boltzmann constant	k_B	1.38×10^{-23} J/K
		8.62×10^{-5} eV/K
Stefan–Boltzmann constant	σ	5.67×10^{-8} W/m^2-K^4
Permeability of free space	μ_o	$4\pi \times 10^{-7}$ N/A^2
Permittivity of free space	ε_o	8.85×10^{-12} C^2/N-m^2

Source: Servey, R.A. et al., *Modern Physics,* Thomson, Brooks Cole, Belmont, CA, 2005.

E.2 ENERGY EQUIVALENCE FOR PHOTONS

$$E = hf = k_B T = hc/\lambda$$

433

Energy, E (J)	(eV)	Frequency, f (Hz)	Temperature, T (K)	Wavelength, λ (m)	1/Wavelength, λ⁻¹ (m⁻¹)	(cm⁻¹)
1	6.242×10^{18}	1.509×10^{23}	7.243×10^{22}	1.486×10^{-25}	5.034×10^{24}	
1.602×10^{-19}	1	2.418×10^{14}	1.160×10^{4}	1.240×10^{-6}	8.066×10^{5}	8.066×10^{3}
6.626×10^{-34}	4.136×10^{-15}	1	4.799×10^{-11}	2.998×10^{8}	3.336×10^{-9}	3.336×10^{-11}
1.381×10^{-23}	8.617×10^{-5}	2.084×10^{10}	1	1.439×10^{-2}	69.50	0.6950
1.987×10^{-25}	1.240×10^{-6}	2.998×10^{8}	1.439×10^{-2}	1	1.0	0.01
1.987×10^{-25}	1.240×10^{-6}	2.998×10^{8}	1.439×10^{-2}	1.0	1	0.01
1.987×10^{-23}	1.240×10^{-4}	2.998×110	1.439	0.01	100	1

Source: Booker, G., *Modern Classical Optics*, Oxford University Press, New York, 2003.

Appendix F: Examples on Fresnel Diffraction Done on Mathematica

F.1 GENERAL CASE

Detailed Code: The code starts with defining or calculating the values of the main quantities needed for a Fresnel problem, like wavelength λ, L, Z1, Z2, W1, and W2 described in the first part (lines 1–6). Lines 7–15 form the central part of the body. The phrase(s) in brackets are comments and have no bearing on the program.

Another way that would enable you see the plots, Fresnel spiral, the lines representing EE′, and diffraction, is to add the lowest six lines to the upper part of the code below, from line $\lambda = 550 \cdot 10^{-9}$ to end of line $w2 = z2\sqrt{\dfrac{2}{\lambda L}}$. This makes the more complete code look as follows:

```
(*Determine w1, w2*)

λ = 550 · 10⁻⁹;
L = 5;
z1 = -2.5 · 10⁻⁴;
z2 = 2.5 · 10⁻⁴;
```
$$w1 = z1\sqrt{\frac{2}{\lambda L}};$$
$$w2 = z2\sqrt{\frac{2}{\lambda L}};$$

```
(*Solve Fresnel integrals numerically*)
Cw[t_]:=NIntegrate[Cos[(π w²)/2],{w,0,t}];
Sw[t_]:=NIntegrate[Sin[(π w²)/2],{w,0,t}];
I_p=1/2 ((Cw[w2]-Cw[w1])²+(Sw[w2]-Sw[w1])²);

cornuSpiral=ParametricPlot[{Cw[t],Sw[t]},{t,-2 π,2π}];
line=ListLinePlot[{{Cw[w1],Sw[w1]},{Cw[w2],Sw[w2]}},
  PlotStyle->Red];
(*line2=ListLinePlot[{{-0.5,-0.5},{0.5,0.5}},PlotStyle□Green];*)
 (*Print intensity and display curves*)
Print["The relative intensity is: ", I_p,"×I_u"]
          Show[cornuSpiral,line,line2, AxesLabel->{C_w,S_w}]
============================================================
```

EXAMPLE 9.3

Consider a plane wave of 550.0 nm incident normally on an obstacle of a square aperture 0.500 mm on a side. Determine the intensity of the diffraction at a point lying along the straight line connecting the source and the center of the aperture 5.00 m away. (*Hint*: Results can be obtained by linear interpolation from the Fresnel table (Table 9.1) or from direct integration of integrals in Equations 9.53 and 9.54.)

Solution

i. Brief Code:

```
(*Determine w1, w2*)

λ = 550 · 10⁻⁹;
L = 5;
z1 = -2.5 · 10⁻⁴;
z2 = 2.5 · 10⁻⁴;
w1 = z1 √(2/λL);

w2 = z2 √(2/λL);


(*Solve Fresnel integrals numerically*)
Cw[t_]:=NIntegrate[Cos[(π w²)/2],{w,0,t}];
Sw[t_]:=NIntegrate[Sin[(π w²)/2],{w,0,t}];
I_p=1/2 ((Cw[w2]-Cw[w1])²+(Sw[w2]-Sw[w1])²)
```
----

ii. Another more complete code that provides plots of Cornu spiral and lines E_1 E_2

```
λ = 550 · 10⁻⁹;
L = 5;
z1 = -2.5 · 10⁻⁴;
z2 = 2.5 · 10⁻⁴;
w1 = z1 √(2/λL) ;
w2 = z2 √(2/λL) ;

(*Solve Fresnel integrals numerically*)
Cw[t_]:=NIntegrate[Cos[(π w²)/2],{w,0,t}];
Sw[t_]:=NIntegrate[Sin[(π w²)/2],{w,0,t}];
Iₚ=1/2 ((Cw[w2]-Cw[w1])²+(Sw[w2]-Sw[w1])²);

cornuSpiral=ParametricPlot[{Cw[t],Sw[t]},{t,-2 π,2π}];
line=ListLinePlot[{{Cw[w1],Sw[w1]},{Cw[w2],Sw[w2]}},
PlotStyle->Red];
(*line2=ListLinePlot[{{-0.5,-0.5},{0.5,0.5}},PlotStyle□
Green];*)
  (*Print intensity and display curves*)
=================================================
```

F.2 WORKBOOK

F.2.1 GENERATE THE CORNU SPIRAL

```
ClearAll["Global"`*"] Off[NIntegrate::nlim];

(*Upper and Lower Bounds for the Cornu Spiral*)
l = -5;
u = 5;
(*Upper and Lower Bounds for the Tick Marks*)
lt = -2.5;
ut = 2.5;

(*Define the C(w), S(w) Functions*)
Cw[T_] := NIntegrate [Cos [πw²/2], {w, 0, T}];

Sw[T_] := NIntegrate [Sin [πw²/2], {w, 0, T}];
```

```
(*Define a function that is the ordered pair f(t)={x,y}
  where x,y are given by C(w), S(w) *)
f[t_] := {Cw[t], Sw[t]};

(*Finds a point inward of the curve, orthogonal to the
  curve: for tickmark*)
inward[f_, t_] : = RotationTransform[π/2] [Normalize
  [f ' [t]]];

(*Does the same, but is reversed for text to be interior
  to curve*)
inwardText [f_, t_] : = RotationTransform[π/2] [Normalize
  [f ' [t]]] /; t >= 0
inwardText [f_, t_] := -RotationTransform[π/2] [Normalize
  [f ' [t]]] /; t < 0

(*Makes tickmarks and text*)
smallTickGraphics [f_] : =
  Function[t, {Line[{f[t] -inward[f, t]/150, f[t] +inward[f, t]/150}],
    Text [Style [Abs [t] , FontSize → 8] , f[t] +
    inwardText [ f, t]/35]}];

vsmallTickGraphics [f_] :=
  Function[t, {Line[{f[t] - inward [f, t]/500, f[t] +
    inward[f, t]/500}]}];

largeTickGraphics [f_] : =
  Function[t, {Line[{f [t] -inward[f, t]/150, f[t] +
    inward[f, t]/150}]}];

(*Plots Curve and Tickmarks, Text*)
curve = ParametricPlot[f[t] , {t, 1, u},
  Ticks → {{{-0.8, -0.8}, {-0.7, -0.7}, {-0.6, -0.6},
    {-0.5, -0.5}, {-0.4, -0.4},
    {-0.3, -0.3}, {-0.2, -0.2}, {-0.1, -0.1}, {0, Null},
    {0.1,0.1}, {0.2, 0.2},
    {0.3, 0.3}, {0.4, 0.4}, {0.5, 0.5}, {0.6, 0.6},
    {0.7,0.7}, {0.8, 0.8}},
    {{-0.8, -0.8}, {-0.7, -0.7}, {-0.6, -0.6}, {-0.5,-
    0.5}, {-0.4, -0.4},
    {-0.3, -0.3}, {-0.2, -0.2}, {-0.1, -0.1}, {0,Null},
    {0.1, 0.1}, {0.2, 0.2},
    {0.3, 0.3}, {0.4, 0.4}, {0.5, 0.5},{0.6, 0.6}, {0.7,
    0.8}, {0.8, 0.8}}},
  Epilog → ({smallTickGraphics[f] /@Range[lt, ut, 0.5],
    vsmallTickGraphics[f]/@Range[lt, ut, 0.01],
    largeTickGraphics[f] /@Range[lt, ut, 0.1]}),
  PlotStyle → Black, (*GridLines→ {{-0.8,-0.7,-0.6,-0.5,-0.4,
```

```
        -0.3,-0.2,-0.1,0.1,0.2,0.3,0.4,0.5,0.6,0.7,0.8},
        {-0.8,-0.7,-0.6,-0.5,-0.4,-0.3,-0.2,-0.1,0.1,0.2,0.3,0.4,0.5,
          0.6,0.7,0.8}},
  *) AspectRatio → Automatic, Axes → True, Imagesize → Full] ;
mainCornu = Show[curve, AxesLabel →
    {Style ["C(w) ", FontSize → 18] , Style ["S(w) ",
      FontSize → 18]}, AxesStyle → Black]
```

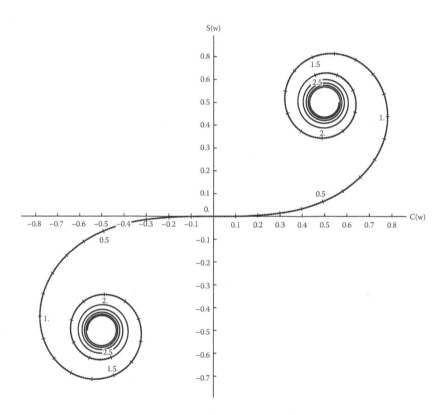

```
w2 = 0.0;
line1 = ListLinePlot[{{-0.5, -0.5}, {0.5, 0.5}},
  PlotStyle → Green];
line2 =ListLinePlot[{{Cw[w2] , Sw[w2]}, {0.5, 0.5}},
  PlotStyle → Red];
Show[mainCornu,line1,line2]
```

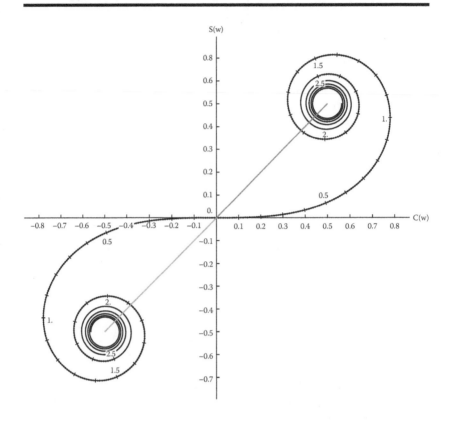

```
w1 = -0.5;
w2 = 0.0;
line1=ListLinePlot[{{Cw[w1], Sw[w1]}, {0.5, 0.5}},
  PlotStyle → Green];
line2 =ListLinePlot[{{Cw[w2], Sw[w2]}, {0.5, 0.5}},
  PlotStyle → Red];
Show[mainCornu,line1,line2]

w1 = -1.2;
w2 = -1.9;
line1 = ListLinePlot[{{Cw[w1] , Sw[w1]}, {0.5, 0.5}},
  PlotStyle → Green];
line2 =ListLinePlot[{{Cw[w2] , Sw[w2]}, {0.5, 0.5}},
  PlotStyle → Red];
Show[mainCornu,line1,line2]

(*Haija, July 7,Text*)
(*Example 9.4,Text*)
(*Plots Curve and Tickmarks, Example 9.4,Text*)

w1 = -2.0;
w2 = -2.5;
```

```
line1 = ListLinePlot[{{Cw[w1], Sw[w1]}, {Cw[w2], Sw[w2]}},
  PlotStyle → Green];
line2 = ListLinePlot[{{0.5, 0.5}, {-0.5, -0.5}},
  PlotStyle → Red];
Show[mainCornu, line1, line2]
```

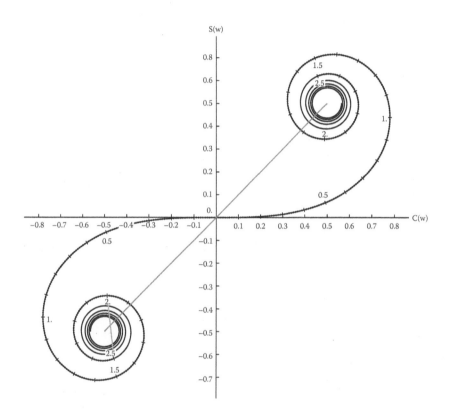

F.3 HOMEWORK PROBLEMS

F.3.1 TEXT EXAMPLE

```
(*Determine w1, w2*)
λ = 550 × 10⁻⁹;
L = 5;
z1 = -2.5 × 10⁻⁴;
z2 = 2.5 × 10⁻⁴;
```

$$w1 = z1 \sqrt{\frac{2}{\lambda L}} ;$$

$$w2 = z2 \sqrt{\frac{2}{\lambda L}} ;$$

```
line=ListLinePlot[{{Cw[w1],Sw[w1]},{Cw[w2],Sw[w2]}},PlotStyle→Red];
Show[mainCornu, line]
```

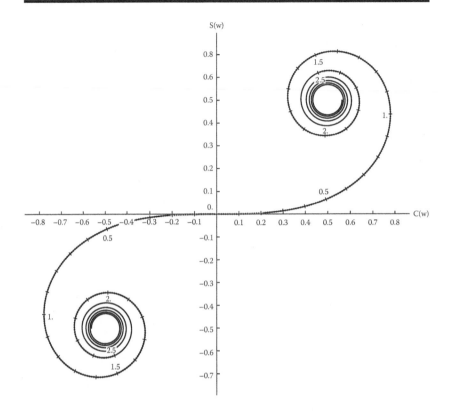

F.3.2 Problem 9.8

$\lambda = 540 \times 10^{-9};$

$\rho = 0.4;$

$r = 0.4;$

$L = \dfrac{\rho r}{\rho + r};$

$z1 = -3 \times 10^{-4};$

$z2 = 3 \times 10^{-4};$

$w1 = z1 \sqrt{\dfrac{2}{\lambda L}};$

$w2 = z2 \sqrt{\dfrac{2}{\lambda L}};$

```
line=ListLinePlot[{{Cw[w1],Sw[w1]},{Cw[w2],Sw[w2]}},PlotStyle→Red];
Show[mainCornu, line]
```

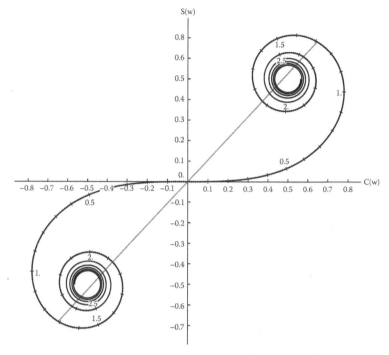

```
Show[mainCornu, ListLinePlot[{{-0.5,-0.5}, {0.5, 0.5}}]]
```

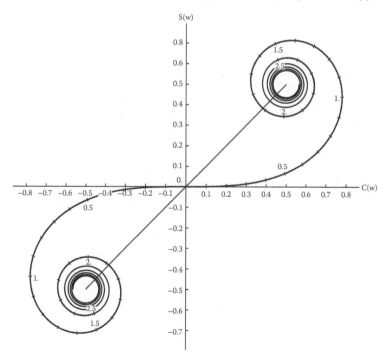

```
w1 = - 2;
w2 =  2;
line1=ListLinePlot[{{Cw[w1], Sw[w1]}, {0.5, 0.5}},
  PlotStyle → Green];
line2 =ListLinePlot[{{Cw[w2], Sw[w2]}, {0.5, 0.5}},
  PlotStyle → Red];
Show[mainCornu, line1, line2]
```

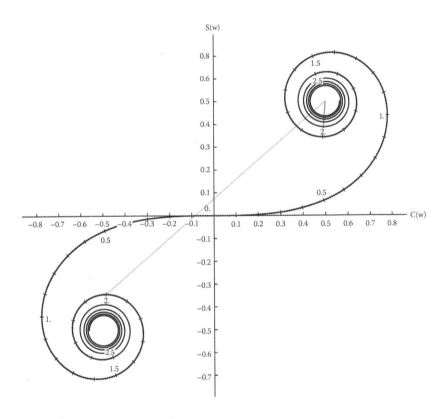

```
(*Haija, July 7,Text*)
(*Example 9.4,Text*)
(*Plots Curve and Tickmarks, Example 9.4,Text*)

w1 = -0.213;
w2 = 0.213;
line1 = ListLinePlot[{{Cw[w1] , Sw[w1]}, {Cw[w2], Sw[w2]}},
  PlotStyle → Green];
line2 = ListLinePlot[{{0.5, 0.5}, {-0.5, -0.5}},
  PlotStyle → Red];
Show[mainCornu, line1, line2]
```

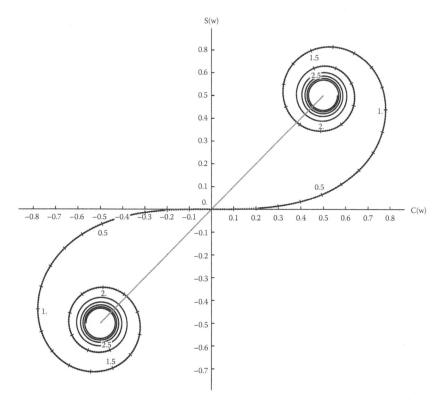

```
(*Haija,July 16,New,Text*)
(*Fig9.14,Text*)
(*Plots Curve and Tickmarks, Example 9.4,Text*)
line1 = ListLinePlot[{{c (-1.2) , S (-1.2)}, {0.5, 0.5}},
  PlotStyle → Green];
line2 = ListLinePlot[{{c (-1.9), S (-1.9)}, {0.5, 0.5}},
  PlotStyle → Red];
Show[mainCornu, line1, line2]
```

Appendix G: Solutions of Selected Examples from Chapter 10 Using Excel

G.1 PROBLEM STATEMENTS

EXAMPLE 1

Consider a dielectric thin film with index of refraction n = 2.0 and thickness h = 75.0 nm surrounded by air from both sides. Determine the film reflectivity and transmissivity for an incident wave of 600.0 nm. What would the value of the reflectivity and transmissivity for the film be if h = 225.0 nm, while all other quantities are kept unchanged.

Solution

First case: h = 75 nm.

We first determine the value in terms of h of the elements of the characteristic matrix as defined in Equation 10.27:

$$m_{11} = \cos\beta; \quad \beta = k_0 nh = (2\pi/600)(2.0)(75.0) = \pi/2.$$

Thus, $m_{11} = 0$, $m_{12} = -\dfrac{i}{n}\sin\beta = (-i/2.0)$, $m_{21} = -in = -2i$, $m_{22} = 0$.

Substituting in formulae (10.28) and (10.29) gives

$$r = E_r/E_o = \frac{(m_{11} + m_{12}n_\ell)n_o - (m_{21} + m_{22}n_\ell)}{(m_{11} + m_{12}n_\ell)p_o + (m_{21} + m_{22}n_\ell)} = \frac{(0 - i/2) - (-2i + 0)}{(0 - i/2) + (-2i + 0)}$$

$$= \frac{1.5i}{-2.5i} = -0.60.$$

Thus, $R = 0.36$

$$t = E_t/E_o = \frac{2n_o}{(m_{11} + m_{12}n_\ell)p_o + (m_{21} + m_{22}n_\ell)} = \frac{2}{(0 - i/2) + (-2i + 0)}$$

$$= \frac{2}{(-i/2) + (-2i)} = \frac{2}{-2.5i} = -0.80i.$$

Thus, $T = 0.64$.
   Note that $R + T = 1.0$, that is, 100%, because the film is perfect dielectric with no loss of energy.
   Second case: h = 225.0 nm.

$$m_{11} = \cos \beta_1; \quad \beta = k_o n h = (2\pi/600)(2.0)(225.0) = \pi/2.$$

Thus, $m_{11} = 1$, $m_{12} = -\dfrac{i}{n}\sin \beta = 0$, $m_{21} = -in\sin \beta = 0$, $m_{22} = 1$.
   The elements having the same values as those in the first case (a) should give the same reflectivity and transmissivity as in (a).
   Thus, $R = 0.36 = 36\%$.
   Thus, $T = 0.64 = 64\%$.
   Note that $R + T = 1.0$, that is, 100%, because the film is perfect dielectric with no loss of energy.

G.2 EXCEL APPROACH

To do the calculations via Excel, the following steps should help.
   Enter the following quantities in Excel column headings A, B, C, ... as comments for the user. (Explanations of symbols are given in parentheses where needed.)

   A → nsub (index of refraction of the substrate in complex form);
   B → nfilm (index of refraction of the film in complex form);
   C → h (film thickness); D → Lambda (wavelength); E → **2*3.14*h/ Lambda;**

F → Complex E (heading indicating complex quantity); G → COMPLEX Beta;

H → cos Beta (cosine of Beta); I →sine Beta (sine of Beta);

J → M11 (renaming cosine Beta);

K → neg i (–i);   L → (–i/n);   M → i (–in); N → M12; O → M21;   P → NUM 1; Q → NUM 2

R → NUM 1 – NUM 2; S → NUM 1 + NUM 2;

T → r (reflection Coeff); U → r* (Complex conjugate of r); V → **R**(Reflectivity Executed)

W → t (transmission Coeff); X → t* (Complex conjugate of t);

T (Transmissivity Executed)

Enter under each column the appropriate values for that column. In this example, the start was from Row 5. Any row as a start is of course left to the user.

A → =COMPLEX(1,0);

B → =COMPLEX(2,0); C → 75; H → =IMCOS(G5); I → =IMSIN (G5);

J → =H5; K → =COMPLEX(0,–1); L → =IMDIV(K5,B5); M → =IMPRODUCT(K5,B5);

N → =IMPRODUCT(L5,I5); O → =IMPRODUCT(M5,I5); P → =IMSUM(J5,N5);

Q → = IMSUM(O5,J5);

R → =IMSUB(P5, Q5); S → IMSUM(P5, Q5);

T → IMDIV(R5,S5); U → r* =IMCONJUGATE (T5); V → **R** → = IMPRODUCT(T5,U5);

W → =IMDIV(2,S5); X → =IMCONJUGATE (W5); Y → = IMPRODUCT(W5,X5);

Bibliography

The following list details a number of sources, mainly books, which were very helpful in guiding our general approach to this text. In frequent instances, some of these were also used in exposing a subtle relevance in specific topics of the book chapters. The nature of our textbook being concise shored us from the advanced details, mathematical or otherwise, that could be pursued in higher optics courses.

1. F.A. Jenkins and H.E. White, *Fundamentals of Optics*, 4th edn., McGraw-Hill, New York (2001).
2. E. Hecht, *Optics*, 4th edn., Addison Wesley, San Francisco, CA (2002).
3. L.M. Pedrotti, L.S. Pedrotti, and F.L. Pedrotti, *Introduction to Optics*, 3rd edn., Pearson/Prentice Hall, Upper Saddle River, NJ (2007).
4. A. Ghatak, *Optics*, McGraw-Hill, New York (2010).
5. G. Brooker, *Modern Classical Optics*, Oxford University Press, Oxford, U.K. (2003).
6. C.A. Benneet, *Principles of Physical Optics*, John Wiley & Sons, Hoboken, NJ (2008).
7. B.B. Rossi, *Optics*, 2nd printing, Addison Wesley, Reading, MA (2002).
8. M. Born and E. Wolf, *Principles of Optics*, 7th edn., Cambridge University Press, Cambridge, U.K. (1999).
9. R.W. Wood, *Physical Optics*, 3rd revised edition, Dover Publications, Inc., New York (1961).
10. C.L. Andrews, *Optics of the Electromagnetic Spectrum*, 2nd printing, Prentice-Hall, Inc., Upper Saddle River, NJ (1961).
11. R.H. Webb, *Elementary Wave Optics*, Dover Publications, Inc., New York (2005).
12. B.E.A. Saleh and M.C. Teich, *Fundamentals of Photonics*, 2nd edn., John Wiley & Sons, Hoboken, NJ (2007).
13. M.L. Boas, *Mathematical Methods in the Physical Sciences*, 2nd edn., John Wiley & Sons, New York (1983).
14. D.J. Griffiths, *Introduction to Electrodynamics*, 4th edn., Pearson, San Francisco, CA (2013).

15. N.A. Riza, *Photonic Signals and Systems: An Introduction*, McGraw Hill, Inc., New York (2013).
16. F.G. Smith, T.A. King, and D. Wilkins, *Optics and Photonics*, John Wiley & Sons, Hoboken, NJ (2007).
17. J.D. Felske and P.F. Roy, Computation of the reflectance and transmittance of an absorbing film on a transparent substrate, *Thin Solid Films*, 109(4), L113–L117, Letters (1983).
18. A.J. Haija, W. Larry Freeman, and R. Umbel, Effective optical constants and effective optical properties of ultrathin trilayer structures, *Physica B: Physics of Condensed Matter*, 406, 225–230 (2011).
19. A.H. Madjid and A.J. Abu El-Haija, Optical properties of very thin, nonideal multilayer superlattice stacks, *Applied Optics*, 19 15, 2612 (1980).
20. G. Zissis and W. Wolfe (Eds.), *The Infrared Handbook*, prepared by The Infrared Information and Analysis (IRIA) Center, Environmental Research Institute of Michigan, for the Office of Naval Research, Department of the Navy, Washington, DC (1978).
21. R.M. Eisberg, *Fundamental of Modern Physics*, John Wiley & Sons, New York (1990).
22. A. Heald and J.B. Marion, *Classical Electromagnetic Radiation*, 3rd edn., Dover Publications, Inc., New York (1995).
23. A. Beiser, *Concepts of Modern Physics*, 6th edn., p. 312, John Wiley & Sons, Inc., New York (2003).
24. T. Thornton and A. Rex, *Modern Physics for Scientists and Engineers*, 2nd edn., p. 375, Saunders College Publishing, Belmont, CA (1999).
25. C. Kittel, *Introduction to Solid State Physics*, 6th edn., John Wiley & Sons, Hoboken, NJ (2005).
26. G.R. Fowles, *Introduction to Modern Optics*, 2nd edn., Dover Publications, Inc., New York (1975).
27. L.M. Krauss, *Fear of Physics*, Basic Books, New York (2007).
28. K. Perry, *Philosophy*, Fall River Press, New York (2015).
29. A.P. French, *Vibrations and Waves*, W. W. Norton & Company, New York (1971).
30. R.A. Servey, C.J. Moses, and C.A. Moyer, *Modern Physics*, Thomson, Brooks Cole, Belmont, CA (2005).

Index

Nonexperimental Social Science: Two Debates (Washington, D.C.: American Educational Research Association and American Statistical Association).

Rosenbaum, P. R. (1995), *Observational Studies* (New York: Springer).

Rose, H. (2001), 'Colonising the Social Sciences?' in H. Rose and S. Rose (eds.) *Alas Poor Darwin: Arguments Against Evolutionary Psychology* (London: Vintage).

Rubin, D. B. (1974), 'Estimating Causal Effects of Treatments in Randomized and Nonrandomized Studies', *Journal of Educational Psychology*, 66: 688–701.

Rubin, D. B. (1977), 'Assignment to Treatment Groups on the Basis of a Covariate', *Journal of Educational Statistics*, 2: 1–26.

Rubin, D. B. (1986), 'Comment: Which Ifs Have Causal Answers?', *Journal of the American Statistical Association*, 81: 961–62.

Rubin, D. B. (1990), 'Formal Models of Statistical Inference for Causal Effects', *Journal of Statistical Planning and Inference*, 25: 279–92.

Rueschemeyer, D. (1991), 'Different Methods—Contradictory Results? Research on Development and Democracy' in C. C. Ragin (ed.) *Issues and Alternatives in Comparative Social Research* (Leiden: E. J. Brill).

Rueschemeyer, D. (2003), 'Can One or a Few Cases Yield Theoretical Gains?' in J. Mahoney and D. Rueschemeyer (eds.) *Comparative Historical Analysis in the Social Sciences* (Cambridge: Cambridge University Press).

Rueschemeyer, D. and Stephens, J. D. (1997), 'Comparing Historical Sequences— A Powerful Tool for Causal Analysis', *Comparative Social Research*, 16: 55–72.

Rueschemeyer, D., Stephens, E. H., and Stephens, J. D. (1992), *Capitalist Development and Democracy* (Cambridge: Polity Press).

Rule, J. B. (1997), *Theory and Progress in Social Science* (Cambridge: Cambridge University Press).

Runciman W. G. (1983), *A Treatise on Social Theory: I—The Methodology of Social Theory* (Cambridge: Cambridge University Press).

Runciman, W. G. (1998), *The Social Animal* (London: Harper Collins).

Rutter, M. (1981), 'Epidemiological/Longitudinal Strategies and Causal Research in Child Psychiatry', *Journal of the American Academy of Child Psychiatry*, 20: 513–44.

Rutter, M. (1994), 'Beyond Longitudinal Data: Causes, Consequences, Changes and Continuity', *Journal of Consulting and Clinical Psychiatry*, 62: 928–40.

Ryan, A. (1970), *The Philosophy of John Stuart Mill* (London: Macmillan).

Salmon, W. C. (1980), 'Probabilistic Causality', *Pacific Philosophical Quarterly*, 61: 50–74.

Sartori, G. (1994), 'Compare Why and How' in M. Dogan and A. Kazancigil (eds.) *Comparing Nations* (Oxford: Blackwell).

Scharpf, F. W. (2000), 'The Viability of Advanced Welfare States in the International Economy', *Journal of European Public Policy*, 7: 997–1032.

Scharpf, F. W. and Schmidt, V. A. (eds.) (2000), *Welfare and Work in the Open Economy* (Oxford: Oxford University Press).

Scheff, T. J. (1992), 'Rationality and Emotion: Homage to Norbert Elias' in J. S. Coleman and T. J. Fararo (eds.) *Rational Choice Theory: Advocacy and Critique* (Newbury Park: Sage).

Scheuch, E. K. (1989), 'Theoretical Implications of Comparative Survey Research: Why the Wheel of Cross-Cultural Methodology Keeps on being Reinvented', *International Sociology*, 4: 147–67.

Schmidt, M. G. (1993), 'Gendered Labour Force Participation' in F. G. Castles (ed.) *Families of Nations: Patterns of Public Policy in Western Democracies* (Aldershot: Dartmouth).

Schmitter, P. C. (1991), 'Comparative Politics at the Crossroads'. Instituto Juan March, Madrid.

Schoemaker, P. J. (1982), 'The Expected Utility Model: Its Variants, Purposes, Evidence and Limitations', *Journal of Economic Literature*, 30: 529–63.

Schumpeter, J. A. (1954), *History of Economic Analysis* (London: Routledge).

Schulze, G. G. and Ursprung, H. W. (1999), 'Globalization of the Economy and the Nation State', *World Economy*, 22: 295–352.

Sciulli, D. (1992), 'Weaknesses in Rational Choice Theory's Contribution to Comparative Research' in J. S. Coleman and T. J. Fararo (eds.) *Rational Choice Theory: Advocacy and Critique* (Newbury Park: Sage).

Searle, J. R. (1993), 'Rationality and Realism: What Is at Stake?', *Daedalus*, Fall: 55–83.

Searle, J. R. (1995), *The Construction of Social Reality* (London: Allen Lane).

Seawright, J. (2005), 'Qualitative Comparative Analysis vis-à-vis Regression', *Comparative International Development*, 40: 3–326.

Sen, A. K. (1977), 'Rational Fools: A Critique of the Behavioural Foundations of Economic Theory', *Philosophy and Public Affairs*, 6: 317–44.

Sen, A. K. (1986), 'Prediction and Economic Theory', *Proceedings of the Royal Society of London*, A407: 3–23.

Sen, A. K. (1987), 'Rational Behaviour' in J. Eatwell, M. Milgate, and P. Newman (eds.) *The New Palgrave: A Dictionary of Economics*, vol. 4, (London: Macmillan).

Seng, Y. P. (1951), 'Historical Survey of the Development of Sampling Theories and Practice', *Journal of the Royal Statistical Society*, Series A, 114: 214–31.

Sennett, R. (1998), *The Corrosion of Character* (New York: Norton).

Shavit, Y. and Blossfeld, H.-P. (eds.) (1993), *Persistent Inequality: Changing Educational Attainment in Thirteen Countries* (Boulder: Westview).

Shavit, Y., Arum, R., and Gamoran, A. (forthcoming), *Stratification, Expansion and Differentiation in Higher Education: A Comparative Study* (Stanford: Stanford University Press).

Shavit, Y. and Müller, W. (eds.) (1998), *From School to Work* (Oxford: Clarendon).

Silver, C. (ed.) (1982), *Frédéric Le Play on Family, Work and Social Change* (Chicago: Chicago University Press).

Silverman, D. (1993), *Interpreting Qualitative Data* (London: Sage).

Simmel, G. (1900/1978), *The Philosophy of Money* (New York: Free Press).

Simmel, G. (1905/1977), *The Problems of the Philosophy of History* (New York: Free Press).

Simon, H. A. (1954), 'Spurious Correlation: A Causal Interpretation', *Journal of the American Statistical Association*, 49: 467–92.

Simon, H. A. (1982), *Models of Bounded Rationality* (Cambridge, Mass.: MIT Press).

Simon, H. A. (1983), *Reason in Human Affairs* (Oxford: Blackwell).

Simon, H. A. and Iwasaki, Y. (1988), 'Causal Ordering, Comparative Statistics, and Near Decomposability', *Journal of Econometrics*, 39: 149–73.

Skocpol, T. (1979), *States and Social Revolutions* (Cambridge: Cambridge University Press).

Skocpol, T. (1982) 'Rentier State and Shi'a Islam in the Iranian Revolution', *Theory and Society*, 11: 265–83.

Skocpol, T. (1984), 'Emerging Agendas and Recurrent Strategies in Historical Sociology' in *idem* (ed.) *Vision and Method in Historical Sociology* (Cambridge: Cambridge University Press).

Skocpol, T. (1986), 'Analyzing Causal Configurations in History: A Rejoinder to Nichols', *Comparative Social Research*, 9: 187–94.

Skocpol, T. (1994), *Social Revolutions in the Modern World* (Cambridge: Cambridge University Press).

Skocpol, T. and Somers, M. (1980), 'The Uses of Comparative History in Macrosocial Inquiry', *Comparative Studies in Society and History*, 22: 174–97.

Slaughter, M. J. and Swagel, P. (1997), *The Effect of Globalization on Wages in the Advanced Economies* (Washington, D.C.: IMF Staff Studies for the World Economic Outlook).

Smelser, N. J. (1976), *Comparative Methods in the Social Sciences* (Englewood Cliffs, N.J.: Prentice Hall).

Smelser, N. J. (1992), 'The Rational Choice Perspective: A Theoretical Assessment', *Rationality and Society*, 4: 381–410.

Smith, H. L. (1990), 'Specification Problems in Experimental and Non-Experimental Social Research', *Sociological Methodology*, 20: 59–91.

Smith, M. (1999), 'What Is the Effect of Technological Change on Earnings Inequality?', *International Journal of Sociology and Social Policy*, 19: 24–59.

Smith, M. (2001), 'La Mondialisation: a-t-elle un effet important sur le marché du travail dans les pays riches?' in D. Mercure (ed.) *Une Société Monde?* (Quebec: Presses de l'Université Laval).

Snow, D. A. and Anderson, L. (1991), 'Researching the Homeless' in J. R. Feagin, A. M. Orum, and G. Sjoberg (eds.) *A Case for the Case Study* (Chapel Hill: University of North Carolina Press).

Snow, D. A. and Morrill, C. (1993), 'Reflections on Anthropology's Ethnographic Crisis of Faith', *Contemporary Sociology*, 22: 8–11.

Sobel, M. E. (1981), 'Diagonal Mobility Models', *American Sociological Review*, 46: 893–906.

Sobel, M. E. (1985), 'Social Mobility and Fertility Revisited', *American Sociological Review*, 50: 699–712.

Sobel, M. E. (1995), 'Causal Inference in the Social and Behavioral Sciences' in G. Arminger, C. C. Clogg, and M. E. Sobel (eds.) *Handbook of Statistical Modeling for the Social and Behavioral Sciences* (New York: Plenum).

Sobel, M. E. (1996), 'An Introduction to Causal Inference', *Sociological Methods and Research*, 24: 353–79.

Sokal, A. (1996a), 'Transgressing the Boundaries: Toward a Transformative Hermeneutics of Quantum Gravity', *Social Text*, 46–47, 217–52.

Sokal, A. (1996b), 'A Physicist Experiments with Cultural Studies', *Lingua Franca*, 6: 62–64.

Sokal, A. and Bricmont, J. (1998), *Intellectual Impostures* (London: Profile Books).

Somers, M. R. (1998), ' "We're No Angels": Realism, Rational Choice and Relationality in Social Science', *American Journal of Sociology*, 104: 722–84.

Sørensen, A. B. (1998), 'Theoretical Mechanisms and the Empirical Study of Social Processes' in P. Hedström and R. Swedberg (eds.), *Social Mechanisms* (Cambridge: Cambridge University Press).

Spencer, H. (1861/1911), *Essays on Education* (London: Dent).

Spencer, H. (1904), *An Autobiography*, 2 vols. (London: Williams and Norgate).

Spirtes, P, Glymour, C., and Scheines, R. (1993), *Causation, Prediction and Search* (New York: Springer).

Spradley, J. P. (1980), *Participant Observation* (New York: Holt, Rinehart and Winston).

Stake, R. E. (1978), 'The Case Study Method of Social Inquiry', *Educational Researcher*, 7: 5–8.

Steinmetz, G. (2004), 'Odious Comparisons: Incommensurability, the Case Study, and "Small N's" in Sociology', *Sociological Theory*, 22: 371–400.

Steinmetz, G. (2005), 'The Epistemological Unconscious of U.S. Sociology and the Transition to Post-Fordism: The Case of Historical Sociology' in J. Adams, E. S. Clemens, and A. S. Orloff (eds.), *Remaking Modernity: Politics, History, and Sociology* (Durham: Duke University Press).

Steuer, M. (1998), 'A Little Too Risky: Does Professor Beck Have a Thesis?' *LSE Magazine*, Spring.

Steuer, M. (2002), *The Scientific Study of Society* (Boston: Kluwer).

Stevens, S. S. (1946), 'On the Theory of Scales of Measurement', *Science*, 103: 677–80.

Stewart, A. (1998), *The Ethnographer's Method* (Thousand Oaks, Calif.: Sage).

Stimson, J. A. (1985), 'Regression in Space and Time: A Statistical Essay', *American Journal of Political Science*, 29: 914–47.

Stinchcombe, A. L. (1968), *Constructing Social Theories* (New York: Harcourt Brace).

Stinchcombe, A. L. (1993), 'The Conditions of Fruitfulness of Theorizing about Mechanisms in Social Science' in A. B. Sørensen and S. Spilerman (eds.) *Social Theory and Social Policy: Essays in Honor of James S. Coleman* (Westport: Praeger).

Stoecker, R. (1991), 'Evaluating and Rethinking the Case Study', *Sociological Review*, 39: 88–112.

Stone, L. (1987), *The Past and Present Revisited* (London: Routledge).

Strange, S. (1996), *The Retreat of the State: The Diffusion of Power in the World Economy* (Cambridge: Cambridge University Press).

Strauss, A. and Corbin, J. (1990), *Basics of Qualitative Research* (Newbury Park: Sage).

Suppes, P. (1970), *A Probabilistic Theory of Causality* (Amsterdam: North Holland).

Surowiecki, J. S. (2005), *The Wisdom of Crowds* (London: Abacus).

Sutherland, S. (1992), *Irrationality* (London: Penguin).

Swank, D. (1998), 'Globalization and the Welfare State'. Eleventh International Conference of Europeanists, Baltimore.

Swedberg, R. (1998), *Max Weber and the Idea of Economic Sociology* (Princeton: Princeton University Press).

Sztompka, P. (1988), 'Conceptual Frameworks in Comparative Inquiry: Divergent or Convergent?', *International Sociology*, 3: 207–18.

Tåhlin, M. (1993), 'Class Inequality and Post-Industrial Employment in Sweden' in G. Esping-Andersen (ed.) *Stratification and Mobility in Post-Industrial Societies* (London: Sage).

Tawney, R. H. (1912), *The Agrarian Problem in the Sixteenth Century* (London: Longmans).

Tawney, R. H. (1941/1954), 'The Rise of the Gentry, 1558–1640' (with a 'Postscript') in E. M. Carus-Wilson (ed.) *Essays in Economic History* (London: Arnold).

Taylor, R. (2002), *Britain's World of Work—Myths and Realities* (London: Economic and Social Research Council).

Teune, H. (1997), 'Stories, Observations, Systems, Theories', *Comparative Social Research*, 16: 73–83.

Therborn, G. (1993), 'Beyond the Lonely Nation-State' in F. G. Castles (ed.) *Families of Nations: Patterns of Public Policy in Western Democracies* (Aldershot: Dartmouth).

Tilly, C. (1997), 'Means and Ends of Comparison in Macrosociology', *Comparative Social Research*, 16: 43–53.

Tooby, J. and Cosmides, L. (1992), 'The Psychological Foundations of Culture' in J. H. Barkow, L. Cosmides, and J. Tooby (eds.) *The Adapted Mind* (New York: Oxford University Press).

Toulmin, S. (1970), 'Reasons and Causes' in R. Borger and F. Cioffi (eds.) *Explanation in the Behavioural Sciences* (Cambridge: Cambridge University Press).

Treiman, D. J. (1975), 'Problems of Concept and Measurement in the Comparative Study of Occupational Mobility', *Social Science Research*, 4: 183–230.

Tsebelis, G. (1990), *Nested Games: Rational Choice in Comparative Politics* (Berkeley: University of California Press).

Tylor, E. B. (1889), 'On a Method of Investigating the Development of Institutions: Applied to Laws of Marriage and Descent', *Journal of the Royal Anthropological Institute*, 18: 245–56, 261–69.

Udéhn, L. (2001), *Methodological Individualism: Background, History and Meaning* (London: Routledge).

Ultee, W. (1991), 'How Classical Questions were Enriched' in H. Becker, F. L. Leeuw, and K. Verrips (eds.) *In Pursuit of Progress: An Assessment of Achievements in Dutch Sociology* (Amsterdam: SISWO).

Urry, J. (2000), 'Mobile Sociology', *British Journal of Sociology*, 51: 185–201.

Usui, C (1994), 'Welfare State Development in a World System Context: Event History Analysis of First Social Insurance Legislation among 60 Countries, 1880–1960' in T. Janoski and A. M. Hicks (eds.) *The Comparative Political Economy of the Welfare State* (Cambridge: Cambridge University Press).

Van den Berg, A. (1998), 'Is Sociological Theory Too Grand for Social Mechanisms?' in P. Hedström and R. Swedberg (eds.) *Social Mechanisms* (Cambridge: Cambridge University Press).

Vanneman, R. and Cannon, L. W. (1987), *The American Perception of Class* (Philadelphia: Temple University Press).

Verba, S., Nie, N., and Kim, J. (1978), *Participation and Political Equity* (Cambridge: Cambridge University Press).

Vidich, A. J. and Lyman, S. M. (1994), 'Qualitative Methods: Their History in Sociology and Anthropology' in N. K. Denzin and Y. S. Lincoln (eds.) *Handbook of Qualitative Research* (Thousand Oaks, Calif.: Sage).

Voas, D. (2003), 'The So-so Construction of Sociology', *British Journal of Sociology*, 54: 129–37.

Wagner, P. (2000), 'The Bird in Hand: Rational Choice—the Default Mode of Social Theorizing' in M. S. Archer and J. Q. Tritter (eds.) *Rational Choice Theory: Resisting Colonization* (London: Routledge).

Wallerstein, I. (1974–89), *The Modern World System*, 3 vols. (New York: Academic Press).

Walsh, W. H. (1974), 'Colligatory Concepts in History' in P. Gardiner (ed.) *The Philosophy of History* (Oxford: Oxford University Press).

Watkins, J. W. N. (1957), 'Historical Explanation in the Social Sciences', *British Journal for the Philosophy of Science*, 8: 104–17.

Watkins, J. W. N. (1963), 'On Explaining Disaster', *The Listener*, 10 January: 69–70.

Watkins, J. W. N. (1970), 'Imperfect Rationality' in R. Borger and F. Cioffi (eds.) *Explanation in the Behavioural Sciences* (Cambridge: Cambridge University Press).

Weakliem, D. L. and Heath, A. F. (1994), 'Rational Choice and Class Voting', *Rationality and Society*, 6: 243–70.

Weakliem, D. L. and Heath, A. F. (1999), 'The Secret Life of Class Voting: Britain, France and the United States' in G. Evans (ed.) *The End of Class Politics?* (Oxford: Oxford University Press).

Weber, Max (1892), *Die Verhältnisse der Landarbeiter im ostelbischen Deutschland* (Leipzig: Schriften des Vereins für Sozialpolitik).

Weber, Max (1903–06/1975), *Roscher and Knies: The Logical Problems of Historical Economics* (New York: Free Press).

Weber, Max (1908/1975), 'Marginal Utility Theory and "The Fundamental Law of Psychophysics"', *Social Science Quarterly*, 56: 21–36.

Weber, Max (1922/1968), *Economy and Society* (Berkeley: University of California Press).

Weir, M. and Skocpol, T. (1985), 'State Structures and the Possibilities for "Keynesian Responses" to the Great Depression in Sweden, Britain and the United States' in P. Evans, D. Rueschemeyer, and T. Skocpol (eds.) *Bringing the State Back In* (New York: Cambridge University Press).

Whelan, C., Layte, R., and Maître, B. (2002), 'Persistent Deprivation in the European Union', *Schmollers Jahrbuch: Journal of Applied Social Science Studies*, 122: 31–54.

Whelan, C., Layte, R., and Maître, B. (2004), 'Understanding the Mismatch between Income Poverty and Deprivation: A Dynamic Comparative Analysis', *European Sociological Review*, 20: 287–302.

Whyte, W. F. (1943), *Street Corner Society* (Chicago: Chicago University Press).

Wickham-Crowley, T. P. (1992), *Guerrillas and Revolution in Latin America* (Princeton: Princeton University Press).

Williams, B. (2002), *Truth and Truthfulness* (Princeton: Princeton University Press).

Williamson, O. E. (1985), *The Economic Institutions of Capitalism* (New York: Free Press).

Williamson, O. E. (1994), 'The Economics and Sociology of Organizations: Promoting a Dialogue' in G. Farkas and P. England (eds.) *Industries, Firms and Jobs* (New York: Aldine De Gruyter).

Williamson, O. (1996), *The Mechanisms of Governance* (New York: Oxford University Press).

Willis, P. (1977), *Learning to Labour* (Farnborough: Saxon House).

Winship, C. and Morgan, S. (1999), 'The Estimation of Causal Effects from Observational Data', *Annual Review of Sociology*, 25: 659–706.

Wolf, F. M. (1986), *Meta-Analysis: Quantitative Methods for Research Synthesis* (Beverly Hills: Sage).

Woolrych, A. (2002), *Britain in Revolution, 1625–1660* (Oxford: Oxford University Press).

von Wright, G. H. (1971), *Explanation and Understanding* (London: Routledge).

von Wright, G. H. (1972), 'On So-called Practical Inference', *Acta Sociologica*, 15: 39–53.

Wrong. D. (1961), 'The Oversocialized Conception of Man in Modern Sociology', *American Sociological Review*, 26: 183–93.

Wrong, D. (1999), *The Oversocialized Conception of Man* (New Brunswick, N.J.: Transaction Books).

Yin, R. K. (1994), *Case Study Research* (Thousand Oaks, Calif.: Sage).

Zaret, D. (1978), 'Sociological Theory and Historical Scholarship', *The American Sociologist*, 13: 114–21.

Zelditch, M. (1971), 'Intelligible Comparisons' in I. Vallier (ed.) *Comparative Methods in Sociology* (Berkeley: University of California Press).

Zellner, A. (1988), 'Causality and Causal Laws in Economics', *Journal of Econometrics*, 39: 7–21.

Znaniecki, F. (1934), *The Method of Sociology* (New York: Farrar and Rinehart).

Zonabend, F. (1992), 'The Monograph in European Ethnography', *Current Sociology*, 40: 49–54.

STUDIES IN SOCIAL INEQUALITY

STUDIES IN SOCIAL INEQUALITY

Printed and bound by CPI Group (UK) Ltd, Croydon, CR0 4YY

13/04/2025

14656449-0004